POROUS BECOMINGS

Edited by Andreas Bandak and Daniel M. Knight

POROUS BECOMINGS

ANTHROPOLOGICAL

ENGAGEMENTS *with*

MICHEL SERRES

DUKE UNIVERSITY PRESS
Durham and London 2024

Designed by Courtney Leigh Richardson
Typeset in Portrait and Changa by Westchester Publishing Services

Library of Congress Cataloging-in-Publication Data
Names: Bandak, Andreas, editor. | Knight, Daniel M., editor.
Title: Porous becomings : anthropological engagements with Michel Serres /
edited by Andreas Bandak and Daniel M. Knight.
Description: Durham : Duke University Press, 2024. | Includes bibliographical
references and index.
Identifiers: LCCN 2023031720 (print)
LCCN 2023031721 (ebook)
ISBN 9781478030287 (paperback)
ISBN 9781478026051 (hardcover)
ISBN 9781478059318 (ebook)
Subjects: LCSH: Serres, Michel. | Anthropology—Philosophy. | Culture—
Philosophy. | Science—Philosophy. | BISAC: SOCIAL SCIENCE / Anthropology /
Cultural & Social | PHILOSOPHY / Movements / Critical Theory
Classification: LCC B2430.S464 P67 2024 (print) | LCC B2430.S464 (ebook) |
DDC 301.01—dc23/eng/20231205
LC record available at https://lccn.loc.gov/2023031720
LC ebook record available at https://lccn.loc.gov/2023031721

Cover art: *Sponge.* Courtesy Shutterstock/Louella938.

Για τους καθημερινούς φιλόσοφους του χρόνου, των παράσιτων και του ιλίγγου. (DMK)

لأولئك الذين يتصلون. إلى ملائكي. ولكل الأصدقاء والمغتربين. (AB)

CONTENTS

More than most, this book is a product of conversation. At a time when the world was in lockdown and social interaction took its place as nostalgic reflection on good times past, conversation simmered between two editors, a bunch of authors in far-flung living rooms and makeshift offices, and a French philosopher of science whose life had fallen agonizingly short of experiencing the next big global cataclysm.

The fire was lit under the boiling pot of brimming energy driving this particular knotty conversation when we, the editors, met on a transatlantic flight from Amsterdam to Vancouver in November 2019. Back then, in the heady days of plenty, little did we know that the Vancouver AAA meetings would be one of the final opportunities for large-scale face-to-face interaction for more than two years. In March 2020, lockdown hit us both hard. Homeschooling, online teaching, our kitchen tables and children's bedrooms hosting staff meetings, hastily rearranged workshops, and, eventually, hours of talk about Michel Serres.

Serres, our companion on our individual academic journeys for almost two decades, became the third person in our COVID-era relationship. We pondered: Why had more anthropologists not delved into Serresian meandering through time and disciplines? Why should anyone dedicate more ink to yet another French white male philosopher? Why had we now turned to Serres to shine a light through our collapsing pandemic world? In Serresian style, the conversations twisted and turned, became tangled and shot off branches in weird and wonderful directions.

We agreed that Serres somehow spoke to "the anthropological project"—a deliberately porous category, if ever there was—but we could not quite pin down how or why. Perhaps this was the point; Serres drifts excitedly in pursuit of the rainbow, in search of that pot of gold that is knowledge. He does so by following his interlocutors beyond time and place and asks you to strap in for the ride (Leibniz, Plato, the Troubadour, Hermes, Lucretius, Jules Verne). But the inability to box Serres the polymath, the time traveler, the poet, means that his work often appears merely as a footnote in anthropological texts or, sometimes, is reduced to a sound bite of his most widely known ideas (the parasite, background noise, the natural contract).

Many of the contributors to *Porous Becomings* had, sometime, somewhere, already engaged with Serres. Some had written stand-alone articles that pivoted on one or more of Serres's core concepts. Others had flirted with an idea but not pressed further into the Serresian cavern. For others still, this book represents their first encounter. Contributors come from diverse schools of thought in anthropology: phenomenology, STS, environment and medicine, media and communication, ontology, and transhumanism. Indeed, our definition of anthropology is itself purposefully porous, reflected by the way that at least two authors could claim residence on the blurry boundaries of the (inter)discipline, with twigs, sometimes branches, snaking their way into social psychology, sociology, and the creative humanities. Ethnography is drawn from New York comedy clubs, African mythology, Balkan war debris, (post)colonial bodies, and the cross-disciplinary comparison of key figures in social theory. Each author navigates Serres's oeuvre according to their own burning questions drawn from their respective field sites and filtered through eclectic epistemological lenses.

Navigation is perhaps the best trope to summarize the whole book project. From helping us navigate the COVID-19 pandemic to the routes Serres suggests our authors navigate their fields, navigation is also how we suggest the reader approach this volume. We have established sections based on Serres's core concepts of the parasite and the natural contract, spatial and temporal topologies, and the quest for knowledge and connection, but these are fluid categories. This is simply our brainchild—one of many—for a potential conversation, but each chapter can and could be placed in any order in all sections. As the author of our afterword, Jane Bennett, points out, Serres's mode of thinking resists systemization; he doesn't seek a standard "order of things." This provides the reader with an opportunity to make their own connections, strike up their own conversations between chapters, in a manner that best suits their intrigue.

Serres navigated time and space, spanned figures of thought, by way of topological relations. He transcended structure and boundaries to make connections that helped him simultaneously hold an array of themes that might otherwise be stamped as the property of the natural or social sciences, bound to a foregone era, or contained to a niche philosophical domain. Relationships, conversations, connections free from preconception: this is how we suggest the reader approach this book. The afterword perhaps best captures the spirit of Serres: a conversation with a modern-day polymath, Jane Bennett. As such, the afterword is not meant to summarize the preceding essays, but rather to strike up a polyphonic dialogue on the relationship between a researcher and Serres as together they navigate their own version of the cosmos; we suggest that perhaps the reader might consult the afterword immediately after digesting the introduction to get a fuller sense of our endeavor.

ACKNOWLEDGMENTS

In acknowledgment of conversations that shaped this book, we thank Kim Fortun and Mike Fortun for engaging with early iterations and Charles Stewart and Bjørn Thomassen for comments on draft sections. Charlotte Garcia and Luka Benedičič offered original insight on the introduction, while continuing adventures with Gabriela Manley provoked Knight to scramble down innumerable rabbit holes, and Simon Coleman has continuously inspired Bandak in their engagements on repetitions, proportions, and lateral thinking. Michael Herzfeld provided invaluable advice at the preliminary stages of the project, while Ken Wissoker at Duke University Press has been outstanding in helping drive the volume forward and Ryan Kendall has been a supportive and skilled Assistant Editor. Most importantly, we thank the contributors for their unrivaled enthusiasm, energy, and intellectual provocations while reading and writing together. We hope that, like us, readers will feel their passion and curiosity seeping through the page.

And finally, to geotag our location as two insignificant figures in the eternal flux of time and space:

Andreas Bandak, Copenhagen, Denmark
Daniel M. Knight, Trikala, Greece
JUNE 2022

ANDREAS BANDAK AND DANIEL M. KNIGHT

ANGEL HAIR ANTHROPOLOGY WITH MICHEL SERRES

Serres and the Spirit of Anthropology

One of the foremost scholars and social commentators of his generation, Michel Serres (1930–2019) had an extraordinary gift to effortlessly transcend epistemological boundaries. Speaking to the foundations of what it is to be human, Serres broke the shackles of his trade to build an accessible philosophy of science free from authoritative metalanguage. He became a tutor to a generation of French social theorists and an inspiration to those wanting to better understand the world beyond the conformities of the academy and narrowly defined conditions of disciplinary thought, traversing the constraints of method, scale, and tradition. While friend and student Bruno Latour and fellow French theorists of a similar era such as Gilles Deleuze, Félix Guattari, and Jacques Derrida have become household names among anthropologists over the past two decades, Serres has gone relatively unnoticed, his work often reduced to sound bites in endnotes or, apparently, becoming the property of a handful of devout followers—something Serres would no doubt detest given his contempt for the notion of intellectual ownership. Immersed in his poetics that foster what he calls "conversations" across epistemological and literary genres are remarkable critiques of the human condition, historical organization, religious quest, and environmental relationality.

Through often audacious navigation of the material and the conceptual world, Serres's interrogation of the fundamentals of human existence can propose new avenues for anthropological knowledge. Serres deploys hyphens, branches, analogies, and messengers to bring to the same conversation the diversity of science, society, and ecology. Always considered a maverick, Serres had a propensity to combine creative prose and literary nous with firsthand life experiences and readings of global events that complement the anthropological endeavor. His weaving of autobiography and literary fiction with deep analyses of the natural and social sciences fosters a unique approach to fathoming contemporary global problems that, in spirit, reflects anthropology at its most engaged. Conversations with Serres help further understanding of humanity and the spatial-temporal coordinates of life on Earth—and the Human in the Cosmos—as we push further into a turbulent new millennium. This book harnesses the potential for anthropology to converse with Serres across a kaleidoscope of topics.

Serres's concerns with untangling the human in and of the world map the core domains of the classic anthropological canon—economy, environment, gender, kinship, politics, religion, technology, and so on. Yet he does so without being restricted by disciplinary dogmas and by fearlessly breaching the confines of the scale and duration of human experience (momentary to epochal, even eternal, individual to collective and planetary). This can be quite a hairy ride for the anthropologist, which, however, with perseverance, can lead the reader to a state of exhilaration, of "voluptuous panic" (Caillois 2001, 23) in what they find.

For instance, much of Serres's work is dedicated to identifying relationality through encounters with Otherness—cultures, disciplines, literatures, and spacetimes. To navigate these porous networks of connectivity across space, time, and episteme, Serres employs the figures of Hermes (1968–1980) and angels (1993) to communicate between the multiple realms of Otherness. Messages delivered across nominally bounded genres reveal, for Serres, the interconnected relationality of temporal agency (particularly society located in nonlinear time), humanity's intrinsically violent disposition, and the influence of technology on the natural senses. Meditation on sameness/difference by way of transcending disciplines is explicated in *The Troubadour of Knowledge* ([1991] 1997), where Serres advocates for a pedagogical basis that combines both science's general truths and literature's singular stories to better comprehend Otherness within and without the human.[1] Continuing the interiority/exteriority distinction and expanding the theme of porosity across borders, in *The Five Senses* ([1985] 2015) Serres addresses the circulation of human bodies in information systems that increasingly place humans indoors, detached from their perceptions and senses. Serres champions the reassertion of sensory perception in tackling the myriad challenges of present and future lifeworlds.

One of Serres's seminal texts, *The Parasite* ([1980] 2007), has already nestled in the imagination of anthropologists working on hospitality, trade, and infrastructure (e.g., Kockelman 2010; Candea and da Col 2012; Lowe 2014; Shryock

2019). Here, Serres poses the human condition as akin to a parasite on a host body, with numerous pathways of Becoming (see also Brown 2002, 2013). For anthropology, the parasite may provide insights into social insurgencies staged by minority groups wishing to eat away at dominant cultures to eventually share in the power games of society-making, the piratical plundering and opportunistic symbiotic emergence of new multispecies biospheres in the wake of ecological disaster, and a deep questioning of relations of hospitality that can be scaled from everyday exchange and reciprocity to interrogate human interaction with the planet in the age of the Anthropocene and climate change. Further, as an inadvertent biproduct of human activity, parasitical noise, a "murmuring messiness" (Bennett and Connolly 2011, 155), serves nascent ecosystems in nonanthropocentric routes to Becoming (Bennett 2020). From this pretext, in *The Natural Contract* ([1990] 1995), Serres calls for a new agreement to be drawn up acknowledging the reciprocal violence perpetrated by humans and the Earth—an accord between two entities that recognizes mutual destruction and demands a reassessment of a relationship founded on extraction and shared violence. Indeed, micro-contracts, Serres suggests in conversation with Latour (Serres and Latour [1992] 1995), are required between feuding bodies of knowledge in order to find futural trajectory rather than spiral into perpetual cyclical return, often to states of violent conflict and mutual destruction. In short, problems will not be solved without a "hyphenated" agreement; despite opposing perspectives, there has to be a collaborative approach, perhaps a series of uneasy alliances, to find novel solutions to existential challenges.

In *L'Art des Ponts: Homo Pontifex* (2006, 77), Serres likens the hyphen to a bridge between the "soft empire of signs" and "the hard realms of physics and biology," allowing communication to transcend two planes. The result is the production of newness, the harlequin figure that is born of the mixing of the hard sciences and humanities, a diversity perhaps shocking or grotesque but indicative of the experiments required to harness novelty (Serres [1991] 1997). The hyphen allows alignment and conjunction of disparate parts, indicating a branching out of expertise into new collaborative domains (centrifugal) but also the drawing together of units of knowledge (centripetal) that do not obviously fit together. The bridging of ideas and domains facilitates movement in physical space and notional time. Scaling the individual and collective, the traditional bounded community and the planetary, "the academy" and "the people," anthropology has a key role to play as both mediator and flagbearer of these hyphenated and hybrid agreements between parasite and host, human and nonhuman, that essentially allow us to participate more fully in the chaos of the contemporary world, moving between concepts rather than becoming boxed in. Symptomatically, Serres sees his philosophy as one of not verbs or nouns but rather prepositions, as each preposition binds, unites, and energizes language and creates webs of relations in and through time and space. Most significantly, prepositions do not create a world merely of objects and things, a world of "marble statuary" ([2019] 2022, 154; see also Serres and Latour [1992] 1995, 101, 106). Anthropology is a natural home to prepositions and the harlequin, full of mischief and diversity but also, perhaps, able to lead the way in a critique of hegemony and what aspects of cosmopolitan existence must be transformed to further collective Becoming. We will return later to the hyphen as central to Serres's method of porous connection.

The chaos of our tenure on Earth is famously tackled in *Genesis* ([1982] 1995). Here, the "background noise" driving humanity, Serres suggests, is fury, violence, disorder, and anxiety rather than any vestiges of philosophical rationale. A multidisciplinary tour de force, *Genesis* speaks to the rapidly expanding anthropological preoccupation with affect whereby the atmosphere, feeling, and aesthetic of life can only partially be explained through logical connections and philosophical reason that might classically be traced back to Rousseau's social contract and Kant's pragmatic anthropology. Much of chaotic human existence is felt, heard, and subconsciously sensed as a continuous resonant accompaniment in the background of everyday practice. In *Genesis*, atmospheres and aesthetics drive the human race forward, even if our propulsion systems choke and splutter on the dusty clouds of the chaos while searching for order and format. The noise drifts in and between phenomena, clots and thins and agitates, posited eloquently by Jane Bennett and William Connolly as "a syrupy material, snotty fluid . . . an emptiness or a fullness, a black absence or a superabundance of presence" (2011, 156; Serres [1982] 1995, 30). The minimal intensification or unannounced blast of noise triggers action and a new branch in the bifurcation process surges forth and novel connections are formed. More on this, too, later. How might anthropologists better capture these resonant affects that point to an intangible background *something*, ebbing and flowing, that indicates life forces in elsewheres and elsewhens beyond the narrative threshold within which our discipline mundanely operates?[2] The world makes sense through parallel and contradictory feelings, the eerily uncanny, affective clouds, "a change in the wind," not simply categories of calculus or narrative, of form and structure. Often, what goes on inside the black box cannot be seen or heard and thus escapes our traditional reach of forming knowledge about the experiential world.

Genesis has been termed "apocalyptic" in asking the reader to think the unthinkable and listen to the piercing shrieks of otherworldly rebirths and Becomings. With an increasing interest in end-of-the-world scenarios and speculative apocalypticism within anthropology, Serres's thinking was well ahead of the curve and offers innovative perspectives on the vertigo-inducing disorder of modern society—what he terms the "turbulence" of a world spinning at different tempos on multifarious axes ([1982] 1995, 109). Yet human beings are inclined, Serres tells us, to search for order in chaos, producing "refined instruments" to apply to "complex and fluctuating" existence ([2004] 2020, 5). Navigating the arborescent network of accidents and obstacles requires instruments of measurement, be they in the form of events and epochs in the discipline of history; units to track flows of credit and debt in economics; margins, chapters, and paragraphs in literature; or the social domains in which anthropology deals. Standardization and formatting are desirable, yet what falls through the cracks, what binds the relationship between orders, is often of utmost fascination and import. Attention must be turned to capturing those resonant *somethings* (Lepselter 2016) that reside in the dark matter between events, mathematical equations, and concepts and in the background noise and silences of narrativized prose.

Concepts are containers that momentarily format or capture fluidity (Shryock and Smail 2018b), but they do not exist in suspended animation, and their edges remain porous. The capture of information in a concept is never complete since the permeable membrane of the container allows for seepage. Concepts cannot contain everything we purport them to have, and as anthropologists, we regularly find ourselves haunted by what remains outside or what gets lost in the transfer between conditions. Serres provides the metaphor of a coffee percolator when discussing how some concepts get stuck in the filtration process; the granules are somehow tangible, they clump together and connect. Other concepts—grains, if you will—pass through the filter with the pressurized flow of the pouring water (Serres and Latour [1992] 1995, 58). Anthropologists tend to be concerned with the modules of social life that are caught in the filter that represent the recorded events and customs of the people we work with. These are often analyzed within the core formats of the discipline itself—say, kinship, religion, economics, politics, gender—which act as units of order and comparison.

Looking at technologies that have brought forth globalized units of ordering information, in *Branches*, Serres again turns to the hyphen of a composite word—*com-putare*—to explain the scramble to order chaos ([2004] 2020, 7; cf. Serres 2012b). Uniting the preposition *cum* with *putare*—from reckon or think and stemming from *putus*, meaning clean and pure—we assume computing allows for objective comparison. Purity, Serres suggests, indicates that information can be committed to law: "When pure, neither things nor humans lie.

Contracts become possible" ([2004] 2020, 8). But repetition of misplaced belief in the *wrong* information, in *deceptive* units of comparison, forecloses potentiality for symbiotic change and instead leads to a vicious circle of pollution and ultimately death. There is no invention, no novelty, no news from accepting a homogenizing world, and Serres enthuses us to search for fresh branches, new hyphenated and hybridized collaborations that flux and flow between concepts. This seems to us to be a crossroads for anthropology: Do we accept our inadequate and antiquated concepts for analysis, shoving fluid material (syrupy noise) into square containers and taping up the leaking creases, or, in our curiosity, seek new branches of hyphenated collaboration in addressing the questions of contemporary and future worlds?

Serres offers poignant readings of myth and history as foundationals that percolate and order Self and Society. In doing so, he returns us to the dual forces of conjoining and branching out as pertinent to the critical rethinking of social (anthropological) assumptions on Becoming. His magisterial *Rome* ([1983] 1991) and *Statues* ([1987] 2014) both attest to Serres's lyricism in proposing how mythistorical traces shape the social reality we might call "modern" more than we may be prepared to readily acknowledge. History and Becoming are two derivative forms, which Serres explores as both conjoining to lay the footings for modern social life but also constantly branching out, bifurcating, often contradicting each other. The powerful sundering and conjoining forces of history and myth, he suggests, necessitate a much broader reading of temporality and historiography than any single disciplinary perspective can facilitate, as post-Enlightenment assumptions on historicism are overridden and overwritten (Knight and Stewart 2016, 13). Splitting and fusion are also nested in religion, which Serres etymologically traces to a dual meaning—*religere* and *religare*, of expelling or binding ([2019] 2022; cf. Ingold 2021; Bandak and Stjernholm 2022). These ostensible contradictions inherent in modernity, historiography, and religion are productive for rethinking the multiplicity of orders, past and present, as they attest to the non-modernity of humans.

From branches to angels and from hyphenation to the frenzied chaos at the birth of the universe, topological connectivity—the distorted fluxes, twists, deformations, and braids of assumed geometric phenomena—is core to Serres's corpus of work, allowing for connection to reside alongside novelty and transformation. A convincing argument for approaching the world through topological connection can perhaps most readily be found in Serres's work on time and temporality, which neatly mirrors the most recent "temporal turn" in anthropology (e.g., Bear 2014; Bryant and Knight 2019; Kirtsoglou and Simpson 2020). In *Conversations on Science, Culture, and Time* ([1992] 1995), Serres argues for a reading of time as topological rather than classically geometrical, where multiple disparate points may be simultaneously superimposed, folded in on each other. Although time may first appear to be flowing in one direction like a river passing beneath a bridge—the Heraclitean view—one often fails to consider the unseen countercurrents running beneath the surface in the opposite direction or the hidden turbulences that churn up sediment (Serres and Latour [1992] 1995, 58). Posing that temporal linearity is a post-Enlightenment Western construction, Serres offers analogies from the natural environment and technological assemblages to provide a critique of historicism that goes beyond the hybrid mythistory of the modern offered in his earlier works. Entwined with personal experiences of wartime violence and the legacy of global events of his lifetime, Serres contends that temporal moments are caught in a filtration process whereby they once again become contemporary when they serve a pedagogical or social purpose. In prose combining science, literature, and autobiography, Serres challenges the reader to question their usually undisputed perception of time and history as neat and linear. In doing so, Serres offers another layer to anthropological concerns with historicity and historical consciousness, such as those tabled by Eric Hirsch and Charles Stewart (2006), rewriting the rulebook on how humans connect with the past through the senses. Becoming in time and history is, for Serres, a matter of nonlinear topological connection where people receive messages through the resonant noise of everything that has gone before. On the theme of temporality, Serres provides a rich conceptual-analogous repertoire to contemplate the porosity of Becoming in time and history.

Ponderings on time and event take a more radical turn in *Branches* ([2004] 2020), where Serres tackles the roots of human history in the form of an origin story to call for epochal change in collective human behavior. His approach to time and eventedness is pertinent to our understanding of the building blocks of modern social life, particularly in a world besieged by a supposedly unprecedented number of crises. For instance, in *Times of Crisis* ([2009] 2014), through medical definitions and classic etymologies of "crisis," Serres considers how the 2008 global financial collapse facilitates a creative choice for the human subject at a fork in the road of historical experience; to change or to repeat, a cyclical temporality of recurrence or an opportunity to invent. Here we are encouraged to think about recent anthropological attention to the need for events to focus social and political action toward change, what Chloe Ahmann (2018) has examined as temporal manipulation to create events out of nothing. Serres's musings on such individual and collective agency in creating and reacting to events foreground human ingenuity in the face of fear and the ability to make decisions at moments of rupture and rapid social change. This is an important branch given the economic, political, environmental, and health crises facing the human race in the early decades of the twenty-first century.

Making Our Intentions Known

Rich in metaphor and breaking free from disciplinary dogmas, Serres's reflections on science, culture, technology, art, and religion provide creative routes into understanding the human condition. In embracing chaos to interrogate format and hegemonic beliefs in the name of exploring diversity and the uncanny, Serres in many ways embodies the spirit of anthropology. This book is not intended to be a review of Serres's insurmountable body of work or the myriad commentaries provided from authors in complementary disciplines. Nor is it a user guide of how to neatly apply Serres to ethnographic analysis. It is rather the start of a new conversation between the philosopher and the discipline of anthropology, hyphenating his world and ours. Relationality and Otherness take on new guises as Serres maps the connections between and across formats and spacetimes. Exploring ethnographies of Becoming inspired by Serres's free-thinking reflections on humanity affords anthropology new hyphenated concepts that take the lead from but then critically destabilize standard disciplinary canons by both expanding and contracting our analytical lens. Anthropology is always trying to develop concepts, but Serres asks us not to make them too rigid or formatted since this inhibits the fluxes in noise that do not fit into the boxes of calculation and measurement. We do need traction to work toward concepts, but the shape-shifting branches that connect scalar domains of life feed off the background noise of our social universe to become its driving force. The concepts are topologically connected by difference, like layers of the harlequin, or through the bounding adventures of Hermes, ever shifting, revealing, piercing rigidity. In truly Serresian guise, the intention here is to ignite "conversation" between anthropologists and a philosopher who has so far remained on the periphery of anthropological thought. In doing so, we might just advance a multidimensional understanding of human life in the face of unprecedented challenges in the twenty-first century.

In short, authors in this volume showcase how Serres can open novel avenues for ethnographic analysis. In many ways, Serres performed an ethnography of philosophy, moving across and between concepts to deliver polyphonic readings of nature and culture. As anthropologists do, Serres mapped the constellations that connect humans, time, technology, and planet Earth. A conversation between Serres and anthropology accentuates the porous Becomings of our research participants as they traverse formats, concepts, and social domains. People operate on the edges of the hypothetical boxes that we claim as units of analysis, at the extremities where the ink disperses on the blotting paper, oozing in multiple directions into the margins, the in-between spaces, and merging with the background noise of what we call "life." Serres helps us better locate our subjects within and between messy trajectories of Becoming.

The collection also straddles a set of wider subclaims, providing subtle undercurrents to many of the chapters. There are resonances throughout of the potential for Serres to impact both anthropology and how the discipline is situated in collaboratively tackling contemporary (and future-oriented) problems. It becomes immediately clear that Serres offers critique on some foundational anthropological assumptions on principles of exchange (Shryock), hunter-gatherer cosmologies (Jackson), hospitality (Lowe), transhuman methodologies (Corsín Jiménez, Povinelli), human-environment relations (Henig, Brown), social structure (Boylston, Nielsen), and modes of comparison (Candea, Pipyrou, Szakolczai). These refreshing new angles propose a rethinking of core texts and alternative interpretations of classic bodies of work.

Looking outward, the volume also hints at what an anthropological reading of Serres can provide the philosophy of science and other disciplines that already engage with his oeuvre, such as English literature, modern languages and culture, and media studies. This is not to say that the contributions here are overtly interdisciplinary; rather, a reader from another school will be able to identify a uniquely anthropological take on Serres that hopefully will add an innovative layer to collaborative knowledge-making. For instance, porosity has been a central theme in recent feminist STS scholarship, representing the merging of STS, feminism, and postcolonial studies. The intersectional approach advocates "fluid, porous and polyvocal" methods to better conceive of "the co-constitution of science and society" (Subramaniam et al. 2016, 407–8). The intention in feminist STS is to break away from simplified categories to complexify hegemonic histories of the present—of which Serres would surely approve—and to puncture nature/culture dichotomies through cross-fertilization (Phillips and Phillips 2010, 3). In proximity to feminist STS, medical anthropology and global health studies have used the concept of porosity as "the relational ecology between bodies and environments, departing from the ambivalent experiences of such relations, mediated by medical and digital technologies, as well as gender, race, and disability" (Iengo, Kotsila, and Nelson 2023, 76). In what have been termed "embodied ecologies," porosity represents the permeability of bodies and often violent substances of the natural and technological world (Tuana 2008; Clarke 2019). Embodied ecologies, fluidity, and the co-constitution of science and technology, in the age of a new natural contract, are debates that run in subtle ways throughout this volume. Porosity, further, emerges as a key term in recent work within the anthropology of religion, where pilgrimage shrines and sacred spaces also are understood to have such porous boundaries allowing for multiple forms of interaction, bodily comportment, and movement. These places can productively be engaged through the lens of porosity to bring out the complex and complicated negotiations within the world of devout and less devout (Coleman 2022; see also Rousseau 2016; Steil 2018). However, one could also see the very figure of the pilgrim as echoing Serres's own figures of messengers, angels, and Hermes, figures that travel and bind together a universe, a world, times, and places.

The intimacy between Serres and anthropology on display in this volume is deliberate, for there are numerous Serres "readers" available in other disciplinary contexts. In the chapters that follow, there is an "accordion effect" as authors move into and out of intimate conversation with Serres and the discipline of anthropology, scaling individual worlds and problems toward planetary and (inter)disciplinary concerns while also sometimes speaking directly to anthropological "insiders." Showcasing the concept-ethnography-Serres hyphenation provides intensity to the ultimate supernova of knowledge creation that is at once aimed at the singularity of anthropology and the wider universe of the humanities and social sciences.

Finally, the fluid scalar work done by Serres introduces the possibility of anthropology addressing issues of global standing, building on the ethnographic method to move between concepts and timescapes. Even when employing Serres as the key to unlock complex local lifeworlds, authors inevitably reference wider issues of Being and Becoming human in the twenty-first century, including accelerated globalization, pollution and climate catastrophe, global health, and decolonization. This reflects Serres's own committed worldliness (in the anthropological sense) and the way in which he dwells immersively in personal archives of experience that ultimately exceed their limits. Serres provides options for scaling contemporary problems through procedural and algorithmic thought. Anthropology can better forge new partnerships and tackle existential questions in conversation with Michel Serres.

Falling for Serres: Ethnographic Explorations

Allow us a moment to change register to consider in more detail what ethnographic analysis working in tandem with Serres might look like. Authors in this collection have had their "Serres moment": that spine-tingling encounter with a line of thought, a particular book, metaphor, or analogy that has inspired them to pursue new and often exciting intertextualities between Serres's ideas and ethnographic practice. Contributors have arrived at Serres through numerous pathways, including prior veneration of the author, post-facto reflection on existing field material, or recognizing something of the Self in Serres's entangled project. To reach a point of conversation with Serres requires patience, a sense of goodwill toward the author, and perseverance as his texts run off on seemingly unrelated tangents and take distorted twists and turns. Moreover, Serres makes puzzling and sometimes downright infuriating assumptions about the reader's prior knowledge on vast subject matters. Yet, for all the abstraction, analogies, and ducking and diving between disciplines and scales, Serres poses unique perspectives on real-world problems that anthropologists find in the fine-grained detail of ethnography. At this point, we believe it pertinent to shift tone and share our own "Serres moment" in the hope that you, the reader, may commence your own conversation.

For Knight, the story begins in the turbulent waters, or rather plains, of Greece in a localized episode of the global financial meltdown. October 2009. A date of rupture, so sudden. The onset of what became known, simply, as the "Greek crisis." One day, the distant corner of the Eastern Mediterranean seemed immune to the worldwide banking collapse, a long way from the troubles of IndyMac, Lehman Brothers, and Goldman Sachs; the next day, the prime minister announced insurmountable public debt and deficit. No money. On the ground, conversations shifted almost as rapidly, from planning for weddings, holidays, and run-of-the-mill social events to pain-filled reflections on crises past. As the months rolled by and Greece received the first of what would become three bailout loans amounting to €326 billion in exchange for implementing crippling structural economic reform measures, talk about hunger, occupation, and violent conflict grew.

When asking a friend in his forties to describe the consequences of economic crisis on his everyday life, he went straight to the Great Famine of the early 1940s to reference his fears and expectations. Kostas believed that he was witnessing a return to a time of starvation, where food was scarce and rationing commonplace. A position supported, he insisted, by the empty supermarket shelves and the queues at petrol stations as wholesalers and transport companies went bust. Without pausing for breath, Kostas skipped to stories of Ottoman landlords in the 1800s, who made the peasant sharecroppers work their fingers to the bone while seizing the produce for themselves. These foreign occupiers were what he was reliving in 2010—just look at the way businesses were being carved up into pieces and sold to international investors. Perhaps inevitably, the narrative string of occupation segued into Nazi rule during World War II and the taking of private lands for the German war machine. What was enforced in the 1940s by military means was now accomplished through economics. Still responding to Knight's initial prompt, Kostas sprang forward to the stock market crash of the late 1990s, when speculative investment became a national obsession. People from all walks of life would buy and sell on unregulated markets, seemingly endless in their prosperity until . . . crash! Lost homes, ravished honor, bankruptcy.

Occupations, periods of hunger and strife, the 1967–74 dictatorship (when his family was displaced to a remote island owing to perceived links to Communism), all these events, Kostas mused, were happening again, all at once. Through what Knight (2017, 37) has called "bouncing around" through the past, Kostas knitted together disparate moments that made his experience of drastic social change make sense. As well as expressing his fears, these excursions also revealed how crisis could be overcome, providing a form of comfort and a sense of futural trajectory. Temporal leaps collapsed the time of Ottoman landlords with the Great Famine, stock market crash, and dictatorship. Each event was a mashup of personal experiences, intergenerational narratives, and nationalized accounts of the past, often hinged on key figures in family history or education textbooks, the messengers bridging time and space. Pressed together, the unsequenced modules of history provided meaning to a life in turmoil. It left the anthropologist feeling seasick.

The embodiment of an assemblage of past crises would quickly become a recurring theme among Knight's research participants in the early years of austerity in Greece. Engrained in this dizzying back-and-forth were accounts of people feeling they were falling through time, expressing disorientation as to where the future lay, and the idea that material objects and artifacts were oozing uncanny history and affect. Childhood photographs from the Civil War (1946–49) years, a dress hand-sewn by a young mother in exile during the dictatorship, a photovoltaic panel indicative of an EU economic recovery plan, flipping a €1 coin between the fingers in a trouser pocket, all transported people on nauseating journeys, almost tearing them at the seams as they searched for a temporal home. "*When* are we?" Kostas rhetorically asked, referring to broken promises of limitless prosperity, modernity, and westernization that had been ruptured and replaced by a sense of spiraling freefall into poverty, peasantry, and existential quandary. How to capture the turbulence of this ethnographic mess?

As the early evening was drawing in one crisp winter's day in 2010, Knight and his partner in crime, Stavroula Pipyrou, found themselves deep in the bowels of Durham University Library preoccupied with some mundane tasks geared toward finishing their doctoral theses. Pipyrou came across a discarded book strewn across a vacated desk. Noticing the title to include the word "time"—the core analytic of Knight's work—she decided to take a look before excitedly and with a knowing smile handing it over. Leaning against the adjacent bookshelf, Knight opened the book to a random page. His world would never be the same, for this was Michel Serres's *Conversations on Science, Culture, and Time* (with Latour [1992] 1995). In analogies of rivers, glaciers, and the weather, Serres critiques how the assumption that time is linear is merely a construction of the post-Enlightenment West that distorts our understanding of events as proximate or distant (Serres and Latour [1992] 1995, 57).[3] He explains through an example of the crumpled handkerchief: "If you take a handkerchief and spread it out in order to iron it, you can see in it certain fixed distances and proximities . . . Then take the same handkerchief and crumple it, by putting it in your pocket. Two distant points are suddenly close, even superimposed. If, further, you tear it in certain places, two points that were close can become very distant" (Serres and Latour [1992] 1995, 60).

The idea of polytemporality—that we are constantly "drawing from the obsolete, the contemporary, and the futuristic" (Bennett and Connolly 2011, 159)—was Knight's "ah-ha" moment where his ethnography made sense. Disparate pasts and futures can become superimposed, be relived, and remain contemporary by way of assemblage in the present. The notion of percolating time—"time doesn't pass, it percolates. This means that it passes and it doesn't pass." It filters, "one flux passes through, while another does not" (Serres and Latour [1992] 1995, 58)—would become the framework for Knight's thesis and his first published paper on "proximity" (2012) and monograph on topological time (2015). The Ottoman landlords, the Great Famine, and the stock market crash were caught in the filtration process and remained socially meaningful as the ebbs and flows, surges and lulls of time and events were woven together in the context of contemporary economic crisis. They remained proximate to lives caught in crisis, connected by a hyphen to each other as building blocks of time and history, but not confined to their own static coordinates.

Perhaps most poignantly, Serres seemed to be echoing the Greek research participants when describing the affective past of his own childhood in a time of "hunger and rationing, death and bombings" (Serres and Latour [1992] 1995, 2). His experience of this "vital environment" was a cumulation of global history, family narratives, and his own memories of the Spanish Civil War, the blitzkrieg, concentration camps, reprisal attacks after the liberation of France, and, at the age of fifteen, the bombing of Hiroshima. Scaling the influences of global events on individuals and collectivities not obviously connected was precisely what Knight's informants were doing. Commenting on the affective struggle of placing oneself in the violent present of a ruptured world, Serres recounts how he cannot look at Pablo Picasso's *Guernica* owing to its association with the Spanish war of 1936. When he encounters such pictures, he physically feels history seeping from them, as witnesses to terrible events.

> I have never recovered—I don't believe I'll ever recover.... Now that I am older, I am still hungry with the same famine, I still hear the same sirens; I would feel sick at the same violence, to my dying day. Near the midpoint of this century (1900s) my generation was born into the worst tragedies of history, without being able to act.... Even my own childhood photographs, happily scarce, are things I can't bear to look at. They are lucky, those who are nostalgic about their youth ... We suffocated in an unbreathable air heavy with misfortune, violence and crime, defeat and humiliation, guilt ... (such events as) the death camps were echoed by Nagasaki and Hiroshima, which were just as destructive of history and conscience—in both cases in a radical way, by attacking the very roots of what makes us human—tearing apart not just historic time but the time frame of human evolution. (Serres and Latour [1992] 1995, 4)

From that day in Durham University Library, Michel Serres became an *as if* informant, a member of Knight's research team. Feeling hungry with the same famine, not bearing to look at photographs of his childhood, deploying angels and mythical figures to make connections across the normal analytical confines of spacetime, Serres's practically scale-free thought captured the messy realities of fieldwork. This is what people were experiencing, trying to communicate. Since their first encounter, one of the most valuable methods for thinking about the multi-scalar consequences of social change has, for Knight, been by bouncing ethnographic material off Serres.

Over the years, Serres has become a muse, influencing Knight's thinking on the ethnographic field as an arena of noise where we navigate chaos while trying to pay attention to the miniscule detail that provides the foundations for grand narratives. In *Vertiginous Life* (2021), Knight draws on Serres's ideas on turbulence, vortices, and elsewhens—as discussed in *Conversations* as well *Variations on the Body* ([1999] 2012) and *Eyes* (2015a)—alongside the work of Roger Caillois, Søren Kierkegaard, Ernesto de Martino, and Eelco Runia.[4] In an era of social rupture, people teeter on the cliff edge of time and history, experiencing intense temporal disorientation across different scales, aesthetics, and materialities. Knight's argument is that certain occasions, situations, and events are vertiginous, laced with a sense of hyperconsciousness, stuckedness, or constant shuddering movement. There is a zooming in and out from the Self and society where people experience nausea, dizziness, the sense of falling, dissociation with former Being, déjà vu, palpitations, and breathlessness. Lifeworlds are sent careening. As apparent when thinking through the vertiginous, Serres's openness to fluid scales of space and time, alongside his focus on surges and turbulence, lends itself to capturing atmospheres beyond the narrative form.

Serres may help anthropology embrace the incomprehensible, assist in the search for understanding by way of scaling nature and culture, pasts and futures, individual experience, and popular rhetoric. Adrift in interpretations and translations, as anthropologists we deliver messages of our craft, much like Serres's angels, working in tandem with message-laden informants and collaborators to listen to the surges in noise from our field. Serres's writing on nonlinear time, events, and turbulence has helped Knight better comprehend his ethnographic material, but that is not to say that their relationship has been without its hiccups. Trouble brews, for instance, in Serres's readiness to quote Christian scripture at the drop of a hat and often without context or explanation. Particularly in his later work, Serres's brashness in addressing sensitive topics of pollution and erotic desire can be received as distasteful or flippant. On occasion, he seems to come close to endorsing cultural evolution. His central characters are often drawn from Western European notions of civilization and the classics, which begs the question of bodies beyond the Occident. And yet those "Serres moments" open hidden, often vociferous pathways for ethnographic analysis by drawing together seemingly unrelated avenues of enquiry to sprout branches of potentiality that break the shackles of conventional anthropological format.

Enrolled in a course on modern French philosophy, in 2002 Bandak was seeking to engage more seriously with trends in continental thought frequently referred to in his anthropology program. During the course, the class slowly read alongside selected thinkers including Henri Bergson, Jean-Paul Sartre, Maurice Merleau-Ponty, Jacques Derrida, Michel Foucault, and Gilles Deleuze. Each lecture was spent working through a short paragraph from a modern master. The course explored duration, Being and nothingness, perception, and subjectivity as well as difference and repetition. Deleuze was an important companion at this point; Serres, however, was not mentioned. On the side, Bandak got hold of a Danish translation of *Genesis*, which he, much like Knight, stumbled on, this time in the basement of a well-stocked Copenhagen bookstore, which today, alas, has closed.

Genesis was captivating. The prose was at once lucid, poetic, and enigmatic: bifurcations, violence, time, dance, multiplicity, and human potentiality. Serres effortlessly crossed vast territories. A couple of lines in particular stayed with Bandak: "Background noise is the first object of metaphysics, the noise of the crowd is the first object of anthropology. The background noise made by the crowd is the first object of history" ([1982] 1995, 54). Succinctly, Serres seized how diverse disciplines work from noise toward relative degrees of order. *Genesis* also entertained a conversation with the biblical narrative of creation, with Serres's emphasis on chaos, turbulence, and fury. The themes breached by Serres allowed Bandak to appreciate the complex fluctuations of time, history, and Becoming in his own work—a Becoming that is multiple, creative, and often marked by violence and chaos.

A signature course convened by Michael Jackson on existential-phenomenological anthropology in 2003 was influential in connecting the French philosophy in which Bandak had just been submerged with a solid anthropological grounding. While Serres did not make an appearance on the official reading list, there was some reference in Jackson's published material, leading to very rewarding conversations during lecture intervals. In *The Politics of Storytelling* (2002), Jackson mentions Serres's *The Natural Contract* ([1990] 1995) and the coinage of the term "epistemodicé," which Serres sees as replacing the theodicé of former times during the Enlightenment period. Bandak was encouraged to think about hierarchies and axioms of knowledge, which were enhanced through regular exchanges with Jackson. In particular, he was taken with the problem of knowledge that runs throughout Serres's body of work— in *The Natural Contract*, it is how we come to know and act in relation to the climate crisis. The general inspiration from Gottfried Leibniz is evident, and the problem of knowledge directly relates to human reasoning, judgment, and the ability to act. However, where Leibniz grappled with the question of God's good vis-à-vis the problem of evil, Serres deliberately moves the question into the domain of human knowledge, believing there to be no authority to pass judgment on good and evil ([1990] 1995, 23) and even construing it as a social problem not of attributing responsibility to God but rather of claiming that "society does not know what it does" (Serres [2019] 2022, 87).

We here touch on some classic debates in anthropology. Take E. E. Evans-Pritchard's exemplary case of the collapse of a granary in his *Witchcraft, Oracles, and Magic among the Azande* ([1937] 1976), where all sorts of explanations are sought by the bereaved parties. According to Evans-Pritchard, the questions revolve not around the collapse of the granary per se as there are plenty of possible explanations, such as termites that eat away the wooden base. The problem is, rather, why did the granary collapse when there was someone sitting underneath it? Addressing the complicated question of suffering, which also preoccupied Clifford Geertz in *The Interpretation of Cultures* (1973), Serres explicates, in similar terms to Evans-Pritchard, that the theodicé may be experienced differently outside a Christian tradition depending on the hierarchical prioritization of axiomatic knowledge (cf. Geertz 1973, 106, 172). The basic conundrum persists: What are we to make of such situations where our knowledge explains but also inevitably falls short?

Serres does not provide a direct solution to this fundamental dilemma, but he does open the floor to creative explorations of knowledge and epistemology, even admitting that the term "epistemodicé" is somewhat ugly.[5] For Serres, the human capacity to think is hampered by our inability to act: a central problem in the context of the climate crisis. By framing this issue as an epistemodicé, Serres plays on the contested role, or awkward relationship, of knowledge and reason when a critical problem of planetary scale is collectively encountered by humans ([1990] 1995, 23). He pushes our understanding of how knowledge, ignorance, and judgment are entwined in the human condition and the various ways we ethnographically locate willed and unwilled ignorance (Dilley 2007; High, Kelly, and Mair 2012; Bandak 2013, 2022).

Beyond Serres's grappling with the epistemodicé, which has informed Bandak's ethnographic musings on (lack of) knowledge, Serres has also put forward the notion of a "black box" of knowledge ([1982] 1995, 5; Serres and Latour [1992] 1995, 86): a device where one may precisely specify the input on one side and equally surely describe the output on the other but be unable to detail what happens inside:

> To its left, or before it, there is the world. To its right, or after it, travelling along certain circuits, there is what we call information. The energy of things goes in: disturbances of the air, shocks and vibrations, heat, alcohol or ether salts, photons. . . . Information comes out, and even meaning. We do not always know where this box is located, nor how it alters what flows through it, nor which Sirens, Muses or Bacchantes are at work inside; it remains closed to us. However, we can say with certainty that beyond this threshold, both of ignorance and perception, energies are exchanged, on their usual scale, at the levels of the world, the group and cellular biochemistry; and that on the other side of this same threshold information appears: signals, figures, languages, meaning. Before the box, the hard; after it, the soft. ([1985] 2015, 129)

Knowledge is only ever partial, based on observation of inputs and outputs, but never the whole process—something most ethnographers, not least Geertz, would sign up to. We can never see or comprehend the entire system of transformation and must, to a degree, accept that knowledge undergoes complicated compressions, manipulations, and changes of states (what Serres often puts in terms of liquid, solid, and gas ([1985] 2015; [1990] 1995) as it gets knocked around in tension between nature and culture, the hard sciences, and the humanities. Each discipline, culture, and epistemology will produce different multiplicities of knowledge depending on the contents of its own black box.[6] Thus, with knowledge there is always ignorance, and a simple event—such as the collapse of a granary—may be filtered through multiple receptacles of interpretative transformation.

We can, however, pick up on the archaisms, repetitions, and traces of knowledge across time and history, and this is significant for Serres's notion of ichnography ([1982] 1995, 19-20; [1983] 2015, 20-21). The ebbs and flows, lulls and surges, of background noise can be understood through Serres's metaphors of crumpled handkerchiefs and the hidden currents of seemingly placid rivers, as discussed previously, but also in how radioactive fallout dates its objects since the Big Bang. We may recognize knowledge, perhaps even place it in spacetime, although a repetition will not identify as an exact copy owing to the turbulences of nature and culture battering its format across history. All of which predicates epistemologies as multiple, not singular, not at peace, but constantly warring and uneasy.

For Bandak, Serres's critiques of knowledge production and consumption have in important ways molded his conceptual positioning on how we strive toward analysis, particularly in the porosity and possibility of rereading and reweaving (Bandak and Kuzmanovic 2014; Bandak and Coleman 2019). Similarly, Serres has been influential in advocating hearing as a model of and for understanding ([1982] 1995, 61; [1985] 2015). In consonance with recent trends in anthropological research, senses beyond solely vision have gathered importance in the ethnographer not being confined to one register of knowing (Erlmann 2004; Hirschkind 2006; Larkin 2008). Along these lines, Bandak has forged a conversation between the material practices of Syrian Christians and their ways of inhabiting one of the oldest cities in the world, Damascus, and the sonorities of place in the time leading up to the uprising (Bandak 2014a, 2014b). This has led to the opening of entangled temporal registers coexisting alongside the material and sensorial (Bandak and Kaur Janeja 2018).

Twenty years on from their first encounter, Bandak still has plenty of conversations with Serres on displacement, memories and documentation, violence, noise, and the possibilities of retying fragmented traditions. One idea Bandak is currently contemplating is the knot, or complex, as formulated by Serres in *The Troubadour of Knowledge*. The knot, as well as the fold, gives texture to multiplicity; multiplicity does not point to a location or a place, but the knot does ([1991] 1997, 24). The complex understood as a knot designates a group of folds that bespeaks a topological rather than arithmetic archive of knowledge, lending itself to the analysis, or perhaps rather synthesis, of complicated systems such as social relations, exchange networks, and cohabitation. In the manuscript he submitted the day before he passed away on June 1, 2019, which came out later that year under the title *Relire le relié* (2019, translated to *Religion* in 2022), Serres again explores bonds and bindings, this time of religion. Religion, understood as a rereading and retying of bonds, amounts to a space of possibility but also of almost inconceivable upheaval. This seems pertinent in the context of the Syrian war. As such, violence and unfathomable destruction are inherent features of Serres's thinking—and these features did, in his coining, begin Serres's constant grappling with the human limit case of Hiroshima ([2019] 2022, 33). But Serres does not end with violence, rather speculating on questions of grace, saintliness, and indeed peace, ending with "a fervent hope that a way forward may yet be found" ([2019] 2022, 191).

Through doctorates and promotions, marriages, and births, Serres has been a traveling companion: a muse, an as-if informant, and often a problem-solver who has reached a point of veneration through processes of deep reflection on ethnographic material and a deal of self-identification. Put bluntly, both Knight and Bandak find that Serres has the tendency to offer direction that at once is obvious and revelatory. Throughout his body of work, it is the porosity of concepts, the links between disciplines and epistemologies, that strikes us. Serres will not be caged. Boundaries are there to be breached, be they the classic format of historiography, instruments of the body, measurements of physical spacetime, or axiomatic ways of knowing. Serres does not rubbish the concepts, but, rather like a Dadaist, he deconstructs them, throws them in the air, and then takes a running jump at finding an absurd or ironic angle to knit the pieces back together—not in original form and always with melted and melded topological morphologies. He identifies what Michael Carrithers (2012), following Kenneth Burke (1969), termed "subcertainties": "a redirection of perspective that (helps) gain a more detailed ethnographic picture . . . Instead of introducing endless fragmentation, subcertainties bring about concentration and strengthening of a perspective precisely because they highlight its fragility" (Pipyrou 2014, 535). For anthropologists, Serres is asking us to ponder the complex compressions, distortions, and vortexing that might be going on inside that black box.

What also strikes us is that despite his reputation for abstraction and wildly meandering prose, Serres sees the world in staggeringly similar ways to our interlocutors. Their lives and problems are not confined by the methods of a single discipline, neither do they experience the world through bounded concepts or containers of format—as with Greeks who bounce around through time weaving together culturally proximate events to provide an assemblage of meaning to their present. People live through the messages of "angels," whatever their guise, delivered from disparate points in space and time. Particularly, metaphor and analogy form significant aspects of our informants' lifeworlds as direct vessels of knowledge that help in the comprehension of the immediate social milieu. In many ways, the people we work with will have had their "Serres moment," although few, if any, would term it such.

Hyphens and Analogies: Serres's Method of Connections

Shimmering with billions of glorious and timid suns, night resembles Verne's cavern with its dazzling gems and innumerable truths linked together by a thousand related networks. This is where thought sparkles, as softly as flowering pearls. More visible and beautiful than the day and peaceful in any case, the night knows while the day pronounces. Stars shiver as they look while the sun's formidable lucidity blinds us. . . . Like any hunting animal, knowledge has night vision. (Serres 2015a, 22)

Metaphors, analogies, anecdotes, and hyphenated connections form the backbone of Serresian method and provide the mesh for how his philosophy of science transitions toward ethnographic enquiry. His detailed interrogation of the local in considering the global; his entwining of personal, cultural, and historical narratives in stories of Becoming; and his frequent application of anecdotes and analogies mirror the practices of people anthropologists encounter in the field. In Serres's own words, he wishes to describe "a general theory of relations" (Serres and Latour [1992] 1995, 127) that showcases multiplicity and diversity. This task cannot be paradigmatic; at all costs, Serres wants to avoid "umbilical thinking," that is, feeding all ideas through a single line, thus reducing all truths and discourses to a single dogmatic point (Watkin 2020, 38).[7] An umbilical approach to knowledge elevates one model or case to the status of a paradigm that explains the whole (Watkin 2020, 48). Instead, Serres champions an algorithmic method to harness the multiplicity of truths, where each individual context is considered on its own terms while being related to a vast web of others. As with the gems in Verne's cavern, each truth sparkles in its own way, has its own luminosity, yet shines still brighter when networked to a thousand others.

SCALING

This algorithmic approach allows Serres to connect the individual and the planetary, local knowledge and universal problems, in a way that is not axiomatic, determinate, or abstract. In a manner that seems to echo a seminar in Sociocultural Anthropology 101, Serres advocates a "procedural" step-by-step approach where we walk alongside our object rather than descending from on high. The universal and planetary should be addressed bottom-up by respecting all the possible combinations of all modes of thought fragmented in a thousand pieces, not in the polemic of truths and falsehoods. Destabilizing the dichotomy between the individual and universal, procedural thought "restores dignity to the knowledge of description as well as of the individual" (Serres 2012b, 43). Building up to the global through embracing the multiplicity of knowledge at the individual level captures Serres's philosophy of scale: "There are no concepts; there are examples and events, that is all" (Serres, Legros, and Ortoli 2016, 84, in Watkin 2020, 83). Attention is paid to each fragment on its own terms, drawing out often unexpected or disruptive connections through analogy and metaphor, providing a trampoline-like web to springboard toward the contemplation of global conditions.

In an outstanding introduction to Serres's method, Christopher Watkin (2020) explains procedural thought through the example of the dictionary, which is delivered as the definitive authority on language through hundreds of thousands of small entries, all unique but connected: "Procedural thought . . . is free of determinate content, and it prescribes operations, not magnitudes. The algorithmic order of the dictionary is practical, conventional and plural; the order of the declarative text is unitary, organised according to 'temporal succession, announcement, suspense, movements of induction and deduction, the confrontation of dialogue' (Serres and Farouki 1998, xii–xiii). The procedural text shows what it is possible to say, without saying anything in particular; the declarative text leaps from the local to the global by universalising its own approach in an umbilical gesture" (Watkin 2020, 82).

Again, we may think of the harlequin, peeling back layer upon layer of multicolored diversity yet maintaining that "everywhere everything is the same." Serres asks, "How can the thousand hues of an odd medley of colors be reduced to their white summation"? (Serres [1991] 1997, xvii). The harlequin is the personification of layered diversity yet claims universal uniformity.

Watkin (2020) explains: "Serres' point is a chromatic one: blank, universal white is not composed of an absence of colour but of all local, determinate colours; the universal and global are arrived at not by jumping out of the local in a puff of abstraction, but by multiplying local instances and seeking carefully to relate them to each other" (41). The global is thus an ensemble of local relations.

For Serres, the connections between the fragments of knowledge are of utmost import. Borrowing from Philippe Descola's (2005) *Beyond Nature and Culture*, Serres describes himself as "analogistic," searching for partial correspondence or similarity across disciplines, thinkers, cultures, and histories. This is where his figures of Hermes and angels do the relational work, drawing ideas together by way of topological connection (Boylston, this volume). "Thousands and thousands of relations" bridge difference; analogism approaches the world as disparates awaiting relations, a web always being spun rather than a fixed umbilical cord (Watkin 2020, 108). Analogism operates to create relationships between seemingly distinct and distant authors, objects, and theories. Jane Bennett and William Connolly (2011, 165) observe how, in *Conversations*, Serres explains being suddenly struck by "an uncanny resemblance between what Archimedes was saying about fluids and what Lucretius was saying about Athens: both writers told of a 'vortical' structure of generativity at work. Serres noted this resemblance and took it as a call to place these two thinkers into dissonant conjunction, 'to explicate, that is, "to unpleat" the fold that they seem to be sharing.'" Bennett and Connolly astutely pin the method of analogy to Serres's crumpled handkerchief where the noise of the past surges forward to form connections to other spacetimes: "a world of noise, with its inherent tendency towards repetition and redundancy, is a world of fractal similitudes" (2011, 165).

There is no single dominating discipline or truth, no umbilical discourse that can exhaustively describe reality but, rather, Serres insists, a cavern of twinkling truths, connected in untold ways—their "subcertainties" in Carrithers's parlance—not blinded by the light of one pure reality. It is to different forms of connection that we now turn.

CONNECTING

A core trope to approach connection and bridge vastly different bodies of knowledge is the hyphen, and it is here that we wish to dwell a little longer. Whereas bifurcation and branches lead to a focus on the ways ideas and disciplines split and are moving in different directions, the hyphen allows consideration of how dissimilar things are brought together. Connection here differs subtly but importantly from comparison. For anthropologists, this may at first seem provocative given their disciplinary training in *comparative* work (Fabian 1983). The anthropological legacy stems from the "armchair" where our forebearers were keen to measure and relate differences comparatively. Ethnography has, to a large extent, rested on unraveling the specificity of a given locale in relation to its Other. Comparison has been formative for the anthropological project but has also increasingly been critiqued (Abu-Lughod 1993). The impossibility of the comparative method, as recently argued by Matei Candea (2019), does not mean that we are to discard it; however, a reflexive rethinking of comparison is indeed required.

Serres's attention to connection rather than comparison is revealing as a means to draw distinct phenomena into the same orbit. His analogies, metaphors, and allegories cross time and space as well as disciplinary traditions and open pathways beyond the realm of direct comparison. Connection operates not by relating kind to kind or measuring degrees of relative similarity or differentiation but rather by having diverse forms of material talk to each other. If anything, such a methodological focus on connection allows for the coevalness that Fabian found lacking in much classical anthropological theory. Serres possesses a fluid ability to engage characters as diverse as Lucretius and Leibniz and to think with them, allowing them to work through our problems with us, in collaboration. There is the same attention to fluid connection in Serres's thematic choices, ranging from mathematics to art, religion to the senses. As such, Serres advocates against ownership of knowledge, either by specific individuals or as located in singular academic fields (umbilical thinking). Serres epitomizes the bold but also increasingly necessary effort to bring things, phenomena, worlds even, together. This also reveals a fundamental porosity in Serres's endeavor; as our informants transcend concepts, disciplinary methods, the regulations of book-bound history, and scientific epistemologies, so must we as analysts seek equally dizzying routes to further connections.

Porosity is then, first, how scholars move across and between concepts in building their analysis algorithmically. There is always an instinctive correspondence between events or entities otherwise assigned to different categories, such as physics and poetry, mathematics and anthropology, ancient and modern. The work of connections through time and space is articulated in Serres's preoccupation with mediating figures, such as the Troubadour of Knowledge ([1991] 1997) or Hermes (1982b) as messengers working toward bringing together unrelated realms. These figures traverse landscapes of incommensurability, dashing between the sparkling gems of the cavern wall, to form audacious new bonds, provide novel insights, and approach life from unusual angles. Similarly, angels bind universes together, crossing between domains of the human and the divine as messengers. By soliciting these figures, Serres inscribes motion, dynamism, and circulation directly as models of and for thought. Hermes, angels, and the Troubadour are ciphers of movement rather than stasis, and they are harbingers of new possibilities, their very presence bringing symbiotic novelty to each landscape they traverse. The messengers bring detail to the individual cases as part of the base process that facilitates algorithmic trajectories toward grander levels of analysis.

Serres thus straddles territories not in a systematic effort to chart everything but to hyphenate what otherwise would have been unrelated or gone unnoticed. Travel and exploration are central to Serres's imagination, exemplified by his focus on Jules Verne (2003; also 2015a). To Serres, the literary experimentations of Verne bring to life not only forms of travel before they were invented—as with Captain Nemo and Nautilus—but also the very quest for invention itself. Verne, like Serres, has the audacity to traverse spacetime as author, with his characters, and through daring scholarly connections. Serres hyphenates realms of science and literature while dancing with figures of thought, bodies of knowledge, geometrical patterns, rhythmic propensities, and the noise of nature.

Mixing, intermingling, and contact form substrata of Serres's method. Take the figure of the hermaphrodite (1987) or, again, the harlequin ([1991] 1997), who attends to the multiplicities of Becoming. In *The Troubadour of Knowledge*, Serres writes, "I wander in the world and the back worlds, in bold abstraction, landscapes, cultures and languages, social castes . . . my soul exposed in learning things, just as it ventured onto the slope of glaciers and still remains there. To open the door, to pierce the partition, is ultimately to expose oneself to death. A life of experiences forges the passage, short or long, sterile or fruitful, from nothingness to death, while passing through indefinitely dilated joy" ([1991] 1997, 32).

In this one passage, Serres explicitly links porosity—"to pierce the partition" of diverse realms—with fearlessness and death. If people take the decisive and difficult leap through "the narrow door that gives access to civilization and history," to progress and to a future, Roger Caillois (2001, 141) warns, then this "basis for collective existence" can lead to a life of captivity inside the whirlpool, clawing away at body and mind, from which there is no escape. The door to create novel connections is laced with vertiginous danger, potentially death, a theme Serres picks up (with reference to Caillois) in *Variations on the Body* ([1999] 2012). In "wandering through worlds" of daring connection, one may incur anger, put alliances on the line, invite scandal, or even run the risk of death (Foucault 2001; Pipyrou 2016, 7–8; also Pipyrou, this volume).

Serres's figures, from Lucretius and Plato to the harlequin and angels, participate in vast networks of information and communication superhighways.[8] Thinking with Serres through hyphens, metaphors, analogies, and allegories allows us to bring disparate phenomena within the same constellation, bridging and exploring provocative realms. Serres works by way of conjunctions in a discourse of topological ties, ligaments, ligatures "in which knowledge grows not through interminable analysis but through overlapping strands casting shadows on each other" (Watkin 2020, 74).

Conclusion: Porous Becoming

Serres invites his readers on a journey where porosity and connection provide potential answers to scalar problems. It may not be an easy ride, but wonders await in Verne's cavern of gems. Porous becoming on a journey with Michel Serres is twofold. First, a scholar must embrace topological movement across and between hegemonic concepts and bounded formats, identifying relations between disparate and seemingly isolated thinkers, objects, and spacetimes. Second, Serres himself almost inevitably becomes an as-if informant, a muse not just for theorization but for ethnographic enquiry. The people we work with experience life on multiple scales, at various speeds, and by tapping into vaguely connected containers of knowledge. They traverse ethical registers and involve figures of mythistory in delivering messages along their personal and shared pathways of Becoming. Bandak and Knight both had their "Serres moment," quite by chance. Ever since, Serres has become part of their journey, being a scholarly companion pointing toward avenues of potential analysis, reminding them of the tribulations of conformity. He is also an as-if informant. His childhood memories, the politics of his time, his schooling in the natural sciences, his stories of navigating waterways as a bargeman and naval officer, the underlying feeling of enthusiastic allegiance to French culture and pastimes, his passion for art and poetry, his at times open animosity toward Descartes and Plato, the influence of his ancestors known and distant ... Serres himself acts as a hyphen, a conjunction where he is at once an interlocutor and an analyst.

From ethnographic endeavors, ideas surge forth with unpredictable and entangled similarities between anthropological interpretations, concepts developed independently by scholars in seemingly incomparable fields, and inspirations drawn from scattered regions of space and time. Without initially knowing it, in their own locales Bandak and Knight were working alongside and across Serres while following the lead of their ethnographic interlocutors—porous knowledge and violent displacement among Syrian Christians and contorted reinterpretations of the past at times of crisis in Greece. The power of ethnography is to get to the point by itself; the anthropologist is then obliged to make what they will of knowledge by way of multidimensional comparison and analysis, to build a bigger picture of the human. As Serres became contemporary with Lucretius, Leibniz, and Verne, collaboratively passing messages and making connections, so Serres has become our contemporary, our colleague, our companion on this crazy ride to make sense of the world.

For Knight, one striking example of the unwitting entwinement of thought lies in his recent conceptualization of vertiginous life. Drawn from twenty years of field research in Greece pre- and post-economic crisis, Knight started from daily narratives of nausea and dizziness of life experienced in the whirlpool of painful pasts and shattered futures to form a theory of social vertigo. Defined in terms of the *ilinx*—from the Greek for "dizziness" or "whirlpool"—it seemed the obvious direction to build from striking ethnography toward novel social theory. On completing the manuscript, constructed on individual lifestories laced with strands of Sartre and Kierkegaard, it became apparent to Knight that someone had been down the vertiginous avenue before—namely, Michel Serres. In *Variations on the Body*, Serres not only discusses social vertigo but also frames it as *ilinx*. In two of his most recent books, *Religion* ([2019] 2022) and *Eyes* (2015a), Serres also talks of vertigo and seasickness in relation to the material and sensory environment. Far from coincidentally citing commonly referenced passages—such as the handkerchief or the parasite—Knight and Serres had independently interpreted unique sets of sociohistorical material through braided topologies of knowledge to come to the same niche concept, somehow connected. Local messages had been scaled to collective truths along eerily similar trajectories.

The contributors to this volume have come to Serres by pathways of self-recognition and post-facto reflection, each on their own journey. Some are deeply versed in Serres and have long identified the need to provide anthropologically informed addendums to his core theories of contracts and planetary cohabitation (Corsín Jiménez, Povinelli), pollution and safety (Brown), and mythistories (Jackson). Others had encountered Serres only in passing, typically through some of his more popular ideas, and have taken this opportunity to become better acquainted with the ways his work might offer new trajectories to their own material (Lowe, Henig). In some chapters, concepts core to the anthropological discipline are taken through the Serresian percolator, including rethinking Mauss and Derrida on hospitality (Shryock), a comparative assessment of Serres and Gregory Bateson's perspectives on the unity of mind and nature (Szakolczai), a retelling of Lévi-Straussian myths in New York comedy clubs (Nielsen), and a turn toward wisdom and the practice of comparison (Candea). Finally, other authors offer a more methodological angle on how anthropology is predicated on messages, with parallels between ethnographic situations we encounter, the structure of storytelling, and our proximity with figures that surge and fade in our texts (Boylston, Pipyrou). Such a variety of approaches plays into the insider-outsider feel that pulls chapters at their seams, offering various degrees of insider knowledge on intimate relationships with Serres and simultaneously peering both deeply into and beyond the borders of anthropology. The afterword with Jane Bennett is presented in Serres's favorite format of knowledge, a conversation, and may, we suggest, provide the reader a lucid "in" to Serres's work worthy of consultation straight after reading this introduction.

Porosity, in its many forms, has been *the* "in" for our authors in the massive Serres oeuvre. The concept is engaged by all contributors, and it is on *Porous Becomings* that the collection rumples together. Each of the three sections corresponds to a particular "in" to Serres's work. *The Parasite* and *The Natural Contract* are two key works that readers are likely to have encountered or wish to take further; the first four chapters deal directly with this material. The second section bundles chapters on the body and temporality, two further "key concepts" in Serres's work through books like *Conversations*, *The Five Senses*, and *Variations on the Body*. Although, arguably and inevitably, links to similar Serresian concepts expand throughout the volume, we simply offer potential avenues for our readers to delve into Serres. Section 3 is explicitly about methods of proximity and connection that run throughout all Serres's work, with chapters clearly addressing connection as method across a wide body of Serres's publications. We offer these knots of thematic conversations while advising the reader to be audacious in their own adventure, tying together chapters of their choosing.

The beauty of ethnography is there is no "right way," no predetermined boundaries of interpretation. So too with Serres, whose refusal to impose borders on disciplines, thoughts, and spacetimes allows for the porous traversing of channels without shutting down conversations along the lines of arbitrary categories. Rather than descending into unproductive chaos, such porosity facilitates novel connections, providing freedom to find pathways through figures of thought, be they our interlocutors, other scholars, or literary characters. The destination of the texts in this volume may ultimately be similar, but their trajectories of Becoming, the assemblages of knotty connections they engender, is invariably very different.

The ethnographic portraits in the current volume influence how the authors engage with Serres as a figure of thought, allowing them to tend to the social situation at hand. Serres provides a rich bank of ideas from which he asks us to withdraw and make our own connections and encourages us to invite our own friends to the party. But for all the adventure and enterprise that porosity gestures, Serres warns that only the courageous and open-minded need apply—he is skeptical about how conversations often lead only to more conversations, frequently reproducing rather than quelling violence (Serres and Latour [1992] 1995). Overfamiliarity with conversation for conversation's sake leads to the constraint of thought in "a context rigid with possibilities" (Serres and Latour [1992] 1995, 43). The audacious step beyond the comforts of disciplinary and cultural dogma is for the fearless. Anthropology must embrace the ride, the raucousness of porous connection, to bring novel socio-technical assemblages to the collective table. This means breaking from canonical thinking and forging new symbiotic relations with other disciplines, figures, and our collaborators in the field.

Having the boldness to embrace means rescaling the ambition of anthropology, even if some cannot see past the "how dare they?" audacity of the endeavor. But this is a call not to blatantly disregard concepts, genealogies, and belief systems but to hyphenate, not to belittle the power of analogy, and to realize that the relatively small-scale detail of ethnography forms the algorithmic basis for tackling large-scale issues through connection rather than dichotomy. Anthropology has the potential and the power to invent, to innovate, to inform, to pierce the rigidity of formatted knowledge, to be audacious. To deliver on this potential, we must act on the obligation to find new ways to symbiotically cross disciplines and scales. Indeed, symbiosis is so often the way forward for Serres. Responding to Latour's question on managing evil while being "emersed in [its] atmosphere," Serres accepts the invitation to consider the violence of the world as a springboard for action. Citing *The Parasite*, Serres ponders, "What is an enemy? . . . Something to expel, excise, reject, or something we negotiate a contract of symbiosis?" (Serres and Latour [1992] 1995, 194–95). For questions to have answers, a debate needs to produce micro-contracts of understanding rather than more cycles of violence. This also seems to be the call of anthropology: to understand the Other, no matter how divergent, perplexing, or potentially repugnant. A series of hyphenated collaborations engaging with the atomic modules of vital questions that make up all scales of the contemporary will help build a grander picture of collective futures. This is something that anthropology could and should contribute to. To break free of the event horizon, to negotiate the vertiginous edge, to navigate through that sticky dark matter, to find trajectory once again, require courage and porosity. As with Serres's organic lifeform learning from a critical event, it is symbiotic novelty or death. We invite you to the conversation.

NOTES

1 See in comparison Jane Bennett's engagement with Walt Whitman in *Influx and Efflux* (2020).

2 This resonant *something* that points toward affects and energies beyond the narrative is the subject of Susan Lepselter's (2016) work on the American uncanny and Daniel M. Knight's (2021) hypothesis on the vertiginous aesthetic.

3 For discussions on the social implications of the handkerchief metaphor, see Knight (2015, 6), and for an ontological take, see Bennett and Connolly (2011).

4 In *Eyes* (2015a, 35), Serres describes his vertigo of being transported to the ancient caves of Lascaux via virtual reality goggles. Not only is the feeling of physical movement while standing still disorienting, the experience of viewing prehistoric cave art through a futuristic technological interface scrambles "ways of seeing."

5 Hierarchies of epistemological knowledge have recently been explored in the collection "Emergent Axioms of Violence: Toward an Anthropology of Post-Liberal Modernity," published in *Anthropological Forum* (Pipyrou and Sorge 2021).

6 An interesting comparison here is with Walt Whitman's "Song of Myself" discussed by Bennett (2020), where outside influences enter bodies, infuse and confuse their organization, and then exit as something new and transformed. Bennett interrogates what goes on inside the black box of the body to transform and be transformed by nonhuman entities.

7 Linguistics, for instance, cannot be the "supplier of all models" (Serres and Badiou 1968, 26), harking back to the importance of the resonant something of atmosphere and aesthetic as a way of knowing.

8 An attempt to chart Serres inevitably leads one to Marilyn Strathern's (1996) significant work on networks, where she makes clear the importance of cuts as well as flow. Analytically, we are always in medias res, in the midst of things, opening in the middle of the plot with flashbacks/flashforwards, porously absorbing and transcending flows of information. It is thus important not only to attend to the ways things connect but also arrest their movement while words are being put on paper. But perhaps Serres also demonstrates how writing can catapult new insights, which depart and accelerate on the page. Cuts and flows in the mesh of life, or in the network, can also be instruments of insight even if, as Serres implies, we are never in full control but are being crisscrossed by forces greater than ourselves, in constant porous becoming ([1982] 1995).

PART I. OF PARASITES AND CONTRACTS

ALBERTO CORSÍN JIMÉNEZ

1

THREE TALES ON THE ARTS
OF ENTRAPMENT

Natural Contracts, Melodic Contaminations,
and Spiderweb Anthropologies

He who takes by force is not apt to trap (i.e., gentleness and not force arrives at truth).
Baganda proverb. —cited in Radin, *Primitive Man as Philosopher* (2017, 156)

It is not uncommon for readers of Michel Serres to stand in awe at his joyful eru-
dition. His works span prodigiously the history of mathematics, music, fables,
cybernetics, and legal theory, to name but a few. His reading and command of
ancient and modern literatures and sciences never fail to dazzle and inspire.
And yet it is still the case that his thinking remains firmly anchored in the
Western tradition. Seldom does one find Serres venturing into anthropological territories. I don't mean this as a criticism because Serres is nothing if not a
pluralist thinker, whose philosophy is forever open to the possible. Yet I would
like to suggest that there is scope for imagining what a Serresian anthropology
might look like; that is, imagining not what Serres can do to or for anthropology but how anthropology can enrich and rearticulate some of the most fertile
aspects of his thought.

This is the challenge I set for myself in this chapter: to reinvent Serres as an anthropologist, reworking some of his scintillating insights with anthropological eyes. To do so, I take as my cue an object that was dear to Serres and has been central to anthropological thought too: exchange. And in proper Serresian style, I stage my exercise through three tales on the nature of gifts and the gifts of nature: how they balance, how they off-balance, and how they recursively capture/captivate one another.

First Tale: Natural Contracts

The Shepherd and the Lion
"In Aesop's tale a shepherd, much a-fuss,
A-fume at losing many a lamb and ewe,
Decides to catch the thief. What does he do?
Assuming someone of the wolfly race's
Foul inclination, off he goes, and places
Traps roundabout those creatures' lair. But then,
Before he turns to leave the den:
"O sovereign of the gods," says he, "I pray
You let me see my popinjay
Caught here and now before my eyes.
And if I have the final laugh,
I promise you the fattest calf
In solemn sacrifice!" As thus he cries,
Out stalks a lion from the den! Half dead
With fright, the shepherd, once again: "Forget
The calf, good god! An ox! An ox, I said,
Is yours if only you can get
This animal to leave and let me be!

The Lion and the Hunter
There's Aesop's tale. Now for his imitator:
A hunter—something of a perorator,
Boastful, brash—lost his dog: fine pedigree,
Good stock . . . Suspecting that the hound had been
Consumed, and that he now was lying in
A lion's belly, asked the hunter: "Where,
Pray tell, does that fell thief reside? It's my
Intent to punish him forthwith!" "Out there,

Off by the mountain," said a shepherd. "I
Pay him one sheep a month. That's how and why
I'm able to go anywhere I please."
As thus they speak, exchanging repartees,
Voilà! The lion saunters by. Our braggard,
Cringing now, turning pale and deathly haggard:

"Jupiter!" cries, "I beg you show me some
Close place to hide; if not my life is lost!"

Courage is courage when the cost
Is high. "Some call for danger: let it come,"
Our poet says, "and off they fly, struck dumb.""
(La Fontaine 2007, 131–32).

The Shepherd and the Lion is not among the corpus of Jean de La Fontaine's famous fables that Michel Serres cannibalized with his multiplicative genius in *The Parasite*. Its absence is somewhat of a surprise, if one may say so, for the text crystallizes many of Serres's favorite themes as well as being one of the original Aesopica that La Fontaine explicitly rivaled with when counter-penning his own *The Lion and the Hunter*.

Both Aesop's and La Fontaine's fables speak about the depths and asymmetries of exchange. Exchanges once deemed bountiful suddenly turn sour. The arrival of the lion, an uninvited guest, interrupts and parasites on both economies, but it does so with logics of substitution that are different in each case. "The parasitic relation," Serres reminds us, "works on the principle of the lion's share: the one who takes does not give; the one who gives never receives anything" (Serres [1980] 2007, 165). In Aesop's fable, the lion interrupts and transforms what was poised to be a display of cunning and deceit (capturing the wolf with a trap) into a presentation of sacrifice and piousness. In La Fontaine's, the appearance of the lion accelerates an existing economy of predictable exchanges into a spectacle of shameful submission and charity.

The lion's violent apparition upends the tranquility of exchange. "Voilà! The lion saunters by," announces La Fontaine, and the amicability between the "exchanging repartees" is violently partitioned anew. A world of quiet stillness is torn apart by a world of upheaval and commotion. Violence inaugurates a blank slate: "Our braggard," says La Fontaine, "Cringing now, *turning pale* and deathly haggard" (La Fontaine 2007, 131-32, emphasis added). Out of the paleness of violence, out of the "white of our dominance.... History bifurcates again.... Mastery and possession begin" (Serres [1980] 2007, 180). Whiteness surges as a violent power of inauguration. The lion doesn't just walk past; its walking gives birth to a new world of history, a new time.

Worlds are born from lionhearts. "Courage is courage," La Fontaine reminds us, "when the cost is high." If the story of Aesop is contained in the story of La Fontaine, this is in turn continued in that of Serres. In *The Natural Contract*, Serres resumes this fabular story about the origins of law and society by positing violence and war as cosmological forces: "We find ourselves in the same position as our unimaginable ancestors when they invented the oldest law, which transformed their subjective violence, through a contract, into what we call wars. . . . War is the motor of history: history begins with war and war set history on its course" (Serres [1990] 1995, 15, 14).

War is a bulwark against parasitism. It is a system of organized violence that, according to Serres, keeps chaos and disorder in check. War turns lionhearts into soldiers and places them under a general's command. As such, war is a prototype for all social contracts: a thermodynamics of places, corpses, and words; a myth of origins for new geometries, new physics, and new judiciaries (see, for example, Serres [1993] 2017). But there is a limit to how many enemies we can wage war with, how many geometries we can inaugurate, before the world itself goes to war with us.

Much to our chagrin, that day has come. Our lionheart philosophy, the ambitions of mastery and possession have culminated in a war of worlds, where the worldly epistemologies of possessiveness, in the "final balance sheet" (Serres [1990] 1995, 37), measure up and clash against the ontologies of Earth. Such is the argument of *The Natural Contract*. However, this clash is hardly an encounter between exchanging repartees. This is not even a war, for there is no prior covenant and disagreement between the parties. Literally and unilaterally, we are sucking the world dry. Thence the need for a new contract where people and things can reconvene their exchanges, a contract that bestows on nature the status of a legal subject and in so doing clears up a space for transforming the "rights of mastery and property" of parasitism into a new "rights of symbiosis": "What should we give back to the world? What should be written down on the list of restitutions?" (Serres [1990] 1995, 38).

What should we give back to the world? It has often been remarked that Serres's oeuvre is a long and recursive meditation on the capacities of the relation. "Relation precedes being. This is indeed the saying of my philosophy," he states unambiguously in *Hominescence* (Serres [2001] 2019, 225). Yet as Christopher Watkin has observed, these relations do not simply work as analogical connections but favor "federation" and "topological isomorphism" instead, a mode of thinking that operates "under the banner not of cuts and divisions but of knots and folds, in which knowledge grows not through interminable analysis but through overlapping strands casting shadows on each other" (Watkin 2020, 74). These are relations that are therefore in a permanent state of transformation and laminar flow. If anything, they operate under the guise of a "general ontology of inclination" (Watkin 2020, 240), such that relational movements are always limping, falling, interrupting, asymmetrical, overflowing, crippled. The parasite is exemplary of such relational flows (Serres [1980] 2007, 79), forever intersecting and discontinuing, mediating, sticking or bifurcating pre-existing relations anew—a lion toppling and cannibalizing prior exchanges. The parasite is therefore more of an abuser than a user (Serres [1980] 2007, 80, 168), less of an exchanger than a changer. "Change comes from a rupture in equilibrated exchanges. Change is the disequilibrium of exchanges" (Serres [1980] 2007, 182). The language of exchange is a poor fit for the logic of the parasite, where host and guest, war and peace, do not balance out. There's always a relation missing, someone stealing from the exchange (Serres [1980] 2007, 11). The world is full of relational thieves.

What should we therefore give back to the world? There is certainly no hope for exchange because there will always be a thief lurking. We must instead give the world the totality of giving. We need to surrender "the totality of our essence, *reason itself*" (Serres [1990] 1995, 90, emphasis added). We must sacrifice epistemology for the sake of ontology, our reason for our habitat. The nature of the gift must be in/commensurate with (must balance through commutation) the gifts of nature.

Second Tale: Melodic Contaminations

The initiate dives under the water and holds onto a pole driven firmly into the streambed, looking up at the stream's surface a few centimeters from his face. While in this position underwater, already initiated shamans on the bank tip burning jatobá resin—*katepo-egru*—from a gourd-like container roughly hewn from the tree's bark (*katepo-pitu*), allowing this pungent "drink" of liquid fire to drip and strike the water surface just above his eyes. The visual ingestion induces bursts of dream imagery from the *tšibo-erem* or "resin owners," which are multiple: fish, terrestrial spirits, sky people, thunder, the sun. . . . As the jatobá resin fizzles in the water, its aroma attracts numerous aquatic species and stuns them, leaving them to float near the initiate in the dark and cold liquid in a semitorpid state. The first named as arriving is the electric eel, the demiurge Imere's final shamanic instructor, followed by anaconda, stingray, river turtle, piranha, alligator, and four tiny shamanic fish species. Apparently dead but actually *tongnore*, "drunk/comatose," the various aquatic beings fill the inner ears of the initiate with their *tumtankom* or "other-language." . . . The initiate surfaces to breathe and returns under the water in cycles for several hours, until finally emerging from the river, cold, exhausted, and now covered too in a translucent agyuru membrane, likened by the Ikpeng to the epidermal coatings of certain tree species and newborn babies (vernix caseosa). . . . The storm passes and the initiate revives, while the various species in the water also reawaken and disperse. Slowly the noise in the initiate's ears distils into intelligible sound, musicalized language, the novice shaman's first absorption of *wonkin-eremri*, "animal-spirit-music." Through this extreme somatic and sensory closure—with mouth and nose sealed, eyes filled with water and fire, and submerged in an alien element, the subaquatic world—the initiate becomes an instrument for auscultating the usually inaudible or unintelligible language of other species. . . . After initiation, the shaman begins to be able to hear the spirits of various animal and fish species, allowing his first acquisition of *eremri*, "songs/music," a melodic contamination subsequently caught during dreams or overheard while wandering through the forest. The spirits and the shaman become acoustically transparent to each other. The new shaman attains the function-state of a container or trap: a vacuum waiting to be filled. . . . This is not merely a symbolic connection: Ikpeng shamans also make *yukutpot* baskets to be thrown into the river to turn into either fish or predators. But what about the traps in the myth? Invisible in the dark, supposedly empty but emitting the hollow cries of *wonkin* spirits, they have a distinctly haunting air. Like the novice shaman and the Ikpeng bamboo flutes, the traps filter the river to produce sounds, an opening and closing mouth. In this respect they are also just like the supernatural owners of species, who the Ikpeng describe as continually capturing and releasing their progeny through their mouths.—Rodgers (2013, 85–90)

What should we give back to the world? Notwithstanding its radical originality, Serres's answer ultimately partakes of a Euro-American philosophical tradition where law and science must in the last instance hold the world to ontological account. "The whole history of the Greek beginning of the sciences," he specifies in *The Natural Contract*, "tells of the shared and tragically eventful life of these enemy twin sisters, justice and justness (or accuracy), reason that judges and reason that proves. Today's question is: when and how do they become symbionts?" (Serres [1990] 1995, 65). As fascinating as Serres's proposal is, there is however scope for imagining a diplomatic strategy whose terms of exchange with the world are sourced on an anthropological imagination of parasitism different from the genealogies of Western science and law. What may an anthropology of parasitism throw into relief?

The parasite, as Andrew Shryock has convincingly argued, helps complicate and nuance certain aspects of the Maussian tradition of exchange in anthropology. Building on his ethnographic and historical work with Bedouins among the Balga tribes of Jordan, Shryock notes how a desire to keep often overcomes any impulses of reciprocity. As he puts it, "the centrality of giving and taking is rivaled by that of keeping, storing, collecting, preserving, hoarding, and hiding things away" (Shryock 2019, 549). Such practices of collection and hoarding point to the important role that acts of containment and reservation hold in social life. If gift exchange illuminates how social energies are displayed and elicited, hoarding speaks instead about the designs through which they are guarded and directed. Shryock describes such acts of containment in terms of hospitality more amply: "A gift is what we take to the feast, or receive once we arrive. It allows us to come or go, but hospitality protocols tell us when we have arrived, how long we can stay, when we should sit or stand, eat or drink, and on what grounds we should entertain Others or keep them away. One might think of hospitality as a container—an enclosing space that temporarily holds people and things inside . . . a form of human interaction ideally suited to controlling parasitism" (Shryock 2019, 565).

Shryock introduces the notion of hospitality as a counterpoint to the Maussian paradigm of exchange. Whereas hospitality seems destined to the conservation of energies, gift exchange would appear to favor their distribution instead. Such an antinomic structure—between a tendency for preservation and a tendency for dissipation—reinstates, however, an overarching figure of "balance" or "proportionality" between polar forces, which the very notion of parasitism was intended to undermine in the first place. Therefore, while one can certainly conceptualize hoarding as a parasite to reciprocity (and hospitality as a super-parasite to exchange more amply), it is also possible to imagine, as Serres would undoubtedly invite us to do, an ur-parasitical force forever destabilizing any such equations and counter-equivalences.

Such was indeed the spirit of Serres's general ontology of inclination as well as his description of parasitism as a changer of exchanges. In anthropology, the work of Eduardo Viveiros de Castro echoes with surprising effects some of these Serresian insights from an ethnographic perspective. In *Cannibal Metaphysics*, Viveiros de Castro offers a profound analysis of Amerindian materials to outline a general metaphysics of predation that resonates with the fundamental impulse of the parasitic principle. Viveiros de Castro distinguishes between an "exchangeist" conception of exchange and a transformational notion of exchange (Viveiros de Castro 2014, 166). In his own words:

> There is exchange, and then there is exchange. There is an exchange that cannot be called "exchangeist" in the market/capitalist sense of the term, since it belongs to the category of theft and gift: the exchange, precisely, characteristic of so-called gift economies—the alliance established by the exchange of gifts, the perpetual, alternating movement of double capture in which the partners commute (counter-alienate) invisible perspectives through the circulation of visible things: it is "theft" that realizes the immediate disjunctive synthesis of the "three moments" of giving,

receiving, and returning. Because even though gifts can be reciprocal, that does not make exchange any less of a violent movement; the whole purpose of the act of giving is to force the recipient to act, to provoke a gesture or response: in short, to steal his soul (alliance as the reciprocal soul theft). And in this sense, that category of social action called gift exchange does not exist; every action is social as and *only as* action on action or reaction on reaction. Here, *reciprocity* simply means *recursivity*. No insinuation of sociability, and still less of altruism. Life is theft. (Viveiros de Castro 2014, 166–67)

The general ontology of inclination (Serres) and the general metaphysics of predation (Viveiros de Castro) both point to the perpetually syncopated nature of exchanges, the ongoing laminar flow of incompletions and subtractions (thefts) whose intensity lies in the breadth of their parasitic involvements, their animus of captures and double captures, their "becoming-enemy" (Viveiros de Castro 2014, 181).

The image of "double capture" that Viveiros de Castro employs to underwrite the oscillatory disjunctions and counter-alienations of gift exchange is one that will stay with us for much of the rest of the chapter. For a start, the imagery is echoed in the ethnographic vignette on Ikpeng shamanic initiation rituals with which I opened this section. In David Rodgers's account, the aim of the ritual is to turn initiates into a "trap" whose interiority and hollowness slowly become "acoustically transparent" to the "songs/music" of the spirits. As Rodgers carefully emphasizes, this is not merely a symbolic connection. Ikpeng shamans are themselves makers of traps, which they release in the river to become fish or predators. In myth, traps are said to filter the river waters and emit the hollow cries of *wonkin* spirits, whose undulating sounds reflect the opening and closing of a mouth, not unlike how supernatural spirits capture and release their own progeny through their mouths. During their initiation rituals, shamans thus learn to become reverberating chambers for the songs of the spirits, their bodies "melodically contaminated"—letting themselves be captured and released—by the ur-language of cosmological music.

The figure-ground gradation of noise into intelligible sounds and then musicalized language plays also a fundamental part in Serres's philosophy. As early as *Hermès II: l'interférence*, Serres was already noting that background noise is the "universal condition of all exchange" (cited in Watkin 2020, 219), while in *Genesis* he described "background noise [as] the first object of metaphysics" (Serres [1982] 1995, 54). "It is through the voice," he wrote in *The Five Senses*, "that the first act of seduction passes between interlocutors, *sotto voce*, a tension that is rhythmic and musical" (Serres [1985] 2008, 120). These insights echo the importance awarded to acoustical infiltrations and remodulations by the Ikpeng. Yet one might argue, following Serres's very own logic, that there is an "excluded third" in his description, which Ikpeng philosophy helps adumbrate, namely, the role that traps and entrapments play as mediators and conducers of musical prehensions and socionatural transformations. If the shaman operates as a filter trap for spirits, as David Rodgers explains, reverberating in his body "the cosmological origin of animal species from traps in the deep past," this is only because it provides a medium for accessing "the trap-like function of supernatural species anomalies . . . in the present: beings who alternately contain and release their animal progeny (our prey)" (Rodgers 2013, 78). In other words, there is a dynamic entrapment of spirits, shamans, and prey in the ongoing availability of the world as a livable, resounding, and reproductive environment.

The hollowness of traps, the reservoirs of passive and cosmological energies that their concavities index and inchoate (see also Lemonnier 2012, 58), was remarked in passing by Lévi-Strauss in *From Honey to Ashes* when noting how Amerindian myths

> seem to offer us the spectacle of a vast group of transformations covering the various ways in which a tree-trunk or a stick can *be hollow*: it can have a natural or artificial cavity or a longitudinal or a transversal orifice; it can be used as a bee-hive, a trough, a drum, a dance stick, a bark pipe, a clapper, or a cang (a portable pillory). . . . Musical instruments occupy a middle position in the series, between the objects at the two opposite extremes which take the form either of a kind of shelter, such as the hive, or of a trap, such as heavy wooden manacle. And we can say that the masks and *musical instruments are themselves, each in their own way, shelters or traps, and sometimes even both at the same time.* (Lévi-Strauss 1973b, 390–91)

Therefore, in this view, the primal ululations of energy embodied by musical instruments are at once refuges from as well as traps for the "rhapsodic repetitions" (Viveiros de Castro 2014, 213) and syncopated possibilities of worldmaking. Said differently, the intense vibrations of music are both capturing and captive of intensity itself. Importantly, this view allows us to move beyond a reductive view of traps as techno-functionalist devices of capture and appreciate instead the dynamics of entrapment—the double movement of capture and captivation—that underpins the ongoing motions of intensities (Corsín Jiménez 2021).

Third Tale: Spiderweb Anthropologies

When we put 50 slot machines in, I always consider them 50 more mousetraps. You have to do something to catch a mouse. It's our duty to extract as much money as we can from customers.
—Bob Stupak, CEO of Las Vegas Stratosphere, 1995 (cited in Schüll 2012, 29)

The notion that musical instruments are at once traps and shelters may sound somewhat ambiguous and mysterious. Yet the sounds of musical instruments can both repel us or appeal to us; they can prompt us to abandon a bar or lure us into a street corner. While we may think at first that our responses to such sounds hinge on matters of taste or musical preference, it is well known that music conduces our bodies into specific forms of inhabitation—moods that are primed by the material designs of acoustics, ambience, and lively resonance. The materiality of music can make us feel at home, or it can make us seek refuge elsewhere. As Emilie Gomart and Antoine Hennion persuasively argued over two decades ago, music can excite a state of abandonment not unlike that of drug use, such that "one abandons one's being to what seizes it" and enters a passionate "suspension of the self" (1999, 227).

Now the notion that musical instruments can function as traps for the self, that is, that they can function as technologies of captivation and capture, has recently been explored by Nick Seaver in his ethnography with developers of music recommendation systems in the United States. Seaver describes how developers worked within the framework of a behaviorist theory that defined human actors as having "habitual minds with tendencies and compulsions that make them susceptible to persuasion and targets for capture" (Seaver 2019, 5). "Captology" and "hooking" were indeed central metaphors employed by industry participants (Seaver 2019, 3–4), who strove to design predictive metrics where the use of "traces of interactions recorded in activity logs" would anticipate behaviors, "elicit more interactions," and, in the last instance, work as a "trap for capturing fickle users" (Seaver 2019, 10). Such conceptions of trap-design fall prey (pun intended) to techno-behaviorist assumptions that reduce the work of entrapment to a narrow push-and-pull contest between the inclinations of the human person and the temptations inscribed in the machine. There are no parasites here, no melodic contaminations, no relational thieves. Capture is a one-way street: a little cheese, a clear pathway, and a mousetrap at the end of the algorithm.

"Mousetrap" is the word that interior designers working for the casino industry at Las Vegas sometimes offered the anthropologist Natasha Schüll when describing the spatial cues and "sensory atmospherics" they employed to guide and conduce the steady flow of casino patrons to gambling machines (Schüll 2012, 29, 45, 46, 65). Schüll's extraordinary ethnography of the design of casinos as "machine zones" (Schüll 2012, 1 and passim), where chronological time is suspended and patrons enter a "world-dissolving state of subjective suspension and affective calm" (Schüll 2012, 19), offers a chilling portrait of the data-atmospherics of contemporary capitalism. In the casino, machines and environments are designed to trigger a state of "perfect contingency" (Schüll 2012, 171), where the relationship between the machine and the gambler collapses into an experience of perfect synchronicity and pure reciprocity. As an informant explained to Schüll, "I'm almost hypnotized into being that machine. . . . It's like playing against yourself: You are the machine; the machine is you" (Schüll 2012, 173, emphases removed).

However, this form of hypnotic communion between machines and their human users should not be mistaken for the exhilarating symbiosis between things and people imagined by Serres or Donna Haraway (Schüll 2012, 179). There is no cybernetic sublime here. Instead, an idiom of overpowering "entrapment" pops up everywhere: the encounter with machines is "entrapping and ultimately annihilating," triggering an experience of "flow that is depleting, entrapping, and associated with a loss of autonomy" (Schüll 2012, 179, 167). One's attention is always "hooked," "in hold," or "captured," rendering one's perception of self-control over the "choices" one makes "emancipatory and entrapping, annihilatory and capacitating, reassuring and demonic" (Schüll 2012, 209). Therefore, there is no double capture here, no possibility of theft, no ontology of inclination. "When I gamble I feel like a rat in a trap," a gambler told Schüll. "'Yes, I feel like a Rat Person, coming out of my dark hole to surface when the money is all gone,' said another. Rats—along with carrier pigeons, rhesus monkeys, and Pavlov's salivating dogs—made continued guest appearances in gamblers' posts" (Schüll 2012, 105). Rats, pigeons, and salivating dogs: parasites without hosts.

What is captured by the machine? "Addicts of gambling machines," notes Schüll, "invariably emphasize their desire for the uncomplicated, 'clean cut' exchanges machines offer them—as opposed to relationships with other humans, which are fraught with demands, dependencies, and risks" (Schüll 2012, 197). The machine captures the world and, in return, offers the promise of escape and self-annihilation in the zone. Or is it perhaps the self that the machine captures? Perhaps it's not the world that the machine traps but the intricacy of our enmeshment in it. The world and the self are entrapped, and the machine mimetizes and sublimates that entrapment. The machine *as trap* becomes our world. One is hypnotized, as Schüll's informants put it previously, into *being the machine*. Abuse value disguised as use value.

In *Mimesis and Alterity*, Michael Taussig described how the power of the mimetic faculty—the faculty that attributes natural qualities to cultural products—was best explained by imagining the "spell of the natural" as that moment "where the reproduction of life merges with the *recapture of the soul*" (Taussig 1993, 2, emphasis added). The idea that gambling machines capture our selves, or that they can momentarily capture and extricate our selves from our relational worlds, participates of this mimetic tradition. Indeed, Taussig acknowledges the role that automata and machine reproduction have historically played in the consolidation of Euro-American mimetic culture, with the behaviorist and cybernetic traditions echoed in Seaver's and Schüll's ethnographies being exemplary in this regard.

However, the power of capture, as Taussig is at pains to emphasize throughout the book, should never be imagined as residing unilaterally in the hands of one single actor. For example, speaking of Cuna healing practices, where small wooden figurines are used to "capture" evil spirits, Taussig explains how the healer's capacity "to diagnose and cure, to restore souls [the indigenous term here meaning also "double"] depends on out-doubling doubling. Through his wooden figurines activated by his chants bringing forth doubles by means of mimetic magic, he brings forth images that battle with images, hence spirit with spirit, copy with copy, out doubling the doubleness of the world. Until the next time" (Taussig 1993, 128). The power of capture is always haunted by the captivations of power. Every capture is always shadowed by a double capture. What is captured by the machine, then? The power to capture back, the dynamics of entrapment.

The idea that one can design and redeploy an environment as a system of entrapment, much like casino developers are intent on doing, has some intriguing precedents. Such an image was, for example, one of the central metaphors employed by Jacob von Uexküll in his investigations into animal environments (Uexküll 2010). In particular, Uexküll regularly resorted to the image of the spiderweb as an analogy or trope for describing systems of environmentalization at large. For example, when Uexküll describes how animals transform their homes into territories, he compares the structured tunnel systems built by moles to a spiderweb (Uexküll 2010, 103). When he explains the developmental rules that give form to the bat's echolocational radar, he similarly draws on the spider's web to make the point that "neither of them is only meant for one, physically present subject, but for all animals of the same structure" (Uexküll 2010, 167). The spiderweb is the image that best exemplifies

Uexküll's biosemiotic metaphysics, an ecological interface capable of hosting and conveying the infinite pulsations and vibrations through which Nature speaks to itself. As he puts it, "one can recognize the reign of Nature's plans in the weaving of a spider's web" (Uexküll 2010, 92).

One of Uexküll's most memorable images a propos the biosemiotics of spiderwebs concerns the surface tension through which the spider and its prey (for example, a fly) counterpoint their mutual describabilities: "The spider's web is configured in a fly-like way, because the spider is also fly-like. To be fly-like means that the spider has taken up certain elements of the fly in its constitution . . . The fly-likeness of the spider means that it has taken up certain motifs of the fly melody in its bodily composition" (Uexküll 2010, 190–91). As is well known, the melodic counterpoint of the spider-fly would later inspire Gilles Deleuze and Félix Guattari's famous conceptualization of the neighborhood of indiscernibility through which a wasp and an orchid are interlocked in one mutual becoming, "a becoming-wasp of the orchid and a becoming-orchid of the wasp" (Deleuze and Guattari 1987, 10), which Deleuze would later refer to as a form of "*double capture* since 'what' each becomes changes no less than 'that which' becomes" (Deleuze and Parnet 2007, 2, emphasis added).

One final example: The image of double capture returns in the work of Isabelle Stengers as she strives to explain how every social action is in the last instance a "gamble" on the milieus through which we get a hold on our presencing and becoming with others. We are one with parasites, she observes, and it is the dynamic of this parasitical involvement—which she dubs "reciprocal capture"—that we must forever reckon with and comprehend (Stengers 2010, 35). She goes on to gloss the concept of reciprocal capture thus: Speaking of the complex ecology of mimesis, camouflage, and undifferentiation that modulates the prey-predator relation between a caterpillar and a bird, she observes that "the specific 'strategies' of mimetic defense employed by the caterpillar refer to the 'cognitive' abilities of the bird that threatens it, but it seems that for the bird the caterpillar is just one kind of prey among others. The definition of the parasite includes a 'knowledge' of the means to invade its prey, but this prey appears to simply endure the parasite's attack" (Stengers 2010, 36).

The parasite and its prey, Stengers adds, "exist in a way that affirms the existence of their respective other, but the opposite does not appear to be true—at least as far as we know at present" (Stengers 2010, 36). This is why the notion of "reciprocal capture" helps us move away from the predatory unilateralism implied in the bird-and-caterpillar example and invokes instead a larger process where "identities that coinvent one another each integrate a reference to the other for their own benefit" (Stengers 2010, 36). The trope of capture and mutual entrapment is a most apposite one for, according to Stengers, the mode of being that grapples with the conditions of its own existence—that comes into the world through reciprocal capture—is ultimately putting itself and its world at stake. This is a mode of existence that must learn to survive its own self-entrapment: "The production of existence for everything for which existence implies a 'gamble,' a risk, [entails] the creation of a point of view about what, from then on, will become a milieu" (Stengers 2010, 37). In other words, a milieu, the experience of the world as a situated environment, is what existence looks like through the dynamic embodiment of reciprocal capture, through the experience of entrapment.

Serres explains in *The Parasite* how the parasite's inhabiting of a milieu must start with its playing a "game of mimicry" whose aim is the "erasure of individuality and its dissolution in the environment" (Serres [1980] 2007, 202). The parasite doesn't so much inhabit an environment as transform the way the environment becomes, such that the parasite and the environment secrete or ooze themselves into—*porously becoming*—a new symbiotic form: "A parasite makes or secretes tissue identical to that of its host at the location of contact points with the host's body" (Serres [1980] 2007, 202). We may call this a spider-web anthropology: ooze value—becoming-milieu—disguised as abuse value.

Let me start wrapping up my argument at this point. In this chapter, I have worked my way through three tales about the nature of gifts and the gifts of nature, about the violence, rituals, and machinic natures of exchange, in an attempt to "parasite" and interject Michel Serres's own parasitical and relational philosophy with a number of anthropological expositions. Serres's oeuvre is unusually (and unfortunately) meager in engaging with anthropological scholarship, and yet anthropology, I have suggested, can adumbrate and amplify his analyses in ever more expanding and rewarding ways.

Serres's arsenal of conceptual persona—the parasite, the natural contract, background noise, the excluded third, to name but the few I have alluded to in this chapter—are always devised to accentuate and propagate the declinations of matter, information, and energy. They are "left-handed" and "lame," as he described himself in his intellectual autobiography *Le Gaucher boiteux* (Serres 2015b), and in this guise, they are conceptual operators that offer a provocative and welcome corrective to the "exchangeist" and "rebalancing" proposals of much social theory.

Given Serres's predilection for off-balancing exchanges, it seemed only natural to host an "exchange" with his thinking on the subject, precisely, of anthropological exchanges. Such has been one of the aims of this chapter. By creating the conditions for mutually parasiting his work and certain anthropological texts, I have tried to clear a space for appreciating what an anthropological vernacular of parasitism might look like.

This brings me, in turn, to the Serresian question with which I concluded the first tale: "What should we give back to the world?" (Serres [1990] 1995, 38). Can an anthropological vernacular of parasitism help us enrich, qualify, or expand the question about the natural contract's future?

Serres's answer to his own question, you may remember, called for our becoming givers rather than takers, to abandon the urges of possessive reasoning for the gifts of collective inhabitation. As he himself put it in *The Natural Contract*, "Rights of mastery and property come down to parasitism. Conversely, rights of symbiosis are defined by reciprocity" ([1990] 1995, 38). We need therefore to surrender the impulses of parasitism for the fertility of symbiosis; we need to "set aside mastery and possession in favor of admiring attention, reciprocity, contemplation, and respect" (Serres [1990] 1995, 38). We need to clear a space where the world can hold us to account.

However, lest we fall for a romantic sublimation of our ecological passions, the new symbiotic contract, Serres also warned us, offers no straightforward redemption from the influence of parasitical powers. As it turns out, there is no standardized template for the symbiotic contract, which must be renegotiated anew for every encounter, every exchange, every situation. There is no unmediated salvation in symbiosis, for the energies of the parasite haunt us all the same, all the time. Perhaps we would do better to "reformulate this question," Serres explained to Latour, by asking, "What is an enemy, who is he to us, and how must we deal with him?" (Serres and Latour [1992] 1995, 195). There is always a thief, remember, lurking around the corner.

Now "What is an enemy?" is of course a question that reverberates with the energies of the metaphysics of predation and the reciprocal captures of a spiderweb anthropology. It is a question that is always and everywhere situated, a question that porously becomes its own milieu: ooze value disguised as abuse value. In this sense, perhaps what anthropology can offer to Serres's original question is not so much an answer as a reformulation, less a reply than a game of entrapment.

In *The Natural Contract*, Serres frames the question about the natural contract using the discourses of science and law and inevitably imagines answers that are constrained by the contractual languages of "justice and justness (or accuracy), reason that judges and reason that proves" (Serres [1990] 1995, 65). Therefore, there remains an unresolved tension in Serres's vision for the natural contract whose terms of resolution seem bound to the metaphysics of geometrical justice. However, as we have seen, the language of give-and-take is only one of the many languages of exchange documented by anthropologists. One can imagine natural contracts, yes, but one can also imagine melodic contaminations or spiderweb anthropologies.

In this sense, the Baganda proverb that opens this chapter—"He who takes by force is not apt to trap"—offers one possible alternative to the Serresian model. The proverb beautifully illustrates the difference between what Eduardo Viveiros de Castro described as an "exchangeist" conception of exchange and a transformational notion of exchange (Viveiros de Castro 2014, 166), where the alternative to taking by force is not quite reciprocity but rather the complex and gentle entrapment that keeps both in motion. If the former indexes the constraints of a Serresian natural contract, perhaps the latter delineates the contours of an anthropological natural contract.

Finally, what all these languages of exchange have in common, I have tried to show, is their use of the recursive inclinations of capture/captivation to describe the solicitudes of social motion and intensity. In this guise, we may think of them more broadly as exemplars of an arts of entrapment. Whether Serres himself would have fallen for the trap, we cannot know, although it's hard to imagine that he would have not, at least, entertained joining the game of entrapment.

NOTE

My thanks to Daniel Knight and Andreas Bandak for generously commenting on an earlier version of this chapter. Michael Jackson enchanted the piece with his own trickster anthropology. I am particularly grateful to Andrew Shryock for his detailed comments and provocative questions, which, like a Baganda trap, gently invited me to reimagine the piece's conclusion.

2

UNDER THE SIGN OF HERMES

Transgression, the Trickster, and Natural Justice

Between 1969 and 1977, Michel Serres published four groundbreaking books, all written under the sign of Hermes. The passages he opened and explored between disparate disciplines, spaces, and times emboldened my own transgressions of the borderlines between social science and literature. As for my aversion to identity thinking, I found in Serres's shape-shifting style and his intolerance of thought constrained "in a context rigid with impossibilities" (Serres and Latour [1992] 1995, 43) the same exemplary openness I had found in Theodor Adorno's negative dialectics, William James's radical empiricism, and Jorge Luis Borges's *ficciones*. In Serres's view, what really matters is not determining the line that divides professions, epochs, cultures, or genres but pressing "closer to the turbulence preceding the emergence of an intelligible, discursively knowable world" (Serres [1991] 1997, 65).

These celebrations of the transitive over the intransitive find expression in the figure of the trickster, particularly Hermes, whose mythological association with porous boundaries pertains equally to the exchange of goods, the transmission of information, sexual intercourse, interdisciplinarity, artfulness, and alchemy.

When one considers the many tricksters that have emerged in different societies and at different historical epochs—Loki in the Eddas; Coyote among the Mohave; the spider trickster Ture among the Zande; Legba the guardian of crossroads in Dahomey; the Hocąk (Winnebago) Wakdjunkaga; the Togo Hare in the West Sudan; and the Māori trickster hero Maui-tikitiki-a-Taranga, who fished up the North Island of Aotearoa from the sea, stole fire from Mahuika, outsmarted his elder brothers, and invented the cat's cradle whose endless loop of string enabled stories to be told, mythological landmarks and figures depicted, or the forms of houses, weapons, articles of clothing, adze and canoe lashings, and flora and fauna entertainingly described—one is at once struck by their bewildering variety of forms, behaviors, and dispositions *and* by recurring motifs that bring into sharp relief some of the most persistent paradoxes of human existence. As Paul Radin observes, the trickster "knows neither good nor evil but he is responsible for both. He possesses no values . . . yet through his actions all values come into being" (Radin 1972, xxiii). Radin's comment resonates with my ethnographic work among the Kuranko of Sierra Leone, where quotidian movements between village and bush are not only essential to economic and political viability but figure as leitmotifs in Kuranko storytelling, implying that social, personal, and moral existence tends toward entropy unless perennially revitalized by the antinomian forces associated with the wild (cf. Serres 1982b, 71–72). While the village evokes images of binding laws and social organization, the bush evokes images of boundless energy and free agency.

Following Serres, my focus in this chapter is the tension between the selfish and social uses of intelligence and the associated tension between parasitism and symbiosis. Implicit in both these dynamics is the question of social and natural justice.

The Mande Trickster

Because the trickster embodies contrary potentialities, perhaps even a split personality (Jung 1972, 141–42), its indeterminate and amorphous character means that the trickster embodies porous becoming, sometimes marking a boundary and sometimes transgressing it; appearing male in one context but female in another; simultaneously assuming the form of an animal that acts like a person and the form of a person that acts like an animal. Alternatively, the trickster's capacity to uphold order or create chaos, to be selfish or sociable or an agent of both restorative and retributive justice is sometimes projected onto a contentious social relationship, usually between elder and younger brothers: Prometheus ("Foresight") and Epimetheus ("Afterthought") or Apollo and Hermes in Greece, Maui's five older brothers and Maui-tikitiki (Maui the last born) in Polynesia, Hyena and Hare in Mande societies of the West Sudan.[1]

It is generally true of traditional tales that they admit no hard-and-fast distinction between empirical, imaginary, ethical, and social constructions of reality. Kuranko descriptions of the Togo Hare (*lepus capensis zechi*) are no exception. They reflect empirical observation, to be sure. Hare is said to be cunning, quick, elusive, solitary, and hypervigilant. Hunters say that hares are skilled in camouflage and dissembling, often standing stock still or feigning death to avoid detection. But these physical characteristics are translated into personality traits (the hare becomes Fa San, "Mr. Hare") such as quick-wittedness and trickiness. His dupe, Fa Suluku (Mr. Hyena), is, by contrast, largely imaginary, since the spotted hyena (*crocuta crocuta*) is not found in Kuranko country. Portrayed as slow-witted and inclined to walk with its nose to the ground and its eyes turned to where it has been rather than where it is going, hyena is a parody of all authority figures and social norms. While the purblind elder is stuck in his ways and wont to abuse his power, the cunning youngster uses guile and ingenuity to counter corruption. Yet no permanent line can be drawn between actions that serve the social weal and actions that subvert it.

As we shall see, cleverness, fairness, and reciprocity are all inherently ambiguous. Ideally, balanced reciprocity is sought between, on the one hand, the protection a chief is duty-bound to give his subjects or that men must give women and children and, on the other hand, the respect that underlings pay their status superiors. The adage *"Nyendan bin to kile, a wa ta an segi"* describes how the *nyendan* grass used for thatching bends before you as you walk through it and bends back when you return. Thus, greetings, goodwill, assistance, and gifts move to and fro within a community, keeping the paths open, keeping relationships alive, including relations between chiefs (who rule the village) and djinn (who rule the bush), the living and the dead, rulers and commoners, and parents and children. However, reciprocity is double-edged and connotes both the generous giving of life-affirming gifts and the taking of life in acts of revenge and retaliation. Gifts may be poisons.

In my first fieldwork in 1970, I was curious to know what happened if a parent was negligent or abusive. What if a chief or big man ignored or exploited those who looked to him for protection? What if an ancestor or djinn was indifferent to the sacrifices offered to them? And how, I asked myself, did people deal with lapses in the ideally symbiotic relationship between superiors and subordinates, as when a person in authority acted autocratically and parasitically toward those who depended on him for support and succor?

One night, in the village of Dankawali, as if in answer to my questions, an elderly man called Nonkowa Kargbo recounted a story that, considering the ominous effects of President Siaka Steven's shadow state in Sierra Leone, was as relevant to national politics as to everyday village life.

Once upon a time, Hare got himself a yam to plant. He showed it to Hyena and asked for his advice on how to plant it. Hyena said, "First boil it, then peel it, then put it in the ground." Hare did as he was told, but that night Hyena came and unearthed the yam and ate it. So the yam never grew.

It wasn't long before Hare realized he'd been tricked. He decided to take his revenge. Hare pounded some rice flour (*dege*), mixed it with honey, and smeared it over his body. Then he lit a fire and lay down beside it. Then he sent word to Hyena, his elder brother, to say that he was ill and that Hyena should come and examine him and tell him if he was going to live or die.

Hyena came. He said, "What is this stuff all over your body?" Hare said, "It is my sickness. Will you taste it and tell me whether I will live or die?" Hyena licked Hare's body. "Eh, younger brother, this sickness of yours is very sweet!" He kept licking the rice flour and honey from Hare's body and saying, "Young brother, this sickness of yours is very sweet!" Finally, he said, "Younger brother, could you show me how you became so sick?"

Hare said, "All right. But you must go home now and return in the morning. Then I will show you how I contracted this illness. But before you go, elder brother, can you tell me whether I will live or die?" Hyena said, "You will live, and I will come and visit you again in the morning."

That night, Hare washed the rice flour and honey from his body but kept it within reach, so that when Hyena came in the morning, Hare told him that it was the residue of his sickness. Hyena ate it up without a word. Hare then said, "Now, come back again tomorrow and I will tell you how I contracted this illness."

When Hyena returned the next day, Hare said, "Elder brother, I will now show you how I became so ill." Hyena listened attentively. Hare told Hyena to call his sons. Hyena did so. Hare then told Hyena to have his sons collect some firewood. They did so. Hare said, "Elder brother, do you think you will be able to endure it?" Hyena said, "Yes, I will." Hare said, "Well, have each of your sons bring a long pole." This was done.

Hare ordered Hyena to light a big fire and then jump into the flames. "When you cry, 'Get me out, get me out!,' your sons should use their long poles to push you farther into the fire. But when you cry, 'Push me in, push me in!,' then it will be time to pull you out. Do you understand?"

Hyena said, "Yes," and immediately jumped into the fire. When he cried, "Get me out, get me out!" his sons pushed him further into the flames. Finally, when he was good and roasted, he cried, "Push me in, push me in!," whereupon his sons pulled him out. They took him to his house and laid him down there. Hare said he would come to see him in a couple of days.

When Hare came to visit Hyena, there were flies everywhere. Hyena had begun to putrefy. Hare said, "Something stinks around here!" People said, "It is Hyena, your elder brother. He is very ill." Hare then said, "Well, this sickness is just like the boiled yam. My elder brother told me to boil and peel it before planting it in the ground. I did what he told me to do, but he came in the night, dug it up, and ate it. Now, if a boiled yam can grow, then my roasted elder brother will live!" And with that he jumped through the window and was gone. Soon after, his elder brother died.

There are implicit connections here between reversals in meaning (pushing into the fire means its opposite), reversals in status (the younger brother shows he is intellectually superior to his elder brother), and reversals in fortune. As Nonkowa explained, the tension between older and younger brother is an expression of a more general problem inherent in hierarchy itself—a tendency toward parasitism, in which the powerful accrue benefits at the expense of the powerless:

> The elder brother is sometimes inclined to abuse his authority and neglect the welfare of his younger brothers. The younger brother is often made to run errands, fetch water, and summon friends for his elder brothers. But the younger brother may also seek the support, protection and friendship of one elder brother if another fails him, and younger brothers may sometimes outsmart the elders by playing them off against one another. The elders fear the possibility that a younger brother may cause rifts or quarrels among them by telling one that another insulted him or refused him help. By enlisting the support of a sympathetic brother, he can cause dissensions among his elder brothers. That is why the elder should not underestimate the younger and why elders look after the younger ones.

As Nonkowa observes, people in positions of authority do not always exercise that authority wisely or well, and underlings are sometimes driven to redress inequities or injustices by devious means. Nonkowa also suggests that cunning and disguise are weapons of the weak and figure as tactics in everyday life *and* as motifs in fantasy, dream, ritual, and myth, as the ubiquitous stories of Hare-Hyena illustrate.[2] It is also clear that the principle of reciprocity is at once a rationale for exchange (which is mutually beneficial) and a justification for revenge (settling a score, getting even). However, as Nonkowa's story shows, recourse to the antinomian powers of the Hare is justified only when a person in authority has forfeited his right to power through fraud, chicanery, or greed. In restoring the status quo, amoral tactics are approved.

What appears to be a simple and risible tale of sibling rivalry proves to be an allegory of the human condition in which the viability and vitality of the nomos requires openness to the antinomian. But a balance must be struck between rebellion and revolution, since revolution involves an irreversible and often violent transition from a known, ancestral order to a new order that derives its justification from the future, not the past. In Sierra Leone in the 1990s, this struggle for the future took the form of a war waged by youth against gerontocracy as well as large-scale migrations from rural villages to Freetown and from Freetown to Europe and the United States. Conversion to Islam or Pentecostalism (particularly the prosperity gospel) was a further expression of what Charles Piot has perceptively called this "nostalgia for the future" (2010).

The Trickster as Rebel

Not long after the end of the Sierra Leone civil war, I returned to the village of Firawa, where I had been doing fieldwork since 1970. I was accompanied by Sewa Magba Koroma, a friend now living in London. Firawa was Sewa's mother's brother's village, and like me, he had a close relationship with it.

One evening, a young man called Fasili Marah introduced himself to us and abruptly declared that he wanted to go "overseas." His father and mother were both dead, and he wanted to go to America. Would I take him?

I explained to Fasili that he was asking the impossible. Besides, America wasn't paradise. Many people were poor and oppressed there. Sewa also tried to spell out the disadvantages of migration. "People will not welcome you. Even if you find work, you will not earn enough to cover the costs of rent and food, let alone medical treatment if you became ill." Fasili was undeterred.

"I will do anything," he said. "I know it is hard. People have taken the little money I have saved, promising to put me in touch with a white man who can help me. When I went to see the white man, all I got was a packet of biscuits."

"What do you want most?" I asked.

"Clothes," Fasili said. He had borrowed a pair of plastic sandals in order to come to see me. His shorts were held up by a broken belt, and he wore a T-shirt with a faded American Express Card logo printed on it.

"Only clothes?"

"Also a car. And money. Take me with you. I will be loyal. I will work hard. If I do not get abroad, my life is at an end."

"You have no life here?"

"Our lives are in the hands of God. *Koe be altal' lon*, everything comes from God, everything is destined by God."

"Everything? Did God take your parents? Did God make you an orphan? Did God decide you should be poor?"

I hated the unkindness of my questions. And I hated myself when I saw, in the light of the fire, that Fasili was crippled. His shin bones were grotesquely bowed, his feet splayed, his growth severely stunted.

"How old are you, Fasili?" I asked.

"I am twenty."

He had become crippled at age four. He fell from a bench. One minute he was normal, the next he was crippled. His mother told him it was witchcraft, though no witch ever confessed.

"But God is responsible for everything," Fasili assured me. "How we are punished and how our destiny is made. God can make you rich one day and poor the next. You have to be patient. You have to accept that you must wait for God and believe in God."

"But if the RUF [Revolutionary United Front] had come and said you could get anything you wanted—clothes, money, a vehicle—would you have gone with them?"

"Yes, right away. Definitely."

Given people's experience of the years in which their village was sacked and burned, young men and women abducted by the RUF, and many people murdered or maimed, I found it hard to understand Fasili's readiness to seize an advantage without much thought for the repercussions. I was also struck by his readiness to turn his back on the past—on ancestral lifeways, farming, his very homeland—and cast his lot on an uncertain future.

But hadn't I been told that hunger gives no thought to the morrow, that there is nothing more tenacious than hunger? And wasn't Fasili's hunger for a better life, his "nostalgia for the future," so great that moral scruples were an unaffordable luxury?

To change the subject, I asked Fasili if he would tell us his favorite story. Fasili did not need to be asked twice, and he hurried through the tale almost without catching his breath.

"The animals formed a labor cooperative (kere) to dig a well. The hare (Fa San) was asked to pitch in and help. But he pleaded sickness and said he could not work. The animals knew he was lying, and after they had dug their well, they set the leopard to stand guard over it in case Fa San tried to take any of the water he had not earned the right to share. Sure enough, Fa San came to the well, bringing a sack of grasshoppers with him. 'Heh,' he said to the leopard, 'don't you know these people are using you? They've got you standing guard over the well while they eat. Have they brought you any food? I don't think so. But here, see what I have brought you to eat.' And Fa San gave the leopard the sack of grasshoppers and leapt into the water, which he spoiled by washing his dirty clothes in it. The animals, infuriated at having been tricked, set the bush cow to stand guard over the well, then lay in ambush in the bushes nearby. Fa San soon returned to the well with another sack of grasshoppers, confident he could fool the bush cow into letting him have access to the water. But the animals captured him, bound his wrists, and took him to a cotton tree where they intended to tie his legs together to stop him running away. But when the animals began tying Fa San's legs together, he cried out, 'Heh, that is not my leg, it's a root. You're not tying my legs, you're tying the roots together.' And when the animals began tying their rope around a root, Fa San cried out, 'Heh, that's not a root, that's my leg. Don't do that. You're tying my legs together.' In this way, Fa San avoided having his legs bound and was able to slip into a dark space among the roots of the great cotton tree and disappear. The animals now went to an old woman and asked how they might deal with Fa San. The old woman asked for food and then told them what they needed to do. They should get their machetes and cut off his arms and legs. If he cries out, 'You're killing me, you're killing me,' this means that you are actually chopping the roots of the tree, but if he cries out 'You're cutting the root, you're cutting the root,' this means that you are cutting him. And so the animals succeeded in getting rid of the deceitful Fa San."

I was stunned by Fasili's story. It was the first story of Fa San I had ever recorded (and I had, over the years, recorded thirty or forty Kuranko trickster tales) in which Fa San gets his comeuppance. What is more, this violent turning of the tables was reminiscent of the Revolutionary United Front, which used machetes to sever the limbs of villagers who had allegedly betrayed them by voting for a government that enlisted the help of foreign forces to defeat the rebellion. And I did not have to think hard to recall the mango and cotton trees on whose long and buttressed roots the rebels laid the limbs of their terrified victims.

I was also mindful of the fact that the story Fasili had told echoed the trickster story that survived the Middle Passage and entered the New World as one of the so-called Brer Rabbit tales, many of which were collected by Joel Chandler Harris and first published in 1868 and 1869.

Of the story from the New World that most closely resembles the Kuranko tale (itself widely known through the Mande-speaking area of the West Sudan), Charles Long arrives at this arresting conclusion: "Brer Rabbit is not simply lazy and clever; it is clear that he feels that *he has something else to do*—that life cannot be dealt with in purely conventional terms" (Long 1986, 196, emphasis in original). Although Long is interested in linking the trickster to the figure of the Black preacher who "kept alive the possibility of another life" among an enslaved people, what struck me in talking to Fasili and other frustrated young men was the way they oscillated between a patient stoicism, declaring a person's fate to be ultimately in the hands of God, and an impatient and urgent desire for transformation, dramatically conveyed by the alacrity with which Fasili said he would not hesitate to join a rebellion if it would improve his lot. Later in this chapter, I will return to this disenchantment with the social contract (in which the powerful are supposed to care for rather than exploit the powerless) and a growing fascination with spiritual contracts like the prosperity gospel of the Pentecostal churches that promise "supernatural abundance."

"Were the rebels like Fa San?" I asked Fasili. "Did they use tricks to get their way?"

"Yes, all kinds of tricks."

Sewa Magba knew them all. How the rebels would use all manner of cunning to cover their tracks, communicate with one another, and take their enemy by surprise. Identifying their comrades by taping red cellophane candy wrappers over their electric torches. Infiltrating towns, allaying suspicion, disguising their intentions, pretending one thing but doing another. Sewa would not be alive had he not been able to beat them at their own game, outwitting them, making his escape. Not for nothing was he nicknamed Bonké (Okra, i.e. "slippery").

"Could you use the trickery of Fa San to get overseas?" I asked Fasili. "To get money?"

"No," Fasili said. "Trickery doesn't pay."

"And yet," I said, addressing Sewa, "you have used your wits not only to escape the rebels when you were captured but to get out of some tight corners in London."

Sewa agreed. But he had never broken the law, he said.

The line between social intelligence (*hankilime*) and trickery (*aliye*) was not, however, easy to draw. There was a gray zone—a zone of ethical ambiguity—in which it was difficult to decide if trickery was justified in retaliation for being tricked or simply your best chance of survival in a situation where the odds were stacked against you.

Sewa had always reminded me of Fa San. Quick-witted, playful, street smart, smooth-talking, charismatic, and attractive to women, he was not above using his wiles to beat a rap or secure an advantage. But he was not a scoundrel.[3] He honored his ancestors as a source of life to the same extent that he sought to create a life for himself that had no ancestral precedent. As such, Sewa and Fa San are both refractions of the classical figure of Hermes who stands on the boundary between strange and familiar worlds, god of the crossroads, of doors, of trade, and of craftsmanship, giver of good things, whose power comes from his contact with strangers and strange places "on the other side" from whence he brings, through trade, theft, and barter, the very goods without which his own community would perish or pale into insignificance (Brown, 1990, 32–45).

Clearly it is never easy to determine how far a person is justified in going in search of the "something else" of which Charles Long speaks—the lost portion that is owed you, the stolen lifeworld for which you should be compensated, the life or livelihood that you imagine you are due. For in securing a fair deal or some form of natural justice, when does one draw the line between what is rightfully yours and what is not? And for we in the affluent West, who enjoy such privilege and power, when do we draw the line between what we feel we must give "to make a difference" or redress a historical wrong and what we owe ourselves, what we must keep if we are to live? Such was the dilemma I experienced, figuring how I could help people like Fasili in some small way without leaving myself with nothing, and that Sewa experienced with even greater intensity as he struggled every day to meet his family obligations yet keep back something of what he had gained from working in menial jobs in London.

The Ethical Ambiguity of Intelligence

For Homer, intelligence is ethically neutral. That it readily switches between serving selfish and altruistic ends is borne out by Odysseus, whose guile was at once a means of saving his own skin and rescuing his comrades from peril. For Michel Serres, this contrast between egocentric and sociocentric action is theorized as a tension between parasitism and symbiosis. The parasite takes without giving. Its own life takes precedence over the life of its host. As such, it exemplifies the negative reciprocity of theft, rape, and pillage. Symbiosis, on the other hand, implies a reciprocal give-and-take that produces mutual benefit and may be regarded as the ethical basis of human sociality.

When one considers colonial regimes or the asymmetrical relationship between an ethnographer and his or her interlocutors, the question of whether these are parasitical or symbiotic is immediately raised.

Sheltering from the rain in the Kuranko village of Fasewoia almost forty years ago, I fell into a conversation with a group of elders, one of whom casually asked me if I considered Kuranko to be my kinsmen. Mindful of the connotations of the Kuranko term *nakelinyorgonu* (literally, "mother-one-partners"), I shook my head and said no. But the old man had used the term in a moral and tactical sense to imply fellow human beings, and I was reproached. "Was I not aware that Africans and Europeans had the same ancestral parents and that our grandfathers were brothers?"

According to the Fasewoia elders, the first people in the world were *bimba* Adama and *mama* Hawa—ancestor Adam and ancestress Eve. They had three sons. The eldest was the ancestor of the whites, the second the ancestor of the Arabs, and the third the ancestor of the Blacks. The first two sons inherited book learning, but the last-born son—the ancestor of the Blacks—inherited nothing.

This story, which has analogues throughout Africa and makes its appearance in the earliest years of colonial rule, echoes the Hare-Hyena stories in which brother is pitted against brother, except that in these myths of ethnic origins, the younger brother can neither take his revenge nor redress the unjust dispensation.[4] I therefore expressed surprise that the old men should imply that Africans were natively inferior to Europeans, and I asked them to explain why the last-born son was doomed to poverty and illiteracy.

"If you uproot a groundnut," I was told, "and inspect the root, isn't it always the case that some of the nuts are bad and some good?"

Many years later, during a heated conversation with an imam in Firawa, I recounted this story only to have Alhaji Hassan vehemently offer his version of the myth. "There were three calabashes. Allah put the book of inventions under one, the Qur'an under another, and groundnuts under the third one. The ancestor of the blacks would have taken the Qur'an or the book of inventions, but the ancestor of the whites tricked him into taking the groundnuts."

I was aware of a similar story of deception and chicanery on the part of whites from Limba oral traditions, but this was the first time I had heard of this motif from a Kuranko informant, and it led me to ponder the connections between notions of well-being, ethics, and natural justice.[5]

To invoke "nature" here is not to imply an essential and universal dichotomy between culture and nature but to capture a sense that our immediate experience is like the tip of an iceberg whose bulk incorporates the residue of past events and forces lying beyond our empirical grasp. Thus, Paul Ricoeur notes that however one defines politics, it is so deeply rooted in our history and humanity that it is in effect, "almost without origin." By this, Ricoeur means "that there has always been politics before politics; before Caesar, there is another Caesar; before Alexander, there are potentates" (Ricoeur 1998, 98). So it is with ethics. "Before the morality of norms, there is an ethic of the wish to live well" (Ricoeur 1998, 94). That is to say, before the advent of any particular cultural, cosmological, or philosophical formulation, ethics has existed not as a unified, normative body of maxims, obligations, duties, or categorical imperatives but as a set of recurring quandaries and questions, a sense of ethical anxiety or disquiet about the very possibility of achieving a good life or of ever reconciling the ideals we espouse with the existential situations in which we find ourselves.[6] Although the origins of this proto-ethical sensibility cannot be pinned down, either in prehistorical or historical time, it is surely grounded in human sociality—in our awareness that our very existence is interwoven with the existence of others who are always there, as Sartre observes, even when physically absent, "in the form of some reminder, a letter lying on the desk, a lamp that someone made, a painting that someone else painted" (Sartre and Levy 2007, 71). To this thought, one might add Maurice Merleau-Ponty's observation that social existence means that we are never quite at one with ourselves. Our perspectives are never independent of each other; they have no definite limits; each "slips spontaneously into the other's," interweaving, merging, and in effect constituting "a common ground" that makes us "collaborators for each other in consummate reciprocity" (Merleau-Ponty 1962, 347, 353–54). This conception of intersubjectivity as a mutually constitutive process of give-and-take also suggests, as Sartre argues, that a meaningful life cannot consist simply in a passive or slavish submission to the world as one finds it. One must have some sense of being an actor whose speech and actions matter or make a difference to the way things are.

The trick here is to strike a balance between our obligations to the common weal and our obligations to ourselves, between what we take from the world and what we give back, what we preserve from the past and what we initiate in creating a viable future. In this vein, Serres argues that before there is a social contract, there is a "natural contract of symbiosis and reciprocity" that finds expression in our sense that however much nature gives us, we must give that much back in return. While most social tracts, such as the Declaration of the

Rights of Man, constitute the citizen of one's own state as the Legal Subject, excluding all those beyond the pale of reason—so-called savages, the insane, women, criminals, aliens, and animals—the natural contract encompasses all humanity and all life forms. Though we may be socialized to play down our sense of owing something to the world at large, this sense of obligation, Serres suggests, is never completely extinguished in any society or any mind and haunts us (Serres [1990] 1995, 38–39).

Given the accelerating effects of climate change on our planet, Serres's invocation of natural law and enlightened leadership as means of change seems idealistic (Serres [1990] 1995, 30–35), and one may well ask whether eco-activism, Gaia consciousness, and a proliferating literature that celebrates the virtues of the natural world and decries its desecration will reverse the present course of human history.

The question as to whether the natural contract is simply the social contract projected onto the extra-human world and therefore a kind of wishful thinking also entails a critique of the "Amazonian perspectivism" of Viveiros de Castro (2012, 83) that assumes that cosmologies mirror empirical reality.

Consider Leslie Marmon Silko's description of how, in the fall of each year, Laguna Pueblo hunters go into the hills and mountains to find deer. "The people think of the deer as coming to give themselves to the hunters," she writes, "so that the people will have meat through the winter. Late in the winter the deer dance is performed to honor and pay thanks to the deer spirits who've come home with the hunters that year. Only when this has been properly done will the spirits be able to return to the mountain and be reborn into more deer who will, remembering the reverence and appreciation of the people, once more come home with the hunters" (Silko 1986, 9–10).

One may be moved by this account of natural mutuality between humans and deer but wonder whether relations between humans and animals are ever so symbiotic that they preclude the possibility of parasitism.

In her compelling study of San art in Southern Africa, Patrician Vinnecombe argues that the naturalistic polychrome representations left by San hunters on rock shelters in the Drakensberg Range may be "read" as symbolic compensations for the killing of animals essential to San life—animals with whom people felt they had a natural contract, particularly with the eland (in the south) and gemsbok (in the north). According to myths among Khoisan-speaking peoples, the ancestral shape-shifter Kaggen created and reared the first eland. When younger members of Kaggen's family killed his "child," Kaggen felt deep sorrow and bade the killer ritually atone for what he had done. This atonement involved "a ceremony *which brought the eland back to life*" so that now, whenever eland are killed, it is vital that the blood and heart fat from the eland are mixed with the pulverized ochers used to paint the eland's image on a rock face. "It . . . seems to be not improbable," Vinnecombe writes, "that many of the eland paintings, particularly those associated with over-painting and re-painting, are connected with an act of reconciliation and of reparation to atone for killing. By this means, dead eland would have been symbolically re-created in order to replace the life which had been taken, and thus to ensure their continued existence" (Vinnicombe 1976, 180, emphasis in original).[7]

If there is an incentive or impulse in human life to receive and return what has been given to us, there is an equally strong tendency to exploit natural resources, avenge slights, and make good our losses, both real and imagined. In other words, the intersubjective logic of exchange, elucidated by Marcel Mauss and echoed in Serres's notion of the natural contract, is anthropocentric and assumes a false equivalence between humans and animals. Although the San kill eland and feel compelled to symbolically restore the lives they have taken, eland do not kill the San or feel the need to honor any natural contract with them.

The real problem here is not only the projection of human morality onto the extra-human world but the way in which reciprocity is understood. When Mauss invoked the Māori spirit (*hau*) of the gift to elucidate the threefold character of reciprocity (1954, 8–12), he glossed over the fact that the Māori word for reciprocity—appropriately a palindrome, *utu*—refers *both* to the gift-giving that sustains social solidarity *and* to the violent acts of seizure, revenge, and re-possession that are provoked when one party denies or diminishes the integrity (*mana*) of another.

For Marshall Sahlins (1968), reciprocity is not always synonymous with fair dealing and fair play. On either side of this "balanced" median lie two extreme positions that may be characterized as all-giving ("generalized reciprocity") and all-taking ("negative reciprocity"). In the case of generalized reciprocity, the line between self and other is so blurred by empathy, interdependence, and physical intimacy that one could not conceive of life without the other. The trust between mother and child exemplifies this modality, as may the bond between a patriot and the motherland or fatherland. At the other extreme, self and other are so polarized that the very existence of one requires the annihilation of another. The absolute antipathy, paranoid fantasies, and ethnic divisions that underwrite genocidal violence provide an obvious example.

The trickster moves between these extremes, sometimes restoring a social contract that has been broken, sometimes acting as if he was a law unto himself. As in Nonkowa Kargbo's and Fasili Marah's stories, the trickster Fa San seeks retributive, not restorative, justice. In other instances, Fa San uses superior intelligence to redress a social injustice rather than pursue some selfish end (Jackson 1982, chap. 4). Kuranko trickster tales bear witness to the ethical ambiguity of the human condition—not only our capacity for selfishness and selflessness, chaos and order, symbiosis and parasitism, but our difficulty in assigning absolute moral value to either.

Reciprocity and Sacrifice

Meditating on "the first and simplest operations of the human soul," Jean-Jacques Rousseau identified "two principles which are anterior to reason. One of them pushes us forcibly to consider our own wellbeing and our own survival, and the other inspires in us a natural repugnance towards seeing any sentient being, and especially our fellow-men, either perish or suffer" (Rousseau 1962, 37).

The spirit of Serres's natural contract and Rousseau's "natural sentiment" of *pitié* or compassion informs Kuranko myths of totemic origins. In turning to these myths, I want to broach the subject of sacrifice, which is neither an expression of symbiosis nor parasitism but echoes both. As with a parasitic relationship, a sacrifice implies that a life is gained at great cost to another, but like a symbiotic relationship, the life that is given is given freely. Moreover, as a form of reciprocity, sacrifice may at one extreme be a consummate expression of altruism but at the other extreme may be synonymous with violence.

First, let us consider a Kuranko myth (on the origin of the Kuyaté clan's special relationship with its totemic animal) in which sacrifice is construed as a form of symbiosis and altruism (in Kuranko parlance, of *morgoye*, lit. "personhood").

The Kuyaté do not eat the monitor lizard (*kana* or *kurumgbe*). Our ancestor went to a faraway place. There was no water there. He became thirsty; he was near death. Then he found a huge tree, and in the bole of the tree was some water left from the rains. The monitor lizard was also there. The ancestor of the Kuyaté sat under the tree. The monitor lizard climbed into the bole of the tree, then climbed out and shook its tail. The water splashed over the man. The ancestor of the Kuyaté realized there was water there; he got up and drank. He said, "Ah, the monitor lizard has saved my life!" When he returned to his home town he told his clanspeople about the incident. He said, "You see me here now because of the monitor lizard." Since that time the monitor lizard has been our totem (*tane*, lit. "prohibited thing"). If any Kuyaté eats it, his body will become marked and disfigured like the body of the monitor lizard. His *sanakuiye* (clan joking partners) will have to find medicines to cure him.

While this myth explains the origins of a clan's affiliation with a totemic animal, a set of complimentary myths explains the origin of the close relationship between certain clans. While several of the totemic myths involve an animal saving the life of a clan ancestor by offering itself to him as food, the clan myths involve the ancestor of one clan offering his body to another. In both cases, the sacrifice is repaid by establishing a privileged relation of respect (*sanakuiye tolon*) between the descendants of the mythic giver and receiver.

The following narrative by Fode Kargbo of Dankawali recounts the journey of the ancestor of the Kargbo clan from the Mande heartland, during which the life of the status superior—the Kargbo—is saved by the status inferior, who thereafter becomes, in effect, an equal, since his *moral* superiority cancels out the *political* superiority of the other.

What I know is this: my father told me that Bakunko Sise and chief Kama Kargbo left Mande and came to this country. Our ancestor came as a hunter, and the ancestor of the Sise clan came with him. On the journey from Mande they came to a river. Bakunko Sise could change himself into a crocodile so he crossed the river, but our ancestor, chief Kama, could not cross over. He became very hungry. He told Bakunko Sise that he was very hungry, and he asked how he was going to cross the river. Then the Sise ancestor cut off his calf and roasted it and sent it for our ancestor to eat. Then he swam across the river and came and took our ancestor across on his back. Our ancestor seven generations back [*bimba woronfila*, "ancestor seven"] came from Mande. I will now tell you how the crocodile became the totem of the Sise clan. This is what I heard from my elders. The Sise killed a crocodile. He ate it and died. His children decided that they should not eat the crocodile; it therefore became their totem. Then the Kargbo said, "Let the crocodile be our totem as well because the Sise ancestor gave part of the calf of his leg to our ancestor to eat, and he swam over the river and helped our ancestor across." So, the crocodile became the totem of the Kargbo. Our own real totem is the bilakunde [a kind of amphibious reptile]; it is not found in this country. The second is the lei, a bird that eats rice. It lives in this country and eats rice by both day and night. These two are our totems. We inherited the crocodile through the Sise. Because the Sise ancestor cut off his calf, roasted it, sent it to our ancestor to eat, then swam across the river and carried our ancestor to the other side, because of this, our ancestor said that their totem should be our totem. If you notice the Sise and Kargbo calling each other *sanakuiye*, it is because of that journey from Mande. Because the Sise offered himself to us, to be a pathfinder and helpmate to our ancestor.

In Western thought, premodern societies are often characterized as having a distributive morality in which a person's worth is relative to birth, age, and gender and that, moreover, notions of impartial justice first appear in Periclean Athens. As Claude Lévi-Strass points out, however, the totemic myths and clan correspondences in Mande societies effectively prevent the closure of each group and promote "an idea something like that of a humanity without frontiers" (Lévi-Strass 1966, 166), and it is likely that people in all human societies have wrestled with the paradox of plurality—that while people are all different, they are united by certain universal and phylogenetic elements. These unitary elements can, however, be forgotten, occluded, or contradicted. Thus, a society that pays lip service to equality, liberty, and fraternity may at the same time be profoundly flawed by systematic inequalities of wealth, opportunity, and power.

Although human beings will heroically or nobly sacrifice their lives to others, they are also willing to sacrifice human lives to political, academic, or economic ends. Thus, in the past, Kuranko chiefs leaving their country to attend festivals or funerals in other chiefdoms might order the sacrifice of a pale-complexioned virgin girl for the safety of the country, burying her alive in a pit at the chiefdom boundary, her mouth filled with gold and her head covered by a copper container—a practice whose logic is reminiscent of the large-scale sacrifices of young men in Western wars for the defense of a supposedly imperiled state or Aztec sacrifice where the lives of women, captives, and children were ritually fed to the sun, their "vital energy transferred" to the cosmos as a kind of "debt payment to the hungry gods" for the expected regeneration of life on Earth (Carrasco 1999, 179, 148).[8] Serres offers a compelling example of this cosmologic in *Statues* ([1987] 2014, 1–5), where he compares the *Challenger* space shuttle explosion in 1986 to the ancient Carthaginian ritual in which children and animals were placed in an enormous statue of the god Baal and burned alive. In the case of the *Challenger*, rubber O-rings in the booster rockets failed at subzero temperatures, a 1:8 probability of disaster that was foreseen by several engineers but dismissed by NASA management, who thereby sacrificed the lives of seven astronauts to keep their tight schedule, secure their funding, and demonstrate their prowess.

The ethical question of what we take from life and what we restore to life is as pressing for the humanities as it is for science, for even our professions can assume monolithic and soul-destroying proportions. How then can we avoid the seductive power of closed systems and reified concepts so that instead of seeking magical forms of security and authority in language, we follow Serres's example of radical openness to the transitive, mutable, fluid, and ambiguous nature of life itself? The ancient Egyptian personification of justice as weighing human actions on a pair of scales suggests that equity and equilibrium are mutually entailed whether we are addressing intersubjective relations that involve persons, groups, species, or objects assumed to possess subjective qualities of consciousness and will. For Serres, we owe a debt to life itself simply by being alive, hence his questions "What do we give back, for example, to the objects of our science, from which we take knowledge? . . . What should we give back to the world?" (Serres [1990] 1995, 38). Ethnographers might ask what they owe the people who grant them the knowledge on which they build their theories and careers. Are we on the side of the parasite who "takes all and gives nothing," or do we exemplify the principle of symbiosis? Or is it more a matter of oscillating between or striking a balance between these hypothetic extremes, as the trickster does, composing texts "outside of solidarity—in fuzziness and fluctuation" on the grounds that "nature does nothing else, or almost?" (Serres and Latour [1992] 1995, 112).

NOTES

1   "Tikitiki" refers to a topknot of hair (Maui's mother, Taranga, found him stillborn at the edge of the sea and incubated him in the topknot of her hair). "Tikitiki" can also mean "repeatedly going back and forth to bring or take things" and as such is a reminder of Maui's journeys between land and sea, or the worlds of light and darkness (*te ao marama* and *te po*), which recall Hermes's role in leading the souls of the dead from this world to the underworld.

2   This Kuranko tale is reminiscent of a recurring theme in South American myth of the *bicho enfolhado*, in which Fox deceives Jaguar by smearing itself with honey (Lévi-Strauss 1973a, 112), and involves a complex structure of culinary symbolism. The "culinary triangle" of boiled/roasted/rotted marks out a semantic field in which the contrast between the boiled, peeled, buried yam and the roasted, charred, putrescent hyena is mediated by "sweet" foodstuffs that are eaten raw. Consumed in their "natural" state, honey and rice flour are also "cultural" products, the former collected from human-made oblong, woven beehives, the latter prepared by pounding rice in similarly oblong but wooden mortars. The sweet, raw substances thus mediate a transformation in the story from an unjust situation to a situation in which the injustice is redressed.

3   In a compelling account of the ethical gray zone of the *favelas* of northeast Brazil, Nancy Scheper-Hughes describes the *jeitoso* "personality type" as "attractive, smooth, handy, sharp and a real operator" who can con people, beat the system, and even "get away with murder." Like the tactical ingenuity of the "rascals" of Papua New Guinea, the Brazilian trickster is amoral and opportunistic, with shallow loyalties and a thick skin (Scheper-Hughes 2008, 47).

4   For an extended, cross-cultural account of such stories, see Jackson (1998, 108–24).

5   According to the Limba narrative, Africans and Europeans were once brothers. That one became less advantaged than the other was a result of their father's favoritism. He wrote a book containing instructions on how to make money, ships, and airplanes, intending to give it to his dark-skinned son. But his wife smuggled the book to her favorite son—the one with white skin. The dispossessed son ended up with a hoe and a basket of millet, rice, and groundnuts.

> You see us, the black people, we are left in suffering. The unfairness of our birth makes us remain in suffering. That is why they want to send us to learn the writing of the Europeans. But our mother did not agree, she did not love us. She loved the white people. She gave them the book . . . Yesterday we were full brothers with them. We come from one descent, the same mother, the same father, but the unfairness of our birth, that is why we are different. We will not know what you know unless we learn from you. (Finnegan 1967, 253)

6   I follow Sartre's and Merleau-Ponty's situated ethics, on the grounds that Kant's ethics was "too well-ordered to be true; to be, alas, too removed from human life—and death—and probably even culpably remote, morally speaking, given the shattering realities of postwar Europe" (Williams 2000, viii).

7   Vinnecombe's insights are reminiscent of Walter Buckert's argument that hunting rituals involve expiation for the guilt of killing an animal. A classic example is the annual Athenian slaying of the ox (Bouphonia) that was followed by a trial for the murder of the animal, with the axe and knife found guilty and cast into the sea (Buckert 1983, 20).

8   Octavio Paz writes that the real rivals of the Aztecs are not to be found in the East (he first suggests the Assyrians) but in the West, "for only among ourselves has the alliance between politics and metaphysics been so intimate, so exacerbated, and so deadly: the inquisitions, the religious wars, and above all, the totalitarian societies of the twentieth century" (Paz 1985, 307–8).

3

KEEPING TO ONESELF

Hospitality and the Magical Hoard in
the Balga of Jordan

Good and Bad

Fareed is driving me through the Balga countryside of central Jordan. The terrain is his livelihood. He grew up farming it, tending livestock and olive trees, and he now belongs to a growing class of real estate agents who specialize in selling tribal properties to wealthy urbanites. Today, he seems especially interested in new houses, or big ones, being constructed by local Bedouin. These are not Fareed's clients, or his clan. His pithy accounts of how they are paying for their homes are not kind.

"That one," he says, "is from cigarette smuggling."

"That one over there," he adds, "is from selling off inherited land to Palestinians."

"That one, God forbid, said he would build a mosque and instead built himself a castle."

"You knew that one when he was a boy. He sends home big checks from America. He tells us he sells rugs."

"See the new addition on that one? Paid for with money he got from arranging the theft of his brothers' goats and sheep."

Fareed chuckles. He is enjoying this guided tour of supposedly ill-gotten gains. I have heard these digs before. Similar comments are made about Fareed himself, who is widely criticized for selling land to strangers to pay for his own second marriage and household. Anyone who does well financially seems to attract this kind of debunking rhetoric.

"Do you mean to say they are all cheats and criminals?" I ask.

"No. There are good and bad. But you have sat with the old men who tell about 'the age of shaykhs,' when Bedouin ruled this region."

This is shorthand. Fareed and I have recorded and transcribed many hours of Balgawi oral history over the years. Themes of taking and trickery in matters of landed wealth are common in these stories, but my habit when faced with such prompts is to bring up more edifying subject matter.

"Yes. They say people were more generous then. There was more hospitality back then, they say. Like what Abu Hamid told us."

Abu Hamid is Fareed's octogenarian uncle. The old man served me lamb and rice along with stories of how his ancestors had acquired (then lost or gave away) sizeable holdings between the highlands and the Jordan Valley. They were impoverished by their own hospitality, he said; they gave land as wedding gifts, and to bring men to their side in tribal disputes. Strong men could give land and still control it, he said, because "giving it meant it was yours." Fareed sat next to me, translating, asking for his favorite stories.

"Noble talk. And true. But the Hajj also told you about wars. And raiding and plunder. It was the age of the sword and lance. There was no government. The strong ate the weak."

"Yes. My tapes are filled with such talk."

Later in the evening, as we drink tea on Fareed's veranda, we return to his earlier comments about "the age of the sword and lance."

"We are not far from those times," he says. "That was two generations ago, or three."

Like so many Bedouin men, Fareed considers pre-Hashemite times much worse, and much better, than the present day. I am never sure which side of this moral dichotomy he will want to explore.

"I must tell you something," he says, leaning toward me and gently squeezing my hand for emphasis.

"We are still specialists in plunder and pillage. I look at all these big villas and I think of an English word: 'skifinjus.'"

"Skifinjus? I don't understand."

"Ski-fin-jus. It is an English word. Like hyenas. The ones who live off the dying ones, off the . . ."

"Oh, yes. Scavengers!"

"Yes. Your pronunciation is American. Ours is English."

Fareed stares at me. I am meant to process his insight and respond with something equally weighty. I am too slow. Fareed goes on without me, explaining how local Bedouin, his own tribe included, are not good at generating wealth, at investing and building. Instead, they prefer to skim off the profits amassed by others—the point is proven by a tangent lecture on the old custom of raiding peasant villages to the North and West, and extracting protection fees from travelers and merchants who moved through the Balga—and, apparently to this day, his relatives would rather sell off parcels of land, or borrow money from each other and never pay it back, than participate in "the modern economy," as Palestinians do.

I could object to Fareed's analysis in a thousand ways, but I can tell he is enjoying the mood of self-criticism. I decide not to lighten it.

"Are you a scavenger, Fareed?"

"Ha! No. I am worse than that. I am a real estate agent. I sell _____ [name of his own tribal group] land to outsiders so we can buy appliances and cars, and eat meat every day, and pay for big weddings. I wish I did not have to do it."

I try to imagine a plot of tribal land as a dying animal, about to be dismembered and carried off by new owners. Fareed often tells me that one of his greatest fears is that he will have to sell off his father's land.

"When I look out over that valley at night," he says, "I see the lights in _____ [a nearby suburb of Amman]. All those new houses built on lands we've sold. It is like they are getting closer and closer."

We study the landscape together. Thousands of lights flicker in the urban sprawl a few miles to the east. Between us and the city a band of dark countryside is waiting to change hands.

"Do you know what the lights look like to me?" Fareed asks.

"What?"

"The eyes of hyenas."

"Scavengers surrounded by scavengers?"

"God, yes."

The Regnant Frame

Entire worlds of social theory are built on the assumption that people must exchange to survive. We give, receive, and return gifts. We trade information and goods. We visit, feed, marry, and bury each other in a steady flow of transactions. Our reliance on scripted forms of exchange has been relentlessly typologized, and the great migration from reciprocity to redistribution and, eventually, to money-market systems is now enshrined, with all its evolutionary implications, in introductory anthropology textbooks. A contrarian approach, which stresses the importance of having things and holding them, is necessarily implied, but it fits easily within the regnant frame, which is always about exchange. One keeps for giving (Godelier 1999), or while giving (Weiner 1992). One holds out only to give later, or to have one's goods stolen or confiscated by others. It is hardly surprising that analysts would find this framework oppressive. Even Levi-Strauss's classic ode to exchange, *The Elementary Structures of Kinship*, ends with the assertion that one of our strongest impulses is to avoid relinquishing our possessions. "To this very day," he writes:

> mankind has always dreamed of seizing and fixing that fleeting moment when it was permissible to believe that the law of exchange could be evaded, that one could gain without losing, enjoy without sharing. At either end of the earth and at both extremes of time, the Sumerian myth of the golden age and the Andaman myth of the future life correspond, the former placing the end of primitive happiness at a time when the confusion of languages made words into common property, the latter describing the bliss of the hereafter as a heaven where women will no longer be exchanged, i.e., removing to an equally unobtainable past or future the joys, eternally denied to social man, of a world in which one might keep to oneself. (1969, 496-97)

It is hard to imagine a more unlikely last sentence for a book in which loss and sharing form the very substance of human kinship. Yet it rings true. The impossible joys of keeping preoccupy Fareed, who does not want to sell tribal lands, but must. Indeed, if humanity's lingering ambition is to accumulate rather than transact, one wonders why exchange became such a predominate theme in twentieth century anthropology. It matters greatly where the emphasis is laid. A quick catalog of any human political economy will show that the centrality of giving and taking is rivaled by that of keeping, storing, collecting, preserving, hoarding, and hiding things away. Our material cultures are dedicated to storage—the last 12,000 years have seen a profusion of containers of all sorts, and things to fill them with (see Gamble 2007; Kuijt 2009)—and much of our conceptual life is given to holding people and things in place. Kinship systems operate in this way (Lambek 2011), as does the making of rules, of laws and social norms more generally, which could be thought of as a storing up of experience and ethical knowledge (Dresch 2012). The most influential theoretical approaches to giving and taking insist that coercion is needed to overcome our desire to keep. Mauss's classic, *The Gift* ([1925] 1967), is dedicated to the proposition that what appears voluntary, spontaneous, and disinterested—namely, the exchange of gifts—is in fact obligatory and carried out under various regimes of compulsion, including fear of death. The tendency to return a gift, or a simple greeting, requires explanation, and reluctance to participate in these transactions is treated as a moral threat. Ideas of obligation suffuse the literature, as they do social life.

The tension between voluntarism and coercion is oddly central to all talk of exchange. In one of his last (and best) essays, Pitt-Rivers (1992) turned the tables on Mauss, arguing that gift-giving only appears to be obligatory when, in fact, it is voluntary. It is probably safe to conclude that this problem is based on deeper analytical assumptions. Two seem especially potent, and they have troubled theories of gift exchange from the start. The first proposes that modernity is characterized by pervasive social inequality, selfishness, taking more than we give (profit-seeking), and relatively impersonal exchange. The second proposes that the deep human past and all recent forms of the premodern are defined, in varying degrees, by reciprocity, economies of sharing, and gift giving, either between equals or as tribute from lesser to greater persons. If one were to draw this model of human time/space as a figure, reciprocity, then gift-giving, would be the ground (the unmarked, the common field, the past), while "modern forms . . . of contract and sale and capital" ([1925] 1967, 2) would be the object (the marked, the peculiar foreground, the present). The genius of Mauss—which enables us to forgive and even deny that his model was inherently evolutionary—is his ability to do two things at once. He showed us that the gift was dominant in archaic societies, thereby bringing it forward for analysis and typological assessment, simultaneously marking and unmarking it, and arguing that "the same morality and economy are at work, albeit less noticeably, in our own societies" ([1925] 1967, 2). He also made a powerful case for the moral superiority of this shared human past, using it to parochialize economists who thought gift-giving could simply be ignored. And (this is the best part), Mauss used the ancestral humanity of the gift, and of sharing more generally, as a policy-minded argument for the elaboration of social welfare programming in France and elsewhere. In effect, Mauss played the role of a sophisticated shake-down artist, teaming up with primitives and Bolsheviks, the not-so-brutish past and the all-too-revolutionary present, to convince the ruling classes that it was in their best interests to facilitate the redistribution of wealth.

Toward the Good

This interpretation of Mauss is not unusual (see Hart 2007; Fournier [1994] 2006). But I would go further and suggest that Mauss's entire discourse on "the gift" is not an exploration of the gift as it actually existed in archaic societies, in Malinowski's Trobriand Islands, or in Boas's Kwakiutl villages. Instead, to succeed as moral criticism, Mauss's essay relies utterly on the positive connotations of the gift as both a practice and an object known intimately to modern Euro-Americans.[1] The casual cynicism of Fareed, my Bedouin realtor, who sees the tribal past in Jordan as one of violence and taking (along with immense hospitality, of course), is barely detectable in Mauss's famous concluding chapter of *The Gift*. In its place, we find praise for charitable giving, the social justice of King Arthur's round table, and Winnebago elders who give tobacco and compliments to their guests. "I salute you; it is well; how could it be otherwise?" says the Snake Clan chief.

> I am a poor man of no worth and you have remembered me. You have thought of the spirits and you have come to sit with me. And so your dishes will soon be filled . . . It is good that you have partaken of my feast . . . You have helped me and that means life to me. ([1925] 1967, 68–69)

Mauss is clearly swayed by this high-minded oratory, declaring it "wise," and contending that there is no risk in magnanimous giving, which we should view as "a liberty." But his ethnographic signals are curiously mixed. The potlatch, an institution he dissects carefully in *The Gift*, is a venue for competition and social shaming, and Melanesian village feasts are marked by distrust so intense that guests and hosts have been known to slaughter each other. Drawing on Thurnwald's accounts, Mauss offers the following:

> Buleau, a chief, had invited Bobal, another chief, and his people to a feast which was probably to be the first of a long series. Dances were performed all night long. By morning everyone was excited by the sleepless night of song and dance. On a remark made by Buleau one of Bobal's men killed him; and the troop of men massacred and pillaged and ran off with the women of the village. 'Buleau and Bobal were more friends than rivals' they said to Thurnwald. We all have experience of events like this. ([1925] 1967, 80)

Gift giving is inseparable from "events like this"—which most of us will never experience!—but Mauss cannot resist the urge to place exchange on the side of the progressive and good. The moral of the massacre, he concludes, is this: "It is by opposing reason to emotion and setting up the will for peace against rash follies of this kind that peoples succeed in substituting alliance, gift and commerce for war, isolation and stagnation" ([1925] 1967, 80). As if these two realms of possibility were located on separate developmental pathways, not one and the same.

The most trenchant critique of this Maussian worldview was made not by an anthropologist, but by Jacques Derrida, the archetypal "continental philosopher," who startled Mauss-lovers everywhere by arguing, in *Given Time* (1992), that *The Gift* was not really about gifts at all, that it was a book whose author stretched the word *gift* over a panoply of human institutions and practices that, arguably, had nothing to do with that term as Western readers would intuitively understand it. Derrida asserts that a consistent "discourse of the gift" is impossible.

> It misses its object and always speaks, finally, of something else. One could go so far as to say that a work as monumental as Marcel Mauss's *The Gift* speaks of everything but the gift: It deals with economy, exchange, contract (*do ut des*), it speaks of raising the stakes, sacrifice, gift and countergift—in short, everything that in the thing itself impels the gift and the annulment of the gift. All the gift supplements (potlatch, transgressions and excesses, surplus values, the necessity to give or give back more, returns with interest—in short, the whole sacrificial bidding war) are destined to bring about once again the circle in which they are annulled. (1992, 24)

By dismissing Mauss's key term as a false universal, Derrida's critique does to the gift, more or less, what Needham (1971) and Schneider (1984) did to kinship. Yet the importance Derrida ascribes to the "pure gift" as a concept essential to social life, but always impossible in its own terms, eventually led him to create his own Maussian surrogate: "unconditional hospitality." This concept, too, is described as an aporia—it is impossible, forever beyond what we can realize in practice, and self-annulling—and Derrida deploys it much as Mauss deployed the gift, as an elastic analytical construct that can be stretched over multiple institutions and ideas. Candea calls Derrida's hospitality a "scale-free abstraction" (2012, 42), one that applies (perhaps too easily) to diverse relations between hosts and guests, ranging from household encounters with tourists in Corsican villages to refugee resettlement policies devised by government officials in Paris. Moreover, the binding imagery of Derrida's hospitality is oddly akin to Mauss's gift in its recourse to the archaic. It fixates on houses, food, sacrifice, kinship, care-giving, and other demonstrably ancient cultural motifs.

Why should Derrida's *Of Hospitality* (2000), which speaks ultimately to the status of Muslim immigrants in contemporary Europe, rely so heavily on ancient Greek and Biblical sources? Basically, this is another way of asking why Mauss wrote a critique of capitalism that relied so heavily on Kwakiutl and Trobriand ethnography.

In many ways, the gift and hospitality are moralizing concepts founded on what Gombrich (2002), writing of Western art forms, calls "a preference for the primitive," a conscious invocation of the basic and ancestral as criticism of the complex and contemporary. To establish that the preference is itself ancient, Gombrich quotes Cicero:

> How much more brilliant, as a rule, in beauty and variety of coloring are new pictures compared to the old ones. But though they capture us at first sight the pleasure does not last, while the very roughness and crudity of old paintings maintains their hold on us. (2002, 7)

I often see Jordanian Bedouin present their own cultural materials in this way—in modern societies, they embody the primitive, preferred or stigmatized—yet they do so with great ambivalence, refusing to treat "old values," like hospitality, as exclusively virtue or vice (see Shryock 2004). A more useful engagement with *The Gift* is suggested by this ambivalence, which attaches not to the problem of how analytical categories are constructed and applied, but to why certain categories are defined and preferred as primitive. The enduring popularity of *The Gift* among anthropologists is strong evidence that, despite our protestations to the contrary, we still have a strong disciplinary sense of the primitive and, in certain aspects of social life, we prefer it. This is not, I would argue, a tendency we need to correct. It is one we should develop more rigorously and subtly.

What would happen, for instance, if we removed the evolutionary substrates built into Mauss's notion of the gift? What if we assumed that the "negative" and "positive" qualities of exchange were always present in human society—in variable measure—such that "archaic societies" could not offer a uniform guide for moral improvement that, as it were, speaks across the ages? And what if we were to see "premodern" economies as ones in which taking, keeping, hoarding up, and hiding away—and even impersonalized exchange—have a logic as elaborate, as morally necessary, as that of giving to others (Smail and Shryock 2018a)? What if we paid closer attention to the moral ambiguities voiced in Fareed's world-weary discourse, in which predation and generosity, scavengers and magnificently hospitable men, reproduce each other historically and in the present? If "the gift" were not allowed to stand in for the human past, or to be posed as morally superior to taking and keeping, what would be left of *The Gift*? Could Mauss still have written it, and would it be as influential as it is today?

Philosophers and Bedouin

In handling such questions, I have found natural allies in Jordanian Bedouin and European philosophers, whose discourses on moral values often complement each other in fascinating ways. Elsewhere, I have put Abbadi and Adwani shaykhs into conversation with Kant, Derrida, and Benhabib (Shryock 2008), showing how their interests in hospitality and sovereignty are more closely related than most anthropologists would at first suspect. The integrity of social spaces akin to houses, and host/guest relations within houses, explain the resonances. House-like spaces require attention to ideas of protection, of guarded privacy, that have always been problematic for anthropologists enamored of alliance, exchange, and open visions of human mobility (Newell 2018). Filtering these ideas through Middle Eastern ethnographic materials is helpful, since well-established Orientalist traditions have rendered Arab/Muslim societies primitive but not entirely Other and therefore not, for certain anthropologists, preferred. Mauss and Lévi-Strauss largely ignored the region—even the Bedouin could not hold their attention—and this tendency to look away is not surprising. In the Middle East, exogamy gave way to endogamy; complex market economies were matched by ideologies of local household autonomy; and the primitive was never simply marginal, but was instead enmeshed in civilizational structures that ethnographers often found too intricate, too literate or historical, and too familiar morally, to process in the relativistic (much less sympathetic) ways that have elsewhere become the dominant anthropological style (Gilsenan 1990; Dresch 1998; Scheele and Shryock 2019).

To move through this terrain with appropriate care, I will bring my Balagawi hosts into conversation with yet another philosopher, Michel Serres, whose ideas on systems of "oneway relations" methodically reverse the sensibilities (and the substantive conclusions) of *The Gift*. In *The Parasite* ([1980] 2007), Serres develops a complex model of information and energy flows that corresponds in suggestive ways to Fareed's blunt observations on Bedouin economies of predation and scavenging. For Serres, "real production is unexpected and improbable" ([1980] 2007, 4), and it is immediately overwhelmed by secondary consumers. Parasitism, latching onto others and draining their energies, is foundational to social life. Parasites take more than they give, by intent and design; they are lodged in dense networks, "cascades," of exploitative extraction. Humans, Serres argues, are creatures that have parasitic relations with each other and the world around them. They have turned the world into a massive body (indeterminately natural and social) in which they live and on which they feed.

> Man is the universal parasite . . . [and] everything and everyone around him is a hospitable space. Plants and animals are always his hosts; man is always necessarily their guest. Always taking, never giving. He bends the logic of exchange and of giving in his favor when he is dealing with nature as a whole. When he is dealing with his kind, he continues to do so; he wants to be the parasite of man as well. And his kind want to be so too. Hence rivalry. Hence the sudden, explosive perception of animal humanity. (Serres [1980] 2007, 24)

At first glance, this imagery would appear to be a photographic negative of Mauss's romantic vision of reciprocal giving, but Serres's model is more encompassing—all human society is based on parasitism; it is not something that divides along archaic and modern lines—and this ubiquity encourages us to interpret key social institutions as means and responses to parasitism, as byproducts of taking and depletion. In this view, our desire to draw energy from others, to keep others from attaching themselves as parasites to us, or to encourage such attachments, pushes us endlessly toward a handful of interactive strategies. Kinship is one. Hospitality is another. Both relations involve a host, and Serres offers us his strongest imagery whenever he depicts this relationship.

> There is no exchange, nor will there be one. Abuse appears before use. Gifted in some fashion, the one eating next to, soon eating at the expense of, always eats the same thing, the host, and this eternal host gives over and over, constantly, till he breaks, even till death, drugged, enchanted, fascinated. The host is not a prey, for he offers and continues to give. Not a prey, but the host. The other is not a predator but a parasite. Would you say the mother's breast is the child's prey? It is more or less the child's home. But the relation is of the simplest sort; there is none simpler or easier: it always goes in the same direction. This is true of all beings. Of lice and men. ([1980] 2007, 7)

Note the absence here of any notion of return. Although Mauss is never mentioned in *The Parasite*, Serres argues that gift exchange is a structured response to the abuses associated with parasitism, abuses that pre-exist exchange both historically and in the present. It is an intriguing claim. For Mauss, the giving, receiving, and returning of gifts brought protection against "private and open warfare" in archaic societies. For Serres, all forms of exchange begin as attempts to discipline social parasites, who might not intend to kill us, or even to weaken us, but who want to turn us, so to speak, into their endless meal. "That is why the relation of exchange is always dangerous, why the gift is always a forfeit" (Serres [1980] 2007, 80). Parasitism is the flow of life, in which case the urge to "keep to oneself" is not only "unobtainable to social man," as Lévi-Strauss would have it, but is tantamount to social death, to the ultimate loss that, as Fareed insinuated, makes us prey not only to parasites, but to scavengers.

The Soft and the Hard

Models of parasitism (taking and keeping) and gift (giving and returning) are typically portrayed as moral opposites. One is built on a framework of human connection, of kind gestures and a general ethos of caring. The other builds on metaphors that, by comparison, are less appealing, and sometimes repulsive, conjuring up images of rats and lice, blood sucking, theft, infestation, stealth, and the devouring of surplus, both treasure and waste. Yet in my own ethnographic experience, these two languages are used simultaneously and fluently among Balga Bedouin, who speak with unflinching pride of ancestors who robbed peasants, stole from and beheaded Ottoman tax collectors, and plundered the camps and herds of neighboring tribes, insisting all the while that they were, and are, gracious hosts who honor their guests, protect wayfarers and refugees, and pile meat high on their platters. Generally, they see no real inconsistency in these soft and hard versions of their identity. In keeping with Maussian progressivism, however, they realize that the softer image is more suitable to public culture in a modern nation-state, a fact that makes their oral histories hard to represent fully in nationalist contexts.

Elsewhere, I have described in detail the political and economic structures that prevailed in the Balga before it came under Ottoman rule in 1867 (see Shryock 1997). It was a society that, in both oral tradition and written historical accounts, was shaped by predatory, protective, and tributary relations among Bedouin tribes of varying sizes and genealogical constitutions. The 'Adwan tribe, who dominated the Balga for much of the nineteenth century, are portrayed by storytellers as "60 men" whose power was based on a dense web of redistributive and extractive ties. They imposed *khawa* (brotherhood payments) on peasant villages to the North in exchange for protection against other tribes, in effect gaining access to an agricultural economy by force. They took possession of key water resources, granaries, and mills. They purchased a small slave army to fight for them and other slaves to farm their lands. They taxed merchants who entered the Balga, collected protection fees for escorting the annual Hajj caravan through their territory, and monopolized the escort of wealthy European antiquarians, tourists, and explorers who visited the region. With the wealth they derived from these extractive relations, they fortified their alliances with local tribes, feasting them, providing gifts of weapons, sugar, and cloth, and shielding them from their enemies.

In its particulars, the old 'Adwani polity resembled a Serresian cascade of parasitism, a configuration that is not exactly new to social theory. Ibn Khaldun, writing in AD 1378, captured the same logic in his allusion to household pests.

> The premises and courtyards of the houses of the prosperous and wealthy ... who set a good table and where grain and breadcrumbs lie scattered around, are frequented by swarms of ants and insects. There are many large rats in their cellars, and cats repair to them. Flocks of birds circle over them and eventually leave, satiated and full with food and drinks. In the premises of the houses of the indigent and poor who have little sustenance, no insect crawls about and no bird hovers in the air, and no rat or cat takes refuge in the cellars of such houses. (1967, 275)

Such is "God's secret design," Ibn Khaldun assures us; the wealth that produces "swarms of dumb animals" will also produce "swarms of human beings" (1967, 275). The point is to have parasites, as a sign of affluence, but not to be overwhelmed by them. In similar fashion, the 'Adwani economy functioned so as to feed its many parasites and keep them at bay. The meat, tobacco, coffee, guns, wedding gifts, and military support were a high wall between 'Adwani clients and inaccessible sources of 'Adwani wealth. The 'Adwan still boast that they gave gifts but seldom accepted them, that their gifts were signs of mastery, and that no return in kind was ever expected (and would, in fact, imply rivalry). In effect, their power and reputation were resources that the 'Adwan kept to themselves, and the long-term smallness of the tribe is evidence of their reluctance to allow their closest allies to affiliate with them as kin. One of the most controversial topics for 'Adwani historians is determining which families are members of the tribe by patrilineal blood descent, and which are simply "followers" who have acquired 'Adwani status by polite convention.

The Fateful Meal

After the establishment of Ottoman authority in the Balga, and with the arrival in the 1920s of British and Hashemite rulers, the 'Adwan were systematically deprived of their extractive capabilities. Their protection contracts were illegalized, as was slavery and intertribal raiding and warfare, and their political supporters, who continued to eat away their surplus, gradually became less compliant. Even before the Hashemite era, the 'Adwan had begun selling off their lands to merchants in Jerusalem and Nablus. Today, exceptional 'Adwani power is confined to a few shaykhly families whose influence is based on business acumen and close relations with the Hashemite elite. They portray their ties to the royal family as friendship, and political appointments and recognition are exchanged for loyalty to the regime, a relationship of gift exchange that, as I have argued in another essay (Shryock and Howell 2001), belongs to a tradition of "house politics" that many tribal Jordanians see as continuous with pre-Hashemite rule.

When the Amir 'Abdullah arrived in Transjordan in 1921, he was a guest with powerful British patrons. The 'Adwan were among his local hosts, and in 1923 they tried to displace him in an armed revolt that failed. The history of modern Jordan is the story of how the Hashemites moved from the status of guests to that of national hosts, successfully transforming a diverse array of tribal groups, villages, towns, and refugee populations into a political formation that can be meaningfully described (and contested) as "the house of Jordan" and "the big Jordanian family." Among Balgawis, the Hashemites represent a new consolidation of power in a long procession of ruling families, many of whom linger on, eclipsed and domesticated, as part of the Jordanian national elite. The rise and fall of these families is often explained in stories of hospitality. The role of hospitality is literal in these tales: entire political worlds collapse and rise in strategic encounters between guests and hosts, within the physical space of the guest chamber. For the 'Adwan, the climactic tale—which every 'Adwani knows, whether they believe it to be a false rumor or the truth—is one in which 'Abdullah, soon to be named King by the British in 1946, poisoned Shaykh Majid al-'Adwan at a feast. Supposedly, Majid advised 'Abdullah to remain an Amir "because you do not control your own army or mint your own coins." Later that evening, blood flowed from Majid's nose and he died. Shaykh Majid's immediate family considers the poisoning story preposterous, but most rank and file 'Adwanis believe it, and they use it to mark the end of the "age of shaykhs."

What I find most revealing about the story is that no one can agree on who was the host, and who was the guest, at that fateful meal. It is a moment of indeterminate parity, intolerable for both parasite and parasited. As a moral system, hospitality cannot accept the equivalence of guest and host in the same house, in the same moment (Pitt-Rivers 1977, 108–10). The poison, like the feast at which it (might have been) administered, is a narrative re-assertion of this principle.

In their introduction to *The Return to Hospitality*, Candea and Da Col ask us to "imagine what anthropology might look like today if Mauss had chosen hospitality rather than the gift as the subject of his 1924 treatise," and they assure us that the mental exercise is worthwhile because, on "its own merits, hospitality is an even more likely candidate than gift-giving for a foundational anthropological concept" (2012, S1). As I move back and forth between *The Parasite*, *The Gift*, and the Balga, I find myself agreeing with Candea and Da Col's revisionist claim. In a clear case of preference for prop over stage, the massive ethnographic literature on gift-giving routinely downplays obvious conditions: that gifts come to and leave a place; that givers and takers must approach or be received; that the power to offer and accept gifts (and the meaning of gifts) depends on one's status as host or guest; and that all of these determinations require precise forms of movement. A gift is what we take to the feast, or receive once we arrive. It allows us to come or go, but hospitality protocols tell us when we have arrived, how long we can stay, when we should sit or stand, eat or drink, and on what grounds we should entertain Others or keep them away. One might think of hospitality as a container—an enclosing space that temporarily holds people and things inside (Shryock and Smail 2018b)—and gift-exchange as its content.

In all these ways, hospitality is a form of human interaction ideally suited to controlling parasitism, but as Serres realizes, hosting the parasite manages only to produce a new flow of energy to which even more parasites can attach. The best hosts, he writes, are the best parasites. "The logic is unshakable" (Serres [1980] 2007, 111). Playing with the double significance of the French word *hôte*, which means both "host" and "guest," Serres claims that these roles are hard to tell apart, that each manipulates and parasites the other. This insight lies at the heart of Balgawi historical narratives about the rise and fall of shaykhs, men who are routinely generous and great takers of other men's possessions. Their accumulation of power requires that they store part of that power away, keeping a strategic reserve for themselves that is protected from parasites. But how can they do this when, in historical accounts, shaykhs and their families often start their trip to power from very disadvantaged positions? Often, they possess nothing but their horses, the clothes on their backs, and their swords.

A few motifs emerge repeatedly in these stories. First, a powerful family is often said to descend from an ancestor who came to the Balga as an asylum seeker after a conflict in another region. The ancestor is taken in by a local shaykh, who offers to protect him. The initial status of the ancestor is that of parasite and political dependent; he is without family, sustenance, or local allies. His sponsor gives him all three. Loyalty is all he can offer in return. The ancestor then comes into conflict with his sponsor, or must find a new one. Often, the sponsor is depicted as an oppressor. He forces men to hitch themselves to plows, like oxen, and then till his fields; he tries to force marriages on unwilling partners. He is arrogant, suspicious. He is not a good host; in fact, he resembles a parasite, taking without giving, and his clients are in no position to be his host. They grow to resent him.

The story of the 'Adwan is consistent with this formula. Fayiz and Fowzan, the ancestral brothers, arrive from the east, escaping a conflict with their cousins. They are taken in by followers of the dominant regional shaykh. In some traditions, this man is called the Mahfuz al-Sardi; in others, he is called the Amir al-Mihdawi. The (grand)son of Fayiz, Hamdan, serves in the entourage of the Mahfuz, "just to make a living," a status that is clearly subordinate, parasitical, and guest-like. Hamdan dislikes the Mahfuz al-Sardi, whom he considers cruel, but he is too weak to oppose him. How Hamdan reverses this situation is the subject of one of the most famous 'Adwani stories. The compressed account below is a mix of narratives told to me by dozens of old men, but I give preference to versions I recorded from narrators in the Kayid sections of the 'Adwan tribe in 1990.[2]

Keeper of al-Dhabta

The Mahfuz al-Sardi decided to raid the annual Hajj caravan, which he had agreed to protect as it passed through the Balga. This was a great sin. During the division of the spoils, he denied Hamdan a proper share of the booty. Instead, the Amir allowed Hamdan to keep a mangy, hobbled she-camel (called al-Dhabta, "the hobbled one"). The others ridiculed Hamdan, calling him Keeper of the Dhabta. To this day, the 'Adwani battle cry is "Keeper of al-Dhabta" (ra'i al-dhabta).

The others did not know that this worthless camel carried in its saddlebags the gold of the entire Hajj caravan. Hamdan discovered the gold, but he told no one. He hid the treasure away in a cave, in Abdoun. Whenever he needed coins, he would go to the cave and take from his secret supply. Hamdan began to buy away the Mihdawi's followers, one by one, giving them guns and horses, inviting them to feasts, paying for their brides, and refurbishing their tents.

In this way, Hamdan gathered men around him. He was known for his hospitality, for his open hand. After a time, he refused to sit in the Mahfuz's diwan. The Mahfuz asked after him. The others said, "Hamdan now has his own tent, and many followers." One day, Hamdan rode with his entourage, now 40 men, in front of the Mahfuz's tent and did not offer a greeting and did not sit with him to visit. The Mahfuz al-Sardi called out to him, "O Keeper of al-Dhabta, who are all those men with you?" Hamdan said, "I borrowed them from you on loan" (*istagradht-hum grudha*). The descendants of these men are known to this day as "The Borrowed Ones" (*al-grudha*). Hamdan continued to avoid the Mahfuz; he would not visit or greet him, nor would he pay him tribute. The Mahfuz sent four slaves to bring Hamdan to his tent by force. Hamdan killed three slaves, allowing one to survive and return to the Mahfuz to reject his invitation and to report that Hamdan and his men no longer belonged to the Mahfuz and would not serve him or live in his camp.

The Mahfuz al-Sardi was enraged. He declared war on Hamdan.

The stories diverge at this point. Some say Hamdan fought the Mahfuz for one year, then decapitated him on the battlefield. Others say Hamdan fought for twenty-five years and was himself slain in battle, after which his son, 'Adwan, defeated the Mahfuz, killing the old man and driving his sons and allies from the Balga. But in all versions, the hoard of gold discovered in the saddlebags of al-Dhabta sets in motion the dramatic shift from guest to host, from parasite to parasited, from follower to master.

Interference

When I ask old 'Adwanis how their tribe, which is small, came to dominate the Balga for three hundred years, they seldom dwell on the mechanics of gift-giving or the realpolitik of parasitism and predation. No 'Adwani elder has ever pointed explicitly to slave warriors, *khawa*, and control of water and mills as the ultimate sources of 'Adwani power. These factors are always presented as effects, as mere evidence of 'Adwani power. Aside from skill in battle, and the malfeasance of the shaykhs they displaced, the tribal historians I have interviewed find more explanatory potential in the saddlebags of a mangy, hobbled she-camel. Why?

Despite a nagging sense that the story is pure invention—that it is perhaps linked to primordial crimes in roughly the way Fareed assumes a plush villa is related to present day crimes—I have come to the conclusion that the story of al-Dhabta is appealing because of its effectiveness as an interrupter, a concept that figures prominently in *The Parasite*. For Serres, the term "parasite" has a meaning in French that is not available to English speakers. That meaning is "noise" or "interference." The introduction of noise, he argues, disrupts the normal flow of exchange: "Noise gives rise to a new system, an order that is more complex than the simple chain" ([1980] 2007, 14). The gold in al-Dhabta's saddlebags is a kind of "interference," and 'Adwani storytellers describe it as *hazz*, good fortune, or luck. The 'Adwan, they say, are *ahl hazz*, a family of good fortune. This is not to say, simply, that they are lucky. Good fortune is unpredictable, but it is hardly a random occurrence. It is ordained by God, and this puts it outside the realm of what Mauss called "the gifts we give to men"; neither is *hazz* implicated in "the gifts men give to God." No storyteller has ever told me that al-Dhabta was given to Hamdan by God as a reward or in response to Hamdan's request. The 'Adwan are not known for religious devotion; indeed, they are widely portrayed in oral tradition as tough, thuggish shaykhs who were feared by weak and strong alike. Their murderous reputation only adds to the appeal of *hazz* as interference. It is a resource that, for twelve generations, was very hard to take away from the 'Adwan—they kept it to themselves—but it was somehow a source of all the wealth they shared with their followers.

Hazz serves, at least in the idiom of storytelling, as a storehouse as improbable and as inaccessible to others as the gold coins Hamdan stashed away. There is, to my knowledge, no actual cave in Abdoun associated with this magical hoard. No one ever took me to the area where it might have been, unlike the sites of famous tribal battles, which cover the Balqa landscape, or the graves of prominent 'Adwani poets and shaykhs, which I was encouraged to visit and photograph. Did Hamdan ever count his gold coins? No sum, large or small, is ever attributed to his secret share. Did Hamdan wonder why he, of all men, should have received it? Why do we know so much, and so little, about this "noise" at the heart of 'Adwani history?

Laying Up Treasures

Literalism is always a risk. Buried wealth, found gold, ancient coins—they have obvious appeal, and I am often told by Jordanians, who live among Hellenic ruins, that a grandfather went to his grave without telling them where he stashed his treasure. "We knocked down the walls of the old house, but nothing was there." Several men have asked me to bring metal detectors to them as gifts. They are illegal in Jordan. An 'Adwani who lives near an archaeological site swore to me that a Hashemite prince arrived by helicopter to confiscate a golden statue that was unearthed there. A perfect case of parasitical interference! Whatever the status of these missing and stolen treasures, they are now out of circulation, which adds to their allure. They cannot function as gifts (yet), and they have been removed (for now) from parasitical cascades. Beyond the objective value of gold, they have the power of sacred things that have somehow escaped the relentless demands of exchange.

Much of social life is dedicated to protecting valuable things of this kind. My work as an oral historian has brought me into contact with many of them. One of my closest friends in Jordan keeps historical documents in a suitcase and will not show them to anyone. Another wrote a history of his mother's career as a tribal shaykh, helped me translate it into English, then asked me to show the Arabic original to no one in Jordan. Another collects old pictures of 'Adwani shaykhs, which he rarely shows to anyone, lest they ask to borrow them. Old books, genealogies, and tribal poetry are treated in the same way by their collectors, who would rather share them with an outsider, like me, than risk losing them to ordinary circuits of exchange. All of these objects have value precisely because they are adjacent to exchange. They have an alternative value, one akin to land that can be inherited but should never be bought and sold, and thus is sold to strangers who see it differently. Finding, protecting, and channeling valuables of this kind requires human capacities beyond the simple ability to take and give. It requires the ability to "lay up treasures" in places where others cannot get at them—where, in the Biblical idiom, "neither moth nor rust destroys nor thieves break in and steal."

The hoard, magical or mundane, exists apart from mere giving and taking, apart from the ordinary flow of goods and services, and apart from the normal course of political history. It is an advantage, a special burden, and an important alternative to reciprocity and exchange. It is a special challenge to ethnographers because, quite simply, we do not know where the hoard is, but we know it has been present and missing for incalculable stretches of time. Lt. Claude Reignier Conder, the British army engineer who created the first topographical maps of the Balga in the 1880s, noted two beliefs among the Bedouin that he found especially peculiar. "The first is that hidden treasure exists in certain places, and can be discovered by the use of incantations" (1895, 348). Conder admits that stashes of coins, recent and ancient, are found occasionally, but that "the Arabs have exaggerated ideas on this subject, and they suppose treasures to lie hidden in every ruin" (1895, 348). "The second common idea," he writes, "is that the desert was formerly cultivated and full of water," a belief he could not credit at all (1895, 349). What I find intriguing about these ideas is their persistence, their function as a kind of economic history in which wealth is not simply something that circulates among us, or awaits us in heaven, but is something that constantly eludes us now, hidden beneath the earth, in walls and caves, and in memory. It is spent up, like the once fertile and water-filled desert. Or it is misplaced. But evidence of wealth, and the labor and losses and secret gains associated with it, is everywhere. Kept and protected.

Even in Death

In 2003, I flew to Jordan to visit a sick friend, Hamoud al-Jibali. He wanted to tell me a few stories he had been saving for me. "My father had bags of gold that he promised to show me before he died," he said, "but he never did." Hamoud had decided not to take secrets to his grave. Although he was weak and his voice wavered, he told me a story that he had been reluctant to share before "because it seemed like boasting, and as we say, 'the man who praises himself is a liar.'" The story was one Hamoud had heard from another man, many years ago. It was about the generosity of Hamoud's own father, Shaykh 'Abd al-Rahman al-Jibali, and how he had entertained an unexpected party of 150 'Abbadi tribesmen, led by the great Shaykh Krayum Nahar al-Bakhit, who were traveling across country to attend a dispute resolution. The guests filled two tents and sat beneath nearby trees. Shaykh 'Abd al-Rahman slaughtered 40 goats on that day and sent the men off with three donkeys loaded with grapes. The storyteller, Hamoud said, summed up the account as follows:

> I never returned to 'Abd al-Rahman's house after that, but the memory of this event lived in me . . . This is my experience of real generosity in my life. I don't deny that others are generous. But that . . . that I call the real *karam* (hospitality): to host 150 people without seeking prestige. Today they invite five people, and they broadcast it to the world.

Hamoud rested for a moment, having exhausted himself with this Maussian tale.

"That is an excellent story," I said. "Excellent."

"Yes. I'm sure you have heard many like it," Hamoud said. "He was a generous man. I am not finished."

After he took a few more breaths, Hamoud told me "the important part," which seemed to come from another place, from another economy of emotion and memory, the one where precious things are kept for as long as possible, and sharing them is always a great risk. He asked me to write it down word for word, so I would get it right, and so he could recuperate between phrases.

'Abd al-Rahman died on November 8, 1982. On the 40th day, when the time of mourning was over, a man appeared with a herd of 23 sheep that 'Abd al-Rahman had set aside for his own memorial feast. Everyone was amazed. He had not told anybody about these sheep. Even in death, he was able to provide for his guests.

I am still not sure why Hamoud, who had told me so many stories, saved these for last. His mind was intensely focused on matters of wealth and caring for others. He was struggling to insure, while he still had time and strength, that the land he had inherited from Shaykh 'Abd al-Rahman, on which he was planting olive orchards and building a large house, would pass securely to his four daughters, all of whom lived in Sweden. Local relatives were plotting to take control of his property, he told me, and an estranged wife and her brothers were after his cash reserves.

"They eat me now," he said, "and they will eat me when I am dead."

Shaykh 'Abd al-Rahman's reputation as a generous man had been an invaluable resource to Hamoud throughout his life, and I was honored to receive a parting narrative gift that, in effect, came to me from both father and son. It encapsulates two powerful moments in the moral systems I have explored in this essay. The first was an act of radical predation that depleted 'Abd al-Rahman's herds and stores of grain; indeed, one could interpret the sudden appearance of so many armed guests as a raid, not a visit. Over time, this event had evolved into an example of unparalleled hospitality. The second moment, an act of secret storage, a defense against insistent demands to give and share, enabled a posthumous show of respect from a dead host to his living guests. The stories articulate multiple aspects of Bedouin polity and economy, ignoring others entirely. It is notions of host, guest, and house—the component parts of *karam*, of hospitality in safe and sovereign spaces—that emerge, yet again, as the ultimate matters of life and death. Waiting on the threshold between them, Hamoud had seen this truth, and could share it with me.

NOTES

Originally published as Shryock, Andrew. 2019. "Keeping to Oneself: Hospitality and the Magical Hoard in the Balga of Jordan." *History and Anthropology* 30, no. 5: 546–62. https://doi.org/10.1080/02757206.2019.1623793. Reproduced by permission of Taylor and Francis Ltd.

1   Again, this is not an unprecedented conclusion. For an enlightening discussion of similar ideas, centered on Jonathan Parry's classic essay, "The Gift, the Indian Gift, and 'the Indian Gift'" (1986), see Sanchez et al. (2017, 553–83).

2   Fuller, alternative versions of these stories appear in *Nationalism and the Genealogical Imagination: Oral History and Textual Authority in Tribal Jordan* (Shryock 1997). Especially good ones are told by Shaykh Sa'ud Mani' Abu al-'Ammash al-'Adwan (1997, 158–60), and by Hajj Ahmad Yusif al-Waraykat al-'Adwan and Hajj Muhammad 'Abd al-Karim al-Waraykat al-'Adwan (1997, 196–210).

ANDREW SHRYOCK

CHAPTER 3 POSTSCRIPT
Connective Tissue

I wrote "Keeping to Oneself" before I had read widely in Serres, before I had "done" or ingested much of his work. Yes, I am likening his philosophy to a drug. I had not taken enough of it. Or, more to the point, I had taken too much of it at once, overdosing on *The Parasite*, a treatise whose effects are especially potent when mixed with any amount of Mauss's *The Gift*, an elixir on which anthropologists have been blissfully stoned for a century. My initial encounter with Serres pushed me beyond old fixations on exchange and reciprocity toward ideas that were grittier, more ambiguous, and morally darker than anything Mauss could allow. In the end, I found myself in a strange place, contending with things and people who were removed from material flows—by stealth, hoarding, or death—yet were still vital to a moral economy based on giving and receiving hospitality. Aspects of my analysis seemed right, especially the focus on resources stored up near the threshold of death, before and beyond the grave. But I felt the analysis more than I fully understood it, and this sensation went hand in hand with the suspicion that *The Parasite* had intoxicated me. Perhaps I had succumbed to a numbing agent or a masking tissue in the parasitic "ooze" Serres had secreted in his (successful) attempt to enter and alter my way of thinking.[1]

As I took in more Serres, I eventually found his own description of what I felt. It comes in *Branches*, where he describes the transformation of knowledge into comprehension.

> And all of a sudden, by means I can only shed light on by comparing it to digestion, which transforms a piece of bread into active biochemical elements in my body, or to pregnancy, which transforms an oocyte into a fetus, I make this theorem mine. The time it takes is indeterminate: a

fraction of a second or decades; how many times, two or three decades having passed, have I violently felt, from my thighs to my thorax, that this digestion, that this conception were finishing their work and that I was entering into a true comprehension of what I had merely known. ([2004] 2020, 51)

Serres then turns a bodily sensation into a sensing body: "Quickly therefore or little by little, the theorem passes into my head, my eyes, my original perception of my landscape, my genitals even, my active life; I walk in its space, place my hands and feet according to its measure, inhabit and caress its forms in such a way that I recreate it, reinvent it from its foundations; this objective changes into subjective. I no longer know it—I feel it, live it, comprehend" ([2004] 2020, 51).

When this happens, Serres concludes, we move differently through the world, rebuilding it, making it bigger. Invention is possible. In *Statues*, Serres adds the element of intoxication. Certain insights leave us oddly im/mobilized. Indeed, being "under the influence" of this transformative process, being receptive to it, can give its hosts a familiar look: "Naïve, having lost their heads. Enraptured. Ecstatic before what happens in the world" ([1987] 2014, 140).

Serres admits that enlightenment of this kind is exceptional. Beyond comprehension, it comes like revelation, and we cannot assume, and Serres does not claim, that it will come from reading his work. People respond to Serres inconsistently. Some are immune to him; others are left feeling angry and cheated. Serres is not always intelligible. Timing is important. A Serresian argument that seems profound in the morning can seem preposterous at sundown. Indeed, there are passages in Serres that I understood for the first time *while writing this text*. It is hard to find the right occasion for Serres, the right reason. His theory and method are seldom clear; he is averse to both, preferring fables, analogies, imagery, echoes, and, most of all, storytelling. One must devise one's own way of ingesting his work or applying it. Still, after tactical experimentation, I find myself addicted to Serres. All I can say to justify my condition is this: the more Serres I read, the more I see in his work, and the more I can do with it.

Serres has a definite way of writing, of lining up and interweaving a set of images and analogies. He enjoys playing with classical, canonical works. Often, they are literally classics, the stuff of Greek and Roman antiquity and literally canonical, with invocations of biblical motifs so obvious and frequent that his writing seems, at times, theological. The thought-world of Serres is always the West, the (pre- and post-)Christian West. It is not the wider world of anthropology. Yet it is a world bigger and older than anthropology, a world that generated anthropology as a modern discipline and continues to trouble and animate it. When Serres uses the term "anthropology," it is code for things barbarous, rising out of the past, simple or primitive, excluded or marginalized by Enlightenment. He is not quick to separate this terrain from what he calls the "grand narrative" of humanity. "The most archaic anthropology," he says, "buried, forgotten, subterranean, invades the staging of technical and scientific progress" ([1987] 2014, 20). Yet Serres gives little attention to the findings of contemporary ethnography. In *Hermes*, for instance, he filters the problem of giving and counter-giving through the image of the feast in Molière's play *Don Juan*. As he approaches the end of the essay, he offers us this:

> Now open *The Gift*, and you will undoubtedly be disappointed. There you will find match and counter-match, alms and banquet, the supreme law which directs the circulation of goods in the same way as that of women and of promises; of feasts, rituals, dances, and ceremonies; of representations, insults, and *jests*. There you will find law and religion, esthetics and economics, magic and death, the fairground and the marketplace—in sum, *comedy*. Was it necessary to wander three centuries over the glaucous eye of the Pacific to learn slowly from others what we already knew ourselves, to attend overseas the same archaic spectacles we stage every day on the banks of the Seine, at the Theatre Francais, or at a brasserie across the street? But could we ever have read Moliere without Mauss? (1982a, 13)

We land securely in France, in the French canon, where we already were. Does the allusion to Mauss make this location more anthropological or less? Why does *The Gift*, a study of exchange in "primitive" and "archaic" societies, suddenly appear? One hears a trapdoor close behind the reader. Or is the door opening? The entire analytical space is now connected—Melanesians and Molière, the ethnographic and dramaturgical—in a vast terrain marked as much by similarity as by difference. The universal and the particular come together abruptly, as if each is trapped, or caught up, in the other. The same thing happens in *The Parasite*, a book in which *The Gift* is not mentioned even once, but could we ever have read Serres without Mauss? The effect is not one of "coevalness," the small triumph Fabian gained for his small (that is, "disciplinary") conception of anthropology and its subjects. Rather, we are faced with an interconnected vastness in which multiple times and places constitute the human.

As Knight and Bandak note in the introduction to this volume, Serres writes with an audacity few anthropologists would dare to attempt. He pushes his analogies to the breaking point, acknowledges their weaknesses, and then doubles down. The effect can be brilliant. In *The Parasite*, he knows that his usage of the term is unorthodox; actually, by the standards of modern biology, it is wrong! Parasites are small. They live inside other bodies. It is a mistake, then, to view human interactions through the lens of parasitism, which ends with "mollusks, insects, and arthropods" ([1980] 2007: 6). Undaunted, Serres turns the tables: "The basic vocabulary of this science [parasitology] comes from such ancient and common customs and habits that the earliest monuments of our culture tell of them, and we still see them, at least in part: hospitality, conviviality, table manners, hostelry, general relations with strangers. Thus the vocabulary is important to this pure science and bears several traces of anthropomorphism. The animal-host offers a meal from the larder or from his own flesh; as a hotel or a hostel, he provides a place to sleep, quite graciously, of course" ([1980] 2007, 6).

Serres opts for the analogy and its suggestive powers, which derive from the real history of words. The analogy allows him to move through analytical time and space with abandon, finding invariants across centuries and continents, across epistemologies.

In *Branches*, Serres argues that the modern sciences are best understood when seen as another version of the Christian Church, in all its hierarchies and heresies, its dead letters and rejuvenating spirits. Whether this is a strong or weak argument is very much in the eye of the beholder, who will see nothing unless they sense the qualities of Christianity and science that connect both traditions to an interpretive third. It is an analogy or an overlay, not a rigorous comparison. For a discipline like anthropology, stuck in its endless love–hate relationship with comparative analysis, the Serresian mode of analogy—seeing A in B, seeing A as B, keeping all their differences intact—might help us intuit profound likenesses and real historical relations that strict comparison would miss.[2] The tapeworm lives inside its host, but human parasites do not? "We dress in leather and adorn ourselves with feathers," Serres writes. "We live within the flora as much as we live within the fauna. We are parasites; thus we clothe our-selves. Thus we live within tents of skin like the gods within their tabernacles" ([1980] 2007, 10).

What is Serres doing? The power of his analogies rests in their ability to sim-plify and expand. They do not facilitate work across disciplinary boundaries— there is no "interdisciplinarity" in Serres; that is bad marketing—instead, his analogies completely refigure and revalue a discipline's subject matter. This is the radical effect of *Angels*, in which Serres likens modern communications, in-formation flows, and media cultures to a world filled with angels, with message bearers, mediators, linkages. It is playful and in earnest. The problems of a secular age, Serres contends, are connected to—grow out of and reproduce—dilemmas that confounded premodern worlds, in which angels were active and real. The language of angels, as Serres uses it, describes something real in a world nominally without angels but in which the form and function of angels persist, modified and obscured. Is the analogy sound? Can one simply overlay these worlds? Even Serres is unsure. He registers his doubts (or nods to our own) by letting the book's characters express them in conversation. The result is not a new theory of information systems, climate science, or media studies. Serres denies these fields the shaping power of their idiom by dressing their subject matter in a new language, ironically an old language with powers modern science claims to have transcended or replaced. In the appendix to *Angels*, in an interview with himself (!), Serres criticizes and defends his own agenda. "Might these metaphors," he asks, "relieve you, for example, of a treatise on cognitive science or philosophy of language?" To which he responds, "What a relief! Who cares for categories such as these! Their only purpose is to shroud themselves in a terrifying vocabulary in order to protect privilege and corporate interests. Since, as people say, today we are seeing the dissolution of philosophy, we find ourselves once again living in the time of original beginnings. Here glitters a vast and interconnected pool, here flows a torrent of shimmering brilliance, seized at the moment of its commencement" (1993, 296–97).

I am not sure which is more terrifying, disciplinary categories or a "torrent of shimmering brilliance." The figure of the angel, a being or force that enables attachment, that enables movement and flow, is the device Serres uses to trace interconnections, which lead always to starting points. The assemblage is there, but it is the work of analogy, so prominent in everything Serres does, that gives mere connection the spark of annunciation. Serres actually compares this moment to arc welding. "It produces a sudden bright flash of light," he says, "which leaves on two pieces of metal a scar, the color of which goes from blue to cherry red" (1993, 297). If the pools of creation shimmer, it is the fire of these connections that lights them up, and knowledge, which follows relations, is the network of scar tissue that angels have been building from the beginning.

Angels are not exotic creatures. I spent much of my life believing in them, and many people I know still do. Serres captures something essential about angels, and allows me to relate to them in new ways, when he describes them as translators and mediators who flourish in the margins and transition zones of human cultures. In the Abrahamic traditions, angels swirl above us and are usually unseen. They speak to prophets and saints, who are among the few humans blessed with the ability to converse with angels, a process often portrayed in scripture as frightening or painful. The angel Gabriel choked Muhammad three times, after which the Prophet could finally recite the divine revelation in a language intelligible to humans. Angels possess higher knowledge, or they are a link to it (see Boylston, this volume). Where does this analogy take us? What does it allow us to comprehend?

If I see my own work with a kind of Serresian double vision, in which angels pervade it and something like angels—analogous figures and interpretive thirds—interact with me to create comprehension; if I think of my work in this way, I am suddenly overwhelmed by an awareness of how many mediators, translators, and special messengers have instructed me. Is ethnography anything but this? What I call arguments or analysis is made of the blue and cherry-red scar tissue that binds me to my Balgawi friends, and much of that connecting substance was made by a few arc welders—the angels to whom I was a protected person or, alas, a hungry parasite—who spoke to me in special languages, not simply English or Arabic but languages of translation, of revelation. Many of them were themselves travelers. They had spent years in Europe or America. They had become interstitial, hybrid. They could relate to me in diverse registers, from the cosmopolitan to the most intensely local forms of knowledge. If the analogy is parasitism, we fed off each other. If it is angels, we communicated something that could not be heard in ordinary speech.

Serres is fascinated by the churning relationship between life and death, with the flow of energy and information that generates human worlds. In *Statues*, he tells us that one of the first places he visits when he comes to a new city is its graveyards. The history of the place has accumulated there, enshrined on named graves or forgotten in a deep sedimentation of bones. The living city is always in dialogue with the dead one. In Jordan, graveyards are not always impressive places, but the landscape itself is marked by the names of the dead. Entire genealogies can be mapped onto villages and the spaces between them. Local history is a dialogue of dead-living voices; it is filled with messages from the past, spoken and heard now, in territories made of ancestral names. "Oral tradition" is our diminished word for this intricate space/time, and the process of writing it down, itself a translation, is also a disturbance. In a passage that stopped my breath, Serres captures what I thought I was doing and what Balgawis helped me do: "When you're writing or drawing at a desk under the flat light of a lamp, make the page of snow fall over the domain of the dead, the founders or keepers of all the secrets of places; evoke them, help along their return to the sun's blue brilliance. Neither the drawing nor the writing is worth anything if it doesn't reveal those who are veiled beneath this shroud. Their heads bore through the sheet. Without this resurrection, no sign nor language, which come from the sheet and the dead" ([1987] 2014, 39).

I will end with an evocation, a story that seems right for thinking about Serres. He was not afraid to tell tales, and he helps me remember them.

My friend Humoud al-Jibali, whom I wrote about at the end of "Keeping to Oneself," died twice. The first time, he came back. He had finished editing a long account of his mother's life, which we had translated into English together, and his lungs and heart failed. He fell into a coma and was pronounced dead. With family gathered around him weeping, he suddenly revived. He was delirious, but his first act was to call me in America. He left a message. Only later did I realize it was him. The voice on the answering machine was faint, and it spoke in a language I did not understand. I thought it was a wrong number, but it was simply the wrong interstitial language. Humoud was speaking a mélange of Arabic and Swedish, not the English and Balgawi dialect we spoke to each other. He told me that, while he was dead, he had talked to me and to others and that there were more stories he needed to tell me. I had been with him, he said, in whatever place he had gone to, and when he returned to consciousness, he spoke to me in the language—the hybrid, angelic tongue?—he kept for his own family, his Swedish wife and their Balgawi daughters.

"He had to see you again," his oldest child told me after he died again. "He came alive whenever you were around. The work you did together was so important to him."

The connection between us, which produced knowledge, was strongest at the points where Humoud and I had collaborated, where we had made something new that could be finished only in moments of translation, only in hyphenated spaces. Serres knew the importance of these bonds. His work is filled with relations. With transubstantiations. His analogies are connective tissue.

I am reminded of the story of Jacob and the angel. They struggled all night long. They would not let each other go. The angel wrenched Jacob's hip. Jacob would not loosen his grip until the angel blessed him. The encounter left Jacob with a new name and a mangled joint. Blessed, wounded, he moved through the world in a new way.

Neither the drawing nor the writing is worth anything if it doesn't reveal those who are veiled beneath this shroud.

NOTES

1   I thank Alberto Corsín Jiménez for putting this image in my head. He discusses Serresian "ooze value" in his contribution to this volume.

2   To see why this might work, I recommend a close study of Matei Candea's book *Comparison in Anthropology: The Impossible Method* (2019).

CELIA LOWE

4

SERRES, THE SEA, THE HUMAN, AND ANTHROPOLOGY

Like a wave returning to the shore, the relevance of the human is being re-
newed. In contrast with Michel Foucault's "death of man" where an ocean swell
washes away the figure of the human, the Anthropocene has cast the human
back onto the sand and into the center of things. As destroyer of worlds, the
contemporary human is responsible for conditions that make life on Earth precarious. There are the military-industrial complex, the plantationocene, and
the capitalocene responsible for producing atomic fallout, plastic waste, land
degradation, and toxins at a global scale, but there are also the fishers of the
Togean Islands of Indonesia where I did field research in the 1990s who fish in
dugout canoes for reef fish using cyanide and who ignite homemade bombs
in Coke bottles to catch larger schools of fish. At nearly eight billion people,
humanity has become a force of transformation in its own right throughout
every corner of the globe at every scale: extinctions, climate change, zoonotic
disease, plastics, pesticides, war—all are destroying the basis for the habitability
of life on Earth. Anthropologists have historically focused on the specificity of
people within particular communities rather than on the "human" writ large.

We have tended to reject the idea of humanity as a whole or have taken the "whole" itself as a projection of specific and temporally situated communities and conversations (Tsing 2000). And yet, in addition to our anthropological attention to divergent senses of meaning and of disparities between populations, the world itself provides evidence that we need the global scale to think of the totality of material and human processes and to make sense of shared risk and responsibility. From the village to the boardroom, a new relationship with the Earth and a new understanding of our position on a unique and lonely planet are required.

In *The Natural Contract* ([1990] 1995, 3), our friend and interlocutor in this volume, Michel Serres, is interested in humanity and the world as totalities: "Suddenly a local object, nature, on which a merely partial subject could act, becomes a global objective, Planet Earth, on which a new, total subject, humanity, is toiling away." He says, "At stake is the Earth in its totality, and humanity collectively. Global history enters nature; global nature enters history: this is something utterly new in philosophy" (Serres [1990] 1995, 4). And in anthropology! Serres describes the "tectonic" force of humanity while at the same time deconstructing our old social sciences as inadequate to the task at hand: "On Planet Earth, henceforth, action comes not so much from man as an individual or subject, the ancient warrior hero of philosophy and old-style historical consciousness, not so much from the canonized combat of master and slave, a rare couple in quicksand, not so much from the groups analyzed by the old social sciences—assemblies, parties, nations, armies, tiny villages—no, the decisive actions are now, massively, those of enormous and dense tectonic plates of humanity" (Serres [1990] 1995, 16).

Serres uses the marine world to construct his thoughts and understandings of "human" and "world" in the contemporary philosophical and material moment. He often builds his philosophy and understanding of material and human worlds by using models and metaphors taken from the sea and his experiences as a mariner. As a sailor myself who spent seven years circumnavigating the globe and as an anthropologist whose original fieldwork was among "sea nomads" in Indonesia, his metaphors and comparisons appeal to me. Thus, in this chapter I read ethnographically two filmic stories about the ocean and its creatures. The first is a tale of progress found in the biography of mariner Jacques Cousteau, and the other is an exploration of human/animal relationality in filmmaker Craig Foster's story of a yearlong encounter with an octopus. My purpose is to explore contemporary formations of the human and to discuss Serres's reactions, responses, and solutions to conditions of the Anthropocene and what he might have to offer for anthropology. Serres writes, "From the beginning of our culture the *Iliad* is opposed to the *Odyssey* as conduct on land, which takes only people into account, is to the ways of the sea, which deals with the world" (Serres [1990] 1995, 41). For Serres, the sea represents the "world" in contrast with the land, which is centered on the "human."

Diving into Serres

Michel Serres thinks from personal experience. Like anthropological autoeth-nographers, Serres's intellectual world extends his own life experience onto the page: "His thought has been sculpted by the details of his life" (Watkin 2020, 4). It's not that Serres thinks he is so important to the story; indeed, he insists he is not. It's that many of his concepts, images, metaphors, and anecdotes begin from meaningful moments of personal awareness. Born in 1930, he says "from the age of nine to seventeen, when the body and sensitivity are being formed, it was the reign of hunger and rationing, death and bombings, a thousand crimes" (Serres and Latour [1992] 1995, 2). While his better-known contemporaries in French thought cite the political excitement and rebellion of Paris 1968, Serres was moved by the nuclear bombings of Hiroshima and Nagasaki to detest even the metaphorical explosions of political and academic contestation. Serres's thought is embedded in the Anthropocene, a world shaped for him by thermo-nuclear explosion and war and which also makes evident the wholeness of the world through what Joseph Masco calls the "Age of Fallout" (Masco 2015). As Christopher Watkin, an eminent interpreter of Serres, explains it, "The same mathematical science that Serres was so ardently pursuing, with whose progress he was directly complicit, has made possible the devastating bomb. Serres writes of the bomb as a problem which explodes 'on every page of my books', raising questions that science could not answer and driving him to study phi-losophy" (Watkin 2020, 9). Serres claims, "Hiroshima remains the sole object of my philosophy" (Serres and Latour [1992] 1995, 15).

Watkin believes the "body" and the "landscape" shaped Serres's intellectual career, but there are also the sea and its creatures, which form a basis for Serre-sian story and metaphor. While the bomb is an undercurrent for Serres, his re-lationship to water—rivers and seas—overtly influences the surfaces of his writ-ing. Serres was born on the Garonne River, where his father and grandfather were bargemen. "'Family history has it that in the great flood of 1930, when my mother was pregnant with me, she was evacuated from our house by boat from the second-story window. Thus, I had been afloat while still in the womb, and not just in amniotic fluid!'" (Watkin 2020, 5). "My childhood flowed through the riverbed and gave me a riverman's eyes," he writes (Serres 2015a, 61). Serres joined the French navy in 1949, leaving and then returning to become an officer in 1959. From here he draws on stories of being a mariner to add to his riparian observations. One evening on night watch in the middle of a squall, his radar told him abruptly that there was a small island ahead! He would mistakenly run his ship aground, sending two hundred men to their deaths. But that outcome was not to pass. His radar had only picked up the shape of a squall, the heavy rain registering as land on his ocular screen: "My radar's greyish eye returns to his usual calm" (Serres 2015a, 59).

Aboard a fishing vessel on a different voyage to Iceland and Labrador, Serres tells a story of how the sciences and arts confront one another in his experience navigating on the ocean (Serres 2015a, 55–56). On the Atlantic, fishermen perceived the sea directly through currents, surface winds, and the movements of birds and cetaceans. The fishermen could use the "marine map" that was given from nature and built up through perception and time. Serres, on the other hand, used a sextant (his calculations) and celestial bodies through which he produced "angles and numbers" and a location on a paper chart. The first, what an anthropologist might call the indigenous knowledge of a Brittany coast mariner, was an art, while the other, the "angles and numbers," was a science. Yet even Serres's science relied on observations of the stars and had an art to it. Then something happened that took the fisher and the philosopher both away from the practice of observation: the invention of the global positioning system (GPS), "which swept me and the old sea dog away sending us together to the historical open-hopper dredge" (Serres 2015a, 55).

Rather than taking the path of specific or community knowledges, Serres speaks of progressions in analyzing his voyage to Iceland and Labrador: "Every emerging science rejects the preceding one as merely sensory; I became the visual twin of the fisherman, his empiricist brother." While the fisherman's comfortable knowledge of the ocean and its signs and Serres's knowledge of the precise "minute in which to take a bearing as the sun goes down" are swept away by the GPS, the eyes of the world for Serres, the sparkling waves and twinkling firmament, both of which change with "seasons, latitudes, hemispheres, climate, storm and calm before the storm," remain (2015a). Modern human knowledges mistakenly imagine progress and replacement, and yet the human does not replace it all. Instead, the human is one among many elements of the world. Serres views humanity as a "tectonic force," but his adventures aboard the fishing vessel do not support some valences of the term "Anthropocene" that propose that the world-shaping human is all that is left. The human who creates the world does it as partner, parasite, and symbiont with other tectonic forces like ocean and air currents.

The Human in the Anthropocene

To be sure, the history of philosophy is replete with myriad figures of the human. There is the antique figure of Aristotle's *Zoon politikon*, the human as political animal. There is *Homo faber*, the human as maker. There is *Homo ludens*, the human who plays. For antique writers, the human was the creation of a demiurge, a creator responsible for the universe. In Christianity and Judaism, the human is created in God's image. The human gradually became autonomous from God and given qualities that separate us from a deity. In the Enlightenment, Kant's human, for example, was distinct as a bearer of morals, freedom, and rationality and the ability to adopt and pursue our own thinking ends. By the end of the nineteenth century, however, the autonomous human defined by thought, morality, and progress came under suspicion. Nietzsche framed the human as subject to the irrational force of a will to power, and he is credited with the "death of God" leading logically to the death of a universal human in God's image. Likewise, there was Freud's human driven by irrational desires hardly accessible to conscious thought. The human of anti-humanism was a social and historical being, not a creature with capacities over which they had conscious, reasoned control. Late twentieth-century anti-humanists such as Foucault, Derrida, or Deleuze believed that the figure of the universal human did "violence to differences and singularities" (Watkin 2017, 3) and the human was disgraced as universalizing and totalizing. Foucault explored the notion of the human, "an invention of recent date," as the consequence of developments in the fields of life, labor, and language. As institutionally elaborated fictions, transformations to figures of the human would eventually bring about the "death of man."

With God and the universal thinking human under suspicion in modern philosophical figures of the human and in the discipline of anthropology, where do we stand in the Anthropocene, which seems to place the human back at the center of things? Many recent French thinkers including Serres knit together the human and the world in novel ways. Watkin argues that Latour, Meillasoux, Badiou, Malabou, and Michel Serres are not humanist, anti-humanist, or post-humanist. Rather, "what we are witnessing in French thought today is a series of returns to the human after the human" (Watkin 2017, 4). These thinkers are interested in a figure of the human that rejects the impossibility of drawing together matter and meaning or the human and nature. Contrary to the recent history of anthropology, which disdains the universal, Michel Serres describes a new form of universal humanism responsive to the environmental crisis. He develops a figure of the human situated within ecology and objects and within what he calls the "Great Story of the Universe" (Watkin 2015). His philosophy does not have one human of history and meaning-making and a separate world of inert nature and objects.

What could be counted on by previous framers of the human, at least through the later twentieth century, were the air, water, soil, chemical molecules, weather, and climate that form the building blocks of the human as "life form" (as opposed to the human as political, social, thinking, making, playing animal or the human subjected through ideology or institutions). Indeed, scholars who were interested in the human as a biological or physical entity in the late twentieth century tended to be divorced from philosophical thinking on the human, derided as "rank empiricists" particularly within cultural anthropology and allied fields. It was not necessary to imagine the human as integrated with a surrounding biophysical world because the features of that world could be relied on, except in extremis. Today, however, the Anthropocene challenges earlier figures of the human and knits the human as living creature into its surrounding environment. Serres argues that this knitting must take the form of a "contract" lest the force of hierarchy between humans and nature reassert itself. He says that "the Earth speaks to us in terms of forces, bonds, and interactions, and that's enough to make a contract. Each of the partners in symbiosis thus owes, by rights, life to the other, on pain of death" (Serres [1990] 1995, 39). Simultaneously, the Anthropocene introduces new forms of universality that are antithetical to many modern forms of anthropology, a field that emphasizes human difference and specificity. Thinking with Serres pushes anthropology to embrace scalar crossings and porous becomings, both in our concepts and in our conversations with research informants.

The epoch of the Anthropocene is full of rude storms and strip-mined territories but also invisible and barely perceptible material objects and processes lying in wait along a trajectory set to shape our future. From the hidden toxins that will give us cancer someday to the imperceptible greenhouse gases driving climate change, the present and future become apparent through relational interactions with and infestations by minor and micro matter. The human is now inseparable from the hundred or more synthetic chemicals that all of us hold in our tissues. The human is now indistinct from the smogs and smokes that fill our lungs or the temperatures that exhaust us or the storms we must weather. Our skies are filled with so much light and our near space filled with so much space debris (Johnson 2022) that, on land, we can no longer see the stars that inspired the myths of our ancestors (though at sea Serres could still navigate by them). Our agriculture is utterly dependent on the insects, especially pollinators, that we are destroying through pesticide use and habitat loss (Milman

2022). As Serres recognized, it is useless and mistaken to view human agency as inclined toward improvement of the Earth or to view the Earth as the inert repository of resources available for human development. Instead, the human violence, bloodshed, and subjugation that have always shadowed narratives of social improvement and that are so apparent in the splitting of the atom are equaled in our approach to the Earth where despoliation and pollution-as-appropriation submerge ideals of development, enhancement, or productivity.

Serres observes, "The plate of humanity has long disturbed the albedo, the circulation of water, the median temperature, and the formation of clouds or wind—in short, the elements—as well as the number and evolution of living species in, on, and under its territory" (Serres [1990] 1995, 16). "Our collectivities are becoming as powerful as seas and share the same destiny," he writes (Serres [1990] 1995, 20). If the Anthropocene is the age in which the entire biosphere seems to be shaped by human activities, anthropos in the Anthropocene is the figure of the human who, from the molecular level out, is porously imbricated with its own material practices of construction and destruction of the biosphere but also with a nature it does not and cannot successfully manage or control. Serres ridicules the notion of human control and mastery, and he finds more similarities than differences between matter and code. For Serres, the human sits alongside a cosmos that is changed by but does not answer to what is human. This suggests that the anthropologist's attention to disparity and inequality needs to be extended to nature while holding in sight the unifying view of the human as a tectonic force. Serres demands a new "natural contract" ([1990] 1995) to address this configuration in which the human and the world are totalities with which we must grapple.

Jacques Cousteau and *Homo Aquaticus*

Serres identifies the human as a parasite who is killing its host, planet Earth, including its rivers and oceans. The story of another French mariner, Jacques Cousteau, a compatriot and fellow naval officer only somewhat older than Michel Serres, provides an example of how human domination over an enchanting and powerful environment was reframed as the vulnerability of both humans and the planet during a period contemporary with Serres's life. While I knew of Cousteau from my youth, I encountered the details of his story in the film *Becoming Cousteau* (Garbus 2021), a biographical picture about the inventor of the Aqua-Lung (the first self-contained underwater breathing apparatus, or "scuba") and a pioneer of marine conservation. I interpret the film ethnographically as more than a story of Cousteau's achievements; it also explores transformations to the Earth and in the figure of the human within a milieu also inhabited by Serres the mariner.

Jacques Cousteau first entered the ocean as a free diver at the edge of the Mediterranean in the 1930s and experienced its profound beauty. Rather than a philosophical adventurer, Cousteau was a physical discoverer who implicitly grounded his work in an Enlightenment figure of the thinking, creative, and playing human: the human as curious explorer. In the film, he says, "Every explorer that I have met has been driven by curiosity. A single-minded, insatiable, and even jubilant need to know. We must go and see for ourselves." Cousteau is also *Homo faber*, man the maker who controls his environment by using tools. In the winter of 1942, Cousteau and his partner Emile Gagnon invented the scuba regulator from a motor car part. In these early years, nature and culture were bifurcated for Cousteau: there were creative, inquisitive, and vulnerable humans and a powerful yet knowable nature. While the sea was their utopia, in the world of wartime Europe above the surface, nothing made sense. In Cousteau's description of the insanity of war, we can see a specter of anti-humanism as well, how he, like Serres, turned away from war toward the ocean, the realm of the *Odyssey* where every wave is not plowed by humans.

One must think about what most of us knew of undersea worlds before Cousteau in the first half of the twentieth century. The answer is nothing. His film *The Silent World*, which won an Oscar and the Palm d'Or in 1956, was the first to introduce the public to moving images of the undersea ocean in color. Picasso attended the Cannes premier of *The Silent World* and was enchanted by the colors he saw that he claimed were "unknown to man." Cousteau describes his experience below the surface as "a moment of grace." He says, "I slide into the depths aware of living in harmony with an environment very different from the world above. I swim almost effortlessly like the fish I meet. I am an unexpected guest, spellbound by this splendor, this silence, this harmony" (Garbus 2021). Through his attitude of reverence, he imagines that the human fits into these natural worlds without disturbance or disruption. He is like the fish.

By 1962, Cousteau has proposed a time when humans will live continuously under the sea. He says, "It is a succession of carefully planned steps; we are moving into the sea deeper and longer" (Garbus 2021). He imagines the permanent occupation of the seabed and conducts experiments in living at the bottom in the vessel *Diogene*. He sees those living aboard *Diogene* as "serving the future," and for him, the experience creates a new figure of the human that, like Serres, he uses to connect across forms and evolutionary time. "Physically we haven't changed," Cousteau says, "but we are not exactly the same. The dolphins, seals, and whales were once terrestrial. Maybe we are witnessing the gestation of a new man: the water man. To live in this strange world, the man of tomorrow will have to adapt" (Garbus 2021). He calls this new man *Homo aquaticus*.

Yet, precisely at the moment when people were beginning to know the beauty and mystery of the undersea world of Jacques Cousteau, Cousteau realizes that his dream of *Homo aquaticus* is folly, and he begins to understand the human as world destroyer. He says, "The start was curiosity, the enthusiasm about beauty. Then came the period of alert, because we were looking at things that were actually disappearing. And so, my past life as just a mere explorer is over. What we are facing is the destruction of the ocean" (Garbus 2021). Furthermore, Cousteau must include *himself* among the destroyers. Sequences from *The Silent World* show Cousteau's team dynamiting coral reefs to inventory the fish that lived among them, riding on the backs of turtles, brutally killing whales, and spearing sharks. He says, "The world at that time didn't understand the danger to the environment" (Garbus 2021).

In 1977, Cousteau returns to the Mediterranean, where his experiments and explorations under the sea had begun. He narrates, "In only three decades the sea floor has become a desert bleak as the surface of some barren planet. In this submerged desolation, the water temperature seems to rise burning our hands in spite of our gloves. Our eyes are burning, tears flow down our faces blurring our vision, the pain is unbearable. We have penetrated a zone of death, a region where no living thing can long survive" (Garbus 2021). In Cousteau's late twentieth-century turn, he sees evidence, as does Serres, that the rational thinking human was a fiction, and yet Cousteau still possesses the modernist faith that humans can determine the future and fix what they have destroyed. It is under these conditions that he becomes *Zoon politikon*, a global environmental activist and a politician.

Cousteau's rational human is unacceptable both to contemporary anthropologists and to philosophers of the human. Serres writes that "nature according to the Moderns is reduced to human nature, which itself is reduced to reason. The world no longer has any place there" (cited in Watkin 2020, 336). Although Cousteau and Serres inhabited the same milieu and shared the same distaste for war, there are differences between the two men. Serres would not appreciate Cousteau's tool-making as a solution: Serres writes, "You would have trouble finding a single tool whose future flowed in the channel its designer foresaw" ([2004] 2020, 162). In the Togean Islands of Indonesia, I witnessed the scuba regulator Cousteau invented hooked up to a tire compressor to allow cyanide fishers to penetrate the depths looking for grouper fish. Serres would also not find Cousteau's political activism a credible antidote: "I consider ecological ideologies to be the umpteenth instance of the city and city dwellers trans-historical victory over the fields and the woods" (cited in Watkin 2020, 337). Although both French luminaries viewed the human as a parasite killing its host, Serres wanted to establish new relations with the Earth, and this can't be done within Cousteau's old forms of contestatory politics.

The Human as Amphibious Animal

The implicit figure of Cousteau's human acts on nature but is separate from it, even when *Homo aquaticus* is saving nature. Afterall, the undersea world did not exist for most of humanity until Jacques Cousteau showed it to us with his cinematography and gave us a technological path to enter it with our oxygen-dependent bodies. The Anthropocene, on the other hand, is a new time when the human and their surround are conceived of together and can be imagined through the porosity between them. The Earth is at risk due to our efforts at improvement, and we can't use the master's tools, including the concept of the autonomous sovereign human, to extricate ourselves from it. In dealing with planet Earth, Serres claims we need new relations and a new human:

> Today's world is screaming in pain because it is beginning its childbirth labor. At serious risk, we have to invent new relations between humans and the totality of what conditions life: the inert planet, the climate, living species, visible things and invisible things, sciences and technologies, the global community, morality and politics, education and health . . . we are leaving our world for other worlds, possible ones, and will have to abandon a hundred passions, ideas, customs, and norms, brought about by our narrow historical duration. We are entering an evolutionary branch. . . . Either a new human, a citizen of the world, will appear, or humanity will totter. (Serres [2004] 2020, preface)

One way Serres brings human and world together is through the body. In *Variations on the Body* ([1999] 2011, 34), Serres writes, "Whatever the activity you are involved in, the body remains the medium of intuition, memory, knowing, working, and above all invention." While the body remains a foundation for knowing, feeling, or making, the body is not whole or impenetrable but porous. For Serres, the body sustains transcorporeal connections to the world: "The other makes my flesh, their flesh blended with mine: this, this thing right here haunts my body, and this animal too, but this one, the other, above all, enters into my body, one so mixed, so crossbred and penetrated that, lost in the very middle of that great crowd that effaces me, I vanish like a bit of vapor" (Serres [1999] 2011, 57). Anthropologists can be interested in the "great crowd" that is human, nonhuman, and other-than-human rather than the community, the singular individual, or the subject.

A second film I interpret ethnographically problematizes the ocean and its creatures to understand how the body enters into and shares with the world, creating a new space of potentiality. This is the story of cinematographer Craig Foster's yearlong encounter with a single common octopus, *Octopus vulgaris,* in *My Octopus Teacher.* The first time Foster spots the octopus, she is mysterious and hidden, wrapped up in a shield of shells, algae, and other debris. Foster's story unfolds the complexity and interconnection taking place within the octopus's unique world as the two come together in the coastal waters of South Africa. Distinct from Cousteau's global travels aboard the Calypso conquering technology and territory to master the global ocean, Foster immerses himself into the physical spaces and life of this singular creature as he works to invent new relations between himself and his other-than-human companion. I read his work as a Serresian evolutionary branch where a singular person is attempting to establish a new contract with nature.

To experience the octopus, Foster must enter her realm in a way that makes connection possible, and yet his body is vulnerable to temperature, buoyancy, and waves, and he will also need to breathe. He refuses to use scuba equipment to breathe or a wet suit to keep himself warm; rather, he enters the 8.5°C/47°F water simply with fins and a weight belt. For Foster, the wet suit is an unacceptable barrier between him and the water despite the hostile environment; instead of substantive technological interventions, he uses the ocean itself as an instrument of transformation. "When you enter the water," he says, "the cold gives the brain a flood of chemicals and your whole body feels alive" (Ehrlich and Reed 2020). His physical liveliness is dependent on those active chemicals and the temperature of the water. He has only ten or fifteen minutes in the cold when everything feels fine and normal and he can coexist with the octopus and her surroundings. Likewise, he has only minutes below the surface before he needs to breathe. This sets a limit on his octopus encounters, on his contract with nature. When Serres speaks about the capacities of the body, he does not view the human body as weak or unable to accept pain. During Foster's year with the octopus, his body accepts the direct contact with the ocean's harsh conditions and begins to transform: he can hold his breath for an increasingly long time; he admits he looks forward to and even craves the cold; he is psychologically more engaged. Serres says the adventures of the body are like sailing; it is the physical motion of our actions that makes us subjects.

When Foster enters the water, he breaks down separations between subjects and objects, humans and nonhumans refusing human dominion over nature and its creatures along with fantasies of conquest and reason. Foster is not *Homo faber*, man the maker like Cousteau who masters the ocean by inventing new technology. He is not Cousteau's *Homo aquaticus*, "serving the future" through living aboard the *Diogene*. Instead, Foster describes himself as an "amphibious animal," the human who crosses porous boundaries between worlds, illustrating their commonalities as well as the limits of commensuration. In an embodied way, Foster learns about the types of kelp forest, mentally maps the environment around the octopus, and learns about her predator the pajama shark. In doing so, he has broken down barriers between himself and the octopus, between human and animal, and between the human and a surrounding "environment." Throughout the year, he gets closer and closer to the octopus, more integrated into her world, which becomes a shared world as the octopus also merges with Foster. There is a definite moment when the octopus's fear subsides and she also becomes curious and interested in him, though "not taking stupid chances" (Ehrlich and Reed 2020). The Anthropocene both describes and demands this merging of worlds.

In Serres's work, he aims to unite matter and meaning, object and code. In one scene, the octopus reaches out a single tentacle to touch Foster, sending a message across species boundaries. Within human mythmaking, this looked to me like the moment when Michelangelo's God gives life to Adam on the ceiling of the Sistine Chapel. Nevertheless, while all things store and transmit information, codes are not always legible. We don't know what the octopus is actually saying, and yet there is a result. The outstretched tentacle inaugurates trust between Foster and the octopus, leading to a moment when the octopus finally comes all the way out of her den and goes about her business in his presence. Foster believes the octopus to be saying, "I trust you human, and now you can come into my octopus world." This seems like more than anthropomorphism since the octopus from that point on is willing to wrap itself around Foster's hand and rest quietly on his chest. Foster speculates on what she was getting out of the relationship with him. He concludes, "It was stimulating for that huge intelligence," and perhaps it gives a "strange octopus level of joy" (Ehrlich and Reed 2020).

Foster does extend human attributes to the octopus in a way Serres refuses. "A lot of people say that an octopus is like an alien," Foster says, "but the strange thing is, as you get closer to them, you realize that we are very similar in a lot of ways" (Ehrlich and Reed 2020). According to Foster, the octopus has many characteristics we define as human but that through his analysis are extended to this other creature: she has a tremendous intelligence that includes the creativity to deceive her predators and a curiosity about who Foster is. As Foster watches the octopus fling her arms at schools of tiny fish to startle them, he realizes that she is playing with them: *Polypus ludens*—octopus the playful. Watkin argues that Serres does not want to breach the human–nonhuman divide in the way Foster does by extending human characteristics to the nonhuman, however. Instead, Serres insists, "that everything—including human beings—receives, stores, processes, emits information, opposes the exclusivity of human agency by multiplying it, showing it to be one mode of agency among many others, quantitatively notable but not qualitatively unique" (Watkin 2020, 298).

It took going into the water every single day for a year and making himself a part of her environment for Foster to begin to mesh with the octopus. "There is something to be learned from this creature," he says, using his curiosity to learn *from,* not *about,* her in order to take up the octopus as a "who" rather than a type, just like the activists I observed defending zoo elephants in Seattle, Washington, in ethnographic research I conducted with Ursula Muenster on elephant viruses (see Lowe and Muenster 2016). The octopus could transform her color, texture, and pattern, sending information and signaling intention. She could grow horns and walk bipedally. Once, she walked up onto land to escape from a pajama shark, sharing amphibious animality with Foster. What Foster learns from his octopus teacher is part of a larger story of water temperature and flow, octopus intelligence and curiosity, shark predation, weather, human frailty and adaptability, and connectedness. The similarities Foster experiences are not limited to turning the octopus into a human or vice versa. Similarly, the human in the Anthropocene does not have to leave the human at the center of the story, even as *Homo vastator,* man the destroyer. Instead, a new figure of the human for the Anthropocene is a human who exists within a web of a totality multiplying attributes with animals, phenomena, and objects.

In Conclusion

As winter comes to the Weddell Sea in the Southern Ocean, surface ocean temperatures fall below –1.5°C/29°F and the waters begin to freeze (Riddick 2013). Microscopic crystals grow, forming a layer of ice and expelling salt into the water. Brine is denser than sea water, so it slowly sinks two miles to the ocean floor, where this enormous outflow moves toward the equator. The flow becomes part of a global circulatory system stirring the world's oceans over the course of a millennia. This loop is critical to life on Earth; ocean circulations regulate ocean temperatures, which, in turn, steady the air and atmosphere. Earth's climate is a result (in part) of ocean circulation patterns, but the Earth, as we know, is changing. Just as ocean circulations begin with microscopic ice crystals, climate change begins with an infinite number of minor insults to atmospheric composition. The *ketinting*, the single-cylinder gas-powered outboard motor Togean Island fishers use to get to their fishing grounds, and the airplane that flies conservationists to the Togean Islands to speculate on the ways fishers destroy the environment, both drip carbon-filled exhaust into the air one trip at a time. The combined tectonic force of each human addition of imperceptible carbon atoms into the atmosphere is warming planet Earth. In Antarctica, this is causing ice shelves to collapse and sea ice to diminish, which may ultimately reduce the flow of brine, slowing ocean circulation and making the Earth less or uninhabitable. The two forces, one chemical and thermodynamic and the other a result of human ambition, ingenuity, and desire, may eventually come together to bring catastrophe to life on Earth. How can anthropologists place these massive forces into a common frame? What difference does it make that humans, who anthropologists could once study in isolated groups, now comprise what has become a planetary force?

As we return to the "human after the human," we can no longer rely on our environment to support our reproductive capacity, our longevity, our agriculture, our air. The work of Michel Serres points us in several productive directions toward a more planetary consciousness and toward an understanding of the human as one element among many in the larger story of planet Earth and what he calls the "Great Story of the Universe." His examples bring specificity to the tectonic force of the human and to the biological, chemical, and mechanical forces that shape the Earth while working from examples that span multiple times and genres. When he invites the reader aboard his naval ship to look into the eye of his radar, the radar becomes a kind of eye on and of the world. Serres brings together the human technology of radar with the forces of nature, the sea and the squall. In his writing, the radar becomes a naturalistic entity, an "eye," rather than a tool to subdue the Earth like Cousteau's scuba or undersea dwelling. For Serres, the human is not always at the center of the story, nor is the human defined by a limited set of traits, as the maker, the player, or the political animal. For Serres, the human is the animal without essential traits who, for better and worse, can flexibly extend themselves into new environments, like Foster following his octopus into the icy waters of South Africa. Serres's human becomes a subject through bodily engagement like Foster's amphibious animal.

Where the human was once only vulnerable to planetary forces like wave and wind that can make a life precarious, the human is now a force in its own right that has the power to make the planet itself unstable and precarious.

Serres believes humans are "quantitatively notable, but not qualitatively unique" (Watkin 2020, 313) in comparison with other entities who inhabit this planet. As an anthropologist working in an out-of-the-way part of Indonesia, I made heroic and seemingly successful attempts to ignore and explain away the Togean fishers who bombed reefs and fished with cyanide (Lowe 2006). Either I would follow the conservationists who found these practices, and thus these people, despicable, or I must find an argument that could make them the victims of larger forces like capitalism or poor governance and preserve their innocence. To acknowledge their complicity would be to follow the conservationists who had abandoned them to the savage slot. But what if to be human means to be part of Serres's great tectonic force? Perhaps we all are guilty of fouling our own nests, of not seeing alternatives, and of wanting lives made easier through appropriating available resources. The human in the Anthropocene is both a destroyer and simultaneously vulnerable to the power of this environment damaged by me, you, and others.

Serres would find Cousteau's efforts to "improve the Earth" as *Homo aquaticus* or *Zoon politikon* of a piece with the work of the destroyers. As noted, he considers "ecological ideologies to be the umpteenth instance of the city and city dwellers trans-historical victory over the fields and the woods" (cited in Watkin 2020, 337). The contemporary anthropologist today is often, like Cousteau the environmentalist, in the business of identifying who is responsible and pointing a finger. This finger pointing is what we imagine as a fix and displays our critical intellect. Unlike Serres, Bruno Latour, for example, finds it ridiculous to lump humanity together and also wants to point a finger (Latour 2017a). He writes, "The 'anthropos' of the Anthropocene is not exactly anybody, it is made of highly localized networks of some individual bodies whose responsibility is staggering" (Latour 2017a, 6). On a march through New York, he finds himself chanting "behind what I took to be the best banner of all: 'We know who is responsible.'" It is not us, it's them, even though we comprise the ninety-nine percent. Michel Serres refuses to fight these atomic wars that scatter fallout everywhere. For Serres, we, the ones who know who is responsible, are also the city dwellers who have won over the fields and the woods. We need sea stories as an alternative to reflect on ourselves and what we have done in the world. What if it *is* us and our retirement portfolios that mean so many of us incrementally invest in things we despise who are also involved? What if, as Cousteau discovered, we are all among the destroyers? Might it make sense, in addition to our search for divergent meaning and disparities, to examine how we are each a part of the great human tectonic plate and to use ethnography to show how that plate coalesces?

To do this, anthropologists need to grapple with the totality of the global scale and the mutual violence with which humans and the world engage each other. Such a scale does not need to erase difference or specificity. For Serres, each local knowledge has its own truth, which together make up the patchwork of the universal. The great tectonic plate of humanity allows us to appreciate the magnitude of the transformations that have beset us and the ways global history has entered nature while nature has entered global history. Following in Serres's wake allows ethnographers to extend the idea of disparity to nature itself and to see how the human is but one among many.

PART II.

BODIES

IN TIME

ELIZABETH A. POVINELLI

5

VARIATIONS OF BODIES IN MOTION AND RELATION

Fire is dangerous on a ship, it drives you out. It burns, stings, bites, crackles, stinks, dazzles, and quickly springs up everywhere, incandescent, to remain in control. . . . A good sailor has to be a reasonable fireman. —Michel Serres, *The Five Senses: A Philosophy of Mingled Bodies* ([1985] 2008, 17)

Imagine two hundred human beings crammed into a space barely capable of containing a third of them. Imagine vomit, naked flesh swarming live, the dead slumped, the dying crouched. Imagine, if you can, the swirling red of mounting to the deck, the ramp they climbed, the black sun on the horizon, vertigo, this dizzying sky plastered to the waves. Over the course of more than two centuries, twenty, thirty million people deported. Worn down, in a debasement more eternal than apocalypse. But that is nothing yet. —Édouard Glissant, *Poetics of Relation* (1997, 6–7)

Bodies in Portholes

Born in the same month, two years apart, Michel Serres and Édouard Glissant were both sons of the Atlantic. What surprise then that both men began influential books from the perspective of people struggling to survive on ships in peril. Yet, if we are to understand how these men's work might be placed in relation, then we must begin by noting that Serres writes primarily from the point of view of the merchant marines trained to escape peril, while Glissant writes from the perspective of the Black women, men, and children chained into the hull. The crew of which Serres speaks is not manning a slave craft nor directly participating in the slave trade. They are sailing many decades after its formal end. However, since Eric Williams published *Capitalism and Slavery* in 1944, the role that the enslavement of Black people across the Atlantic maritime had in the rise of capitalism is clear. The men Serres describes trying to survive a burning ship do so on seas whose depths hold the oft-occluded history of the Black men, women, and children who made their jobs possible. To paraphrase the title of one of Serres's books, we see two variations of the body in motion, related but not reducible the one to the other.

I doubt Serres would have objected to the way I use the "good sailor" to forge a passageway between the sheer hell of the slave hull and the ice-cold calculations of contemporary maritime capitalism. For Serres, time is not linear and progressive but forks and folds disparate elements from various times into the same space (Serres 1980). In times of bodily crisis, these folds loosen their seams, and the habituated body comes to experience the orthopedic nature of its dispositif. It reaches out blindly into its surroundings knowing it must find a new center of gravity if it is to survive. But as it reaches out, it feels all the other bodies entwined within it. In other words, as the good sailor tries to escape, he finds his body is in motion not merely in the sense of running as fast as it can. His habituated body becomes unwound into millions of other possible bodies as it seeks a new anchor. But what of the other bodies the good sailor meets along the way? Forked and folded into the same desperate passageways are others.

In previous writings, I have asked how philosophies and anthropologies of radical potentiality—scholarship that posits a political or social otherwise that emerges in moments of extreme social indetermination and of radical threshold experiences—fail to differentiate between lives lived as the object of colonial racism and those lived as its beneficiaries (see, for instance, Povinelli 2011, 6–11; see also Povinelli 2021). What are the political stakes of theories of radical potentiality that are anchored in the exhausted relationship between the general and specific, the ontologically given and the socially distributed, universal quantification (All bodies are x, All beings . . . , All human beings . . . , All social relations . . .) and existential quantification (for some . . .)? In this essay, I ask how we might understand the social and political stakes of Serres's attempt to find a universal ground for the variations of the body.

It may seem odd to characterize Serres as a philosopher of the universal. After all, throughout his writings, and especially those texts regularly cited in the English language-based context, such as *The Parasite* and *Conversations*, Serres not merely shatters mental and coherence-based approaches to the human subject but foregrounds the political potential that every biological, informational, and social order produces. Whatever we build—the human being as a built space—is always inhabited by a set of others that the building builds into itself. Each building and its hosts are specific. But, as Christopher Watkin has noted, Serres remains committed to a form of the universal that pivots the geometrical universalism of Descartes to the "algorithmic universalism" of Leibniz. This algorithmic unity "acts less as a transcendental condition of isomorphism and more as a procedural operator that generates local instances of order without any necessary sense of a pre-existing grand unity" (Watkin 2020). It is a proceduralism that is supple, agglutinative, radically open, and positive. For Serres, this form of universalism is one "in which, unlike the traditional one, everything is to be seen, found, constructed and populated, without institutional objects, without already occupied niches defended tooth and nail" (Serres [1993] 2017, 136). Thus, for Serres, there is no Weather but only specific and variable weather conditions (see Serres's reflections on the double meaning of *temps* in *The Natural Contract*). There is no Parasite, the title of his famous book *Le Parasite*. There are only parasitical conditions. Likewise, we might say there are no Bodies but only bodily conditions, bodies-in-motion, bodies in conditions of extreme peril.

This essay probes Serres's understanding of the universal algorithm of bodies-in-motion—bodies-in-variation—from the perspective of the relation of this algorithmic relation, namely, that the algorithmic motion is itself put into motion by the ongoing catastrophic results of the merchant ships that began crossing the Atlantic in the fifteenth century. To use Christina Sharpe's phrasing, in the wake of these ships, certain people, certain bodies—Indigenous, Black, and brown people—chronically inhabit the sort of disequilibrium that interested Serres. The question is not merely, who stands for the All when we look to radical disequilibrium to do the work of the otherwise? Nor is the question merely, what is this All *doing* such that it seems to be necessary for theoretical and social accounting? And to be clear, I am not merely interested in the social person who stands in the place of All but, more importantly, what work the relation between the All of the Some does such that other forms of relationality are excluded. And I am not suggesting that Serres's work should be marked as making sense for white Europeans or the white European diaspora but not for all others. Nor that it makes sense in a certain space, a certain local or locality or scale. The question this essay asks is how do Serres's theories of bodies appear if placed within the relation that Glissant argues emerged during the European invasion of the Earth? Why is Serres interested in the body's relation to its own embodiment rather than bodies in relation to other bodies and all of these bodies in an irreducibly historical and material relation to each other? What is this *common body in relation* if we begin from the perspective of those women, men, and children in near unimaginable conditions in the belly of the slave vessel, who are trying to leave this ship whether it is on fire or not? What is this common body in relation for those for whom their body is the fire?

Birthing Bodies

Serres sees portholes everywhere, the most important being the birthing canal. For him, the birthing canal is the bloody condition that inaugurates and provides the potential for the variations of the body. But anything can be a birthing canal (i.e., it is not *the* birthing canal that interests Serres but a condition of the body that finds itself in a birthing canal). The porthole of a ship on fire creates a birthing canal. A slip off a cliff can as well. A volcano spewing lava turns into one. The list goes on.[1] But a Serresian birth is a condition in which no woman is found; the mother is the near catastrophic background for the small male baby. This is true whether we are discussing a human birth or an adult man, now a sailor, pulling himself out of another terrifying canal (Serres [1985] 2008, 20). There is at "a precise moment, the very moment when the totality of the divided body shouts *ego* in a general toppling movement, I slide out and can drag through the remainder of my body, pull through the pieces that have remained inside, yes, the scattered pieces that have suddenly been blackened in the violent overturning of the iceberg" (Serres [1985] 2008, 20). The body that emerges is celebrated, the birthing canal is backdropped, and the placenta becomes the abject (Kristeva 1982).

Birthing canals are crucial for Serres because they are the site in which the sense of a given body is reduced to corporeal non-sense only to emerge, if it can survive, in a new configuration. This approach to corporeal eventfulness is in keeping with a certain moment in French thought. One can *feel*, for instance, the way Serres's thought leans into Deleuze's *The Logics of Sense* rather than, say, Luce Irigaray's *Speculum of the Other Woman*, the one creating a general philosophy of corporeal immanence and the other asking how another woman might emerge from within the specificities of the patriarchal capture of the current woman. Serres is not trying to find a body to counter a specific historical organization of power—definitely not a gendered one. He seeks and claims to have discovered an All Body. Like the Body-without-Organs, the All Body consists of all the quasi-bodies and quasi-points that compose and circulate through any specifically organized body. It is what is before and within any body that has settled into a rotation around a singular axis. Thus, when we feel *our body*—when I feel *my* body as a set of motions around an equilibrium—I am actually feeling how a set of quasi-bodies have been trained to rotate around a set point, itself composed of all the quasi-points that have been trained to rotate around each other in a certain way (Serres [1985] 2008, 21).

The birthing canal is a crucial site for witnessing the dynamic relationship between this All Body and any particular body. Moments in the birth canal shake the routinization of the rotatory body; suddenly all the other possible rotations flood the body, providing it with radically new forms of knowledge. Serres believes that bodily *events* that occur within birthing canals demonstrate that bodily knowledge predates the cogito or, put another way, that knowledge emerges from bodily sense rather than mental sense. The baby boy struggling to get out of the canal does not use a map. Nor is a map what allows a man to navigate land and ocean terrains. The map can only be drawn after the body has settled into its routine, only after the play of variation has been colonized by a propositional form. But this colonization cannot extinguish the "white noise" and "white light" that suffuse it any more than a trained body can totalize the quasi-objects from which it was amalgamated (Deleuze 2014, 206). Again and again, Serres will reject book learning for corporeal exercises. "No seated professor taught me productive work, the only kind of worth, whereas my gymnastics teachers, coaches and, later my guides inscribed its very conditions into my muscle and bones" (Serres [1999] 2011, 33). Knowledge is located in the posture of the body and new knowledge in the productive possibilities of the body in motion, the body learning new postures. We cannot know the body and its variations if we only *think* about them—we must use the body to find them. Indeed, thought itself is a sort of body, the habituated gathering together of *quasi-cogitatio* into *a thought*. Like the other five senses, thought can find itself in existential peril.

One might be tempted to read Serres's work sociologically. For instance, anthropologists might find this description of Serres's approach to the body resonant with Marcel Mauss's approach in "Techniques of the Body." There, Mauss emphasizes the "techniques of the body in the plural" to cast light on "the ways in which from society to society men know how to use their bodies" (Mauss 1973, 70). Like Serres, Mauss often pulls his examples from contexts in which the body is more or less in peril. Some of his examples come from his war experience—the ease with which the French but not the British can march to the beat of the French bugle, the use of various short and long spades in trench warfare. Other examples come from the social practices of sex and age, walking, climbing, mouth care, and labor efficiency. Like Serres, Mauss argues that bodies do not learn new habits through books—say, instructional manuals for how to use a French short spade—but by retraining the body, a practice that involves repetition over time (Mauss 1973, 71). In short, Mauss would agree with Serres that the bodies we have been given are not the only bodies we can inhabit. The body can "leave behind the domain of the real to enter into potential" because "nothing can withstand training, the ascesis of which repeats rather unnatural gestures" (Serres [1999] 2011, 3, 46). This presupposes, of course, that the body survives the conditions of its training.

One might also be tempted to align Serres's thought to American pragmatics. For instance, we might see some superficial similarities between Serres's critique of the posture of philosophers and William James's. James argues that philosophers, puffing away in their closed rooms with their self-referential abstractions, can never know the world or develop the concepts needed to comprehend it because the concepts that are needed are not in their minds but are distributed across the varieties of human existence. James insists that the mind is formed from the social worlds that put bodies in motion in different ways. Comparing the Harvard and Oxford philosophers Josiah Royce and F. H. Bradley to the clerk John Corcoran, who committed suicide because he was unable to secure employment and thus feed his family and pay his rent in an Upper East Side tenement house, James writes:

> What these people [people like Corcoran] experience IS Reality. It gives us an absolute phase of the universe. It is the personal experience of those most qualified in all our circle of knowledge to HAVE experience, to tell us WHAT is. Now, what does THINKING ABOUT the experience of these persons come to compared with directly, personally feeling it, as they feel it? The philosophers are dealing in shades, while those who live and feel know truth. And the mind of mankind-not yet the mind of philosophers and of the proprietary class-but of the great mass of the silently thinking and feeling men, is coming to this view. (James 2017, 22)

One thing that connects Serres, Mauss, and James is their understanding of the body-in-motion as something different from a body *put* in motion. As with the sailor in peril on his burning ship, every body is the result of older repetitions and their organizational force of habituation.

However, while it is true that Serres uses concrete historically based examples throughout his work and insists that the potential for new variations of the body comes from specific exercises of the body, he does so to conjure bodily motions beyond any specific historical or social condition. Serres is not, in other words, proposing a sociology of the body even as he emphasizes the social formation of bodies. He remains a philosopher of bodies grounded within a philosophy of life—life itself understood as and in relationship to a birthing canal. The ground on which *our bodies* are able to access other possible bodies is not a social distribution but the *body of life* as such. Let me quote at length from Serres's *Variations on the Body*. Discussing the situations of war and peace that drive toward a corporeal equilibrium, Serres notes: "Yet, mysteriously, the body can often thwart the laws of statics. By playing its game off-equilibrium, by confronting its limits . . . it succeeds in establishing another high seat, in the instability. But if it can construct this new state off-equilibrium from the previous equilibrium, it's conceivable then that life itself from the start became established by means of an initial deviation comparable to this one in every respect" (Serres [1999] 2011, 46). While Serres never abandons the importance of social exercises that habituate us and can rehabituate us (techniques of gymnastics, mountain climbing, swimming), he endlessly seeks to press beyond socialization as such in order to touch what he often describes as the animal within us. Serres is endlessly attempting to clarify his project in light of why "the sensualism professed by the Enlightenment" is a radically inadequate approach to the body (Serres [1999] 2011, 68). His sentences endlessly rotate around a grammar that eats semantic order to conjure life as a force always embodied but always more than any particular body.

Bodies know first because bodies are the habitat of life as deviation. Moreover, if Serres often uses commonsense examples of life, by the time Serres writes *Biogea*, life extends its sovereignty over all existence. All flora and fauna emerge in mingled relation to their eco-geological terrains. These eco-geologies dig into the human world, whether in the form of "the howling of wind and the tongue-clacking of waves" that sailors had to overcome to hail each other or the deserts that demand how those who travel across them must comport themselves (Serres [1985] 2008, 119). Everything about Serres's texts—the rhythm of his prose, the repetition of his tropes, the mingling of his life into the pressure of his thought—is meant to give the reader a description of the body in its variation that is, on the one hand, "reasonable and felt" and, on the other, pulsing with the sensation of "places and folds, proximities, penetrations and mixtures" that is life itself ([1999] 2011, 30).

In suggesting how life may have emerged from its own birthing canal, Serres claims to have founded a general economy of the body in motion, a general economy in which all bodies have within them the potential for a new organization. This general economy not merely traffics in gendered stereotypes, but it also occludes two conditions crucial to contemporary social politics and political ecologies. On the one hand, Serres all but evacuates the historical conditions of the birthing canal itself, focusing on the body in it. And on the other, he focuses on bodies in extremis rather than relations of humans whose normal conditions are lived as the object of quotidian and suddenly intensified moments of social violence and those who can be shocked by finding themselves in the same.

Bodies in Extremis

The body in motion is not a body undergoing a moment of unpleasantness, say, when a land lover becomes desperately seasick on the undulating deck of a ship. As I have been indicating, the variations of the body that interest Serres are moments of deadly conditions such as when human bodies are attempting to escape a ship engulfed by a petrol explosion. Serres underlines the anxiety, the pain, and the terror of the body in the process of becoming another body. But what would Serres's algorithmic universalism look like if we weren't interested in the body in relation to itself but bodies in Relation, namely the thought of Édouard Glissant and other critical race theorists? Let me begin with Frantz Fanon.

If Glissant begins his first major philosophical reflections from the perspective of the slave ship's hull, Fanon opens *Black Skin, White Masks* with a quote from Césaire's *Discours sur le Colonialism*—"I am talking of millions of men who have been skillfully injected with fear, inferiority complexes, trepidation, servility, despair, abasement" (Fanon 1967, 7). Fanon begins here to insist on the relational psychodynamics of colonial anti-Blackness. For my purposes here, let us consider anti-Black colonialism as a birthing canal that gave birth to Janus-faced twins. On the one hand are those persons addressed as Black within anti-Black colonialism. For Fanon, Black persons are surrounded by the racist gaze of an Other. From within this Other's hatred, a pathological form is injected into the marrow of their bones. At times, *Black Skin, White Masks* suggests that this anti-Black gaze is something a Black Antillean only experiences when confronted by a white person in continental France. In this reading, an Antillean child in Martinique is nurtured in a very different economy of the gaze than a Black person in Paris. This person would be more like those western African men and women ripped out of their original corporeal equilibriums, suddenly facing the vicious cliff of being nothing for the white Other than a source of profit.

But Fanon slowly unpacks the depth of the anti-Blackness that surrounds the Antilles and seeps into and conditions the bodies that Black Antilleans assume. The very nature of their relation to language, to how one speaks and assesses speech, reflects a morphology of culture. "To speak means to be in the position to use a certain syntax, to grasp the morphology of this or that language, but is means above all to assume a culture, the weight of a civilization" (Fanon 1967, 18–19). This civilization, as Césaire noted, is the Western civilization that emerged out of the decivilizing brutality of colonialism. This civilization does not establish relations among people; it dominates and bends all toward its will to accumulate (Césaire 2001, 31–34). Even on their own island, Fanon claims, an Antillean, feeling imprisoned, leans toward Europe and "breathes in this appeal of Europe like pure air" (Fanon 1967, 21). If French slave boats brought their ancestors to the French Caribbean from West Africa, the Antillean seeks to alleviate the "amputation of his being" by sailing to France. Only there can he recover his limbs. The voyager sees the shared nature of this belief: "In the eyes of those who have come to see him off he can read the evidence of his own mutation" (Fanon 1967, 23). The unnatural gestures are not something one finds oneself suddenly confronting from the outside. The unnatural gestures are the body that you have been given, and on this body, the Black person "has to wear the livery that the white man has sewed for him" (Fanon 1967, 34). James Baldwin, in "Notes on a Native Son," though with different consequences, also emphasizes the gross deformations of the body that anti-Blackness creates for both whites and Blacks, but not in the same way.

> One is always in the position of having to decide between amputation and gangrene. . . . Amputation is swift but time may prove that the amputation was not necessary—or one may delay the amputation too long. Gangrene is slow, but it is impossible to be sure that one is reading one's symptoms right. The idea of going through life as a cripple is more than one can bear, and equally unbearable is the risk of swelling up slowly, in agony, with poison. And the trouble, finally, is that the risks are real even if the choices do not exist. (Baldwin [1955] 1963, 101)

I am, of course, merely repeating what critical race theory has said and said again and again. One need only remember the debates within feminist theory around the pleasures of the cinematic gaze. In answer to questions among feminist critics about why women enjoy cinema if it is structured by the male gaze (Mulvey 1973; Koch 1982; Doane 1987), the Black feminist critic bell hooks argued that one had to begin by the racial organization of the gaze as such. hooks argued that Black women had to look back not only at the anti-Blackness of the cinematic imaginary but at the change in reality of the differential organization of bodies as such. The goal of the critical thinker is not to find the unorganized body that courses through all bodies but to find the specific bodies that are distributed in order for some to accumulate value and others to be the sources of that value and the depositories of harm and to find "those margins, gaps, and locations on and through the body where agency can be found" (hooks 1992, 116). It would be wrong to dismiss the effort, the energy, the sheer will to be otherwise that characterizes the assertions of agency that "by claiming and cultivating 'awareness' politicizes 'looking' relations" and thereby "learns to look a certain way in order to resist" (hooks 1992, 116). Here hooks returns to Fanon to remind us how "power is inside as well as outside," how the glances of the anti-Black Other "fixes" the Black body, how the fragments of the self must be put back together (hooks 1992, 116). Likewise, Zakiyyah Iman Jackson argues that Octavia Butler's science fiction "tarries with dislocation and loss of identity not simply as defeat or the loss of tradition but also as a processual opening to unforeseeable, emergent modes of belonging and existence" (Jackson 2020, 123). This tarrying with dislocation and identity is, as Mathias Nilges notes, less about the universal value of embracing change and more "about the struggle with the necessity of *having* to do so" and "the psychological struggle that arises out of confrontation with change" (Jackson 2020, 123–24; Nilges 2009, 1137).

On the other side of anti-Black colonialism are white subjects and their bodies. Lucas Van Milders has described this side of the colonial birthing canal as consisting of the white subject's constant, continual need to fabricate objects in and relations between the world in order to keep the fantasy of his mastery over it (Van Milders 2022). This isn't the Kantian world of necessary human epistemic mediation but the hallucinations that arise from a social relation built on racism and colonialism. This form of psychosis emerges within the racist colonial relation, but its content and dynamics are different from the content and form within the terrorism of this system. This is what makes Fanon's reworking of the Lacanian *trotte bébé* so important. Lacan used this metaphor to explain the extimate nature of the imaginary. "Unable as yet to walk, or even stand up, and held tightly as he is by some support, human or artificial (what, in France, we call a '*trotte-bébé*'), he nevertheless overcomes, in a flutter of jubilant activity, the obstructions and, fixing his attitude in a slightly leaning-forward position, in order to hold it in his gaze, brings back an instantaneous aspect of the image" (Lacan 1977, 1–2). Fanon focused on who can lean forward—or toward whom? The supports are the ankle chains digging through flesh into bone. Slightly leaning forward or jubilation out of place can get a Black man killed. Thus, disequilibrium within Black bodies is not merely de-territorialized forces circulating through the organized body but also that one's own original body is, for Fanon, borrowed from a society structured by its anti-Blackness. So let me say, alongside Fanon, that there is no mirror stage. There are multiple mirrors staged in such a way to refract the kind of gaze from which different kinds of persons can find and organize their bodies. Don't look at the police in the eye. Or do, but know what's in the eyes looking back at you.

Not only don't these eyes meet for different reasons, the very nature of the world they see is fundamentally different. The Black ego is not hallucinating the world. She is inhabiting it. She must constantly find a way of inhabiting it as a quotidian attack that can intensify suddenly into a deadly one. For Glissant, Serres, Fanon, and, most recently, Christina Sharpe, the ship was a gigantic womb of anti-Blackness with two uteruses and two birthing tracks sorted through who could own and be owned (Sharpe 2016). Thus, the persons were subjected to different practices of survival because they were, on the one hand, in different relations of property (those humans who owned other humans, those humans attempting not to be owned) and, on the other hand, not in relation at all, namely, that the very concept of property, let alone humans as property, was not shared. Because of this, for Fanon, if one holds up a mirror to the imaginary stage, one does not find the other but the gangplank, one leading into the unimaginable conditions of the slave quarters and the other the decks reserved for the crew.

The sailor's obsession with his survival—the way the sailor has been trained to survive—is an effect of the hallucination he inhabits. With this we can return to the good sailor and the human beings who refuse to simply be his cargo.

Bodies on Fire

For Serres, the good sailor is not a good fireman by nature. He is *trained* in how to survive the blaze. After tortuous exercises, his body knows how to take over in moments of peril before his mind thinks. This training is made up of endless drills—climbing down dark, "vertical wells, descending endless ladders, inching along damp crawlways, to low underground rooms in which a sheet of oil would be burning"—which fashions a disposition of survival deep in the sailor's corporeal tissue (Serres [1985] 2008, 17). These specific habituations of the body alter or redirect earlier ones—the sailor learns not to rush toward closed doors but to make of himself a snake that can slither along the ground.

The source from which the sailor learns to reorganize his senses does not come from abstract explications of texts—not from book learning. He is able to *learn* how to survive the fires engulfing his floating world by remembering-as-experiencing the "quasi-objects" (quasi-body) that preceded and constructed the object-body he is now struggling to save by reorganizing. These quasi-objects alternate between habit and invention—for what got the sailor into this mess and for what will get him out of it (Serres [1999] 2011, 75–87; Delueze and Guattari 1983, "the body without-organs" and "the body-with-organs"). In other words, for Serres, every body is the effect of previous trainings of quasi-bodies, an amalgamation produced by habituations of affective touching, looking, turning, listening. While trained to be together in this or that manner, every body also remains what he calls a body in variation, namely, all the other possible bodies that could have been habituated. Thus, the good sailor is a sailor at one moment, a fireman at another moment, a mammal or amphibian at still others. Because the good sailor is trained to have multiple variations of his body available to be put to use when necessary, Serres sees him as a perfect example of an algorithmic approach to corporeality. The sailor is a set of conditions organized around the algorithmic function of survival.

The obsession with the sailor's body in relation to itself is itself an effect of the long slow catastrophic event of racial colonialism, events and quasi-events that distributed the arts of death and survival. What if we looked at the (algorithmic) relationship between sailors and cargo in another form of merchant capitalism—the slave vessel on fire or sinking? We notice that the various bodies trapped on a sinking slave ship were *invested* with very different organic matrixes—they are given very different sets of training. One matrix is meant to organize a human into an agent, to give that human an *agential body*, to experience itself as that which matters and matters forth. In a Lacanian sense, this human is introjected with an "orthopedic body," which through the "armour of an alienating identity" is meant to functionalize the subject and protect his limbs, to work in moments of social and physical equilibrium and disequilibrium (Lacan 1977, 4). In the slave trade, it is the European sailor who has this kind of armored body—has a body that assumes a certain quotidian equilibrium and has been given the means to survive moments of disequilibrium. He has been actively trained to secure and loosen portholes—to know how to escape them even when the rusted flanges refuse to unscrew (Serres [1985] 2008, 18). He knows how to do this because, when he finds himself in the churning smoke, half inside and half outside the ship's hull, half drowning and half burning, the sailor does not ask, "Who was 'I'"? ([1985] 2008, 19) in order to *return* to an *original* firstness—"an instance of that kind of consciousness which involves no analysis, comparison or process whatsoever" as a "quality which consists of nothing else" (Peirce 1994, 152). He reaches into his body and finds that he can access other possible bodies. This reaching, or experiencing, is not thinking, is not a cogito. The good sailor feels that because someone created his body ("*this, my body*") to thrive in one condition, there are other bodies that could be created to survive other conditions. The sailor can experience, feel, and reach for one or another condition because the feeling he has of the bodily qualia subtending his "I" has been carefully, if not consciously, produced by others invested in maintaining the condition of his body. Does the good sailor feel these other bodies inside him, or does he feel the social world that is invested in investing his corporeal qualia with value no matter its variations of form—or how it finds a way of reestablishing its equilibrium? Surely the latter. How can we forget that not all people assume the socially dominant world wishes to imbue them with a balance at any time or any condition, let alone in moments of peril? The socially dominant world not only is built in a way that establishes disequilibrium as a way of life for them but, when ships are on fire, will throw them overboard first.

Even these bodies on fire are not on fire in the same way, but only in relationship to their historical conditions. Rather than merely focusing on the suffering of Africans and Black Americans in the horrific conditions of enslavement, scholars like Saidiya Hartman have long called for considering relations of care and politics (Hartman 1997). Those Africans forced to walk across the ramp into the depth of these fetid ships had been given sturdy bodies by their families. These families provided them, as children, a gaze from which they would build an integrated body. They looked into eyes, they felt hands and chests, they heard and smelled a form of addressivity that was affectively inflected with care, joy, happiness, irritation no doubt when those who held them wished for some sleep, but pulsing throughout with the feeling of an investment in the body they were borrowing. This socially invested body was subjected to new "training" as it was moved through the overland routes of the slave trade, then across the ramp and onto the ship, and when the ship caught on fire or the hull was breached, often into the depths of the mid-Atlantic. This training was the "torment of those who never escaped" the hull, who went "straight from the belly of the slave ship into the violet belly of the ocean depths," who were meant to survive only enough to be sold at good price and then worked to death (Glissant 1997, 7). The ship might make it back full of seawater, but these African women, men, and children would not. Death,

Jessica Millward argues, weaving Jason deCaires Taylor's *Vicissitudes* sculptures with Sowande Mustakeem's *Slavery at Sea*, was an active protagonist whether the boat was sailing clear skies or capsizing in violent storms.

No surprise that the West African women and men related to their sea training in a very different way than did the sailors to their own. The sailor was taught how to be a good fireman, to lie flat to the ground where a thin layer of oxygen might save him. So-called surgeons were hired to keep the enslaved alive, but enslaved men and women knew that this investment in their bodily health was merely so that they would survive along a commodity chain. The enslaved body must be kept alive to be sold into whatever condition followed the purchase. Surgeons cared for their bodies the ways they cared for cattle. Those who refused to eat were force-fed. Those who tried to commit suicide were more firmly chained down.

But, still, everywhere and at all times was the threat of an insurrection, *fugitive planning*, through blade or fire (Moten and Harney 2013). The men, women, and children confined in the hulls found linguistic means of communicating across their many languages, not merely to survive the passage but to refuse the image of themselves as now reflected in the eyes of the sailors. Those who lived to reach American shores continued to find "provisional ways of operating within the dominant spaces; local, multiple, and dispersed sites of resilience that have not been strategically codified or integrated; and the nonautonomy and pained constitution of the slave as person" (Hartman 1997, 61). This "politics at a lower frequency" is the noise then built into the American grammar (Gilroy 2014; Spillers 1987). In other words, the world as disequilibrium created a different form and relation to equilibrium and disequilibrium as such.

We could quote Serres in such a way that would align his approach to these fugitive plannings and politics at a lower frequency. After all, Serres notes that the power of the noise of these frequencies comes not because it "occupies the center" but because it "fills the environment" (Serres [1980] 2007, 95). For instance, Serres might have argued that invading colonial merchants told themselves that they are working themselves to death so that they "can finally be at home when everything is clean" and without the pests the house itself built (Serres [1980] 2007, 92). But for Serres, the problem is not that nothing can be made clean. The point is not only that the house built into itself the parasite that the owner then is driven mad by. The houses of modernity, civilization, capitalism were built by those assigned the role of parasite on lands dispossessed from other "pests." And, equally true, the slave merchant and colonial assassins blind their eyes and stubbornly, forcefully refuse to know their own pestilence. The voices of merchants claim to be the prophets of a world historical General Equivalence (capital, money) even as they are the pestilence ravaging the Earth, making a desert across which only they can be heard (Serres [1985] 2008, 119). They claim to rule by Universal reason and Truth, but they become Universally True only by silencing all other forms of reason and relation. In other words, the ant is not the only one who "produces parasites in eliminating others" (Serres [1980] 2007, 92). Colonizing bodies were trained to seal their ears, connect their minds to the alimentary tracks, so that they could eat and defecate without experiencing how their bodies were internally related to the horrific destruction of other bodies and lands. As a pest that produces pests so as not to know itself, these colonizing bodies increased to deafening levels the noise in the system, drowning out all speech, forcing bodies to lean into each other to have any chance of communicating, exploding eardrums.

To be sure, the bodies of both slave merchants and colonial invaders find themselves stuck in the horrifying assemblage of emerging and actualized capitalism. But the merchants see themselves as *becoming* "charcoal black" while they assign a negative value to the *being* of blackness to those whom they wish to keep in the furnace. Anti-Blackness is the cause of the fire that then chars bodies in times of danger. Sailors may find themselves in total blackness as storms disable the sturdiest ship, but the enslaved are tied to capitalist ships by real or metaphorical ankle chains because they are enclosed in anti-Blackness. The skipper can choose to go down with what has become mingled with himself. But by that time, he will have done everything in his power to make sure the Black persons he is bringing to sale do not have the same powers of choice. This is what we witness when we stand alongside these two regimes of training as the slave ships burn at sea: a body trapped in a porthole and bodies that meet there without creating an equivalence.

Bodies in History

Serres was always a philosopher of history as movement, a philosopher who placed his own history of movement at the center of his thought. He laces his texts with his continual career deviations—from naval academy to scholar academy, from mathematics to philosophy on the chance encounter of the work of Simone Weil, who he saw as an incarnation of the messenger god Hermes. He, like her, opened the fixed point of their corporeal equilibrium to open themselves to the struggles of bodies inside the containers of rigid spaces—one's own body, factories, ships, laboratories. Serres heard these containers screaming the noise of their own crumpled histories. Not silent, clean, or perfect shapes. They were riddled with the parasitic nature of their own building—and the parasites had parasites all the way down. He would use many metaphors for the dynamics of the building which makes any *thing*. The history of things—even history qua history—is not a collection of facts placed in a linear timeline. It is instead a "crumpled handkerchief" that folds the space of time into the archaic relations of bodies (Herzogenath 2012). In conversation with Serres in *Artforum*, Paul Galvez noted how this way of conceptualizing history within bodies applied to technology: "A car is an object that seems to be new, but that in fact is an assemblage of different technologies produced in vastly different times—the prehistoric wheel, the midcentury automatic transmission, the GPS" (Galvez 2013).

Let us linger for a moment on not only this abstract automobile but also the automobile that organizes the affects and senses of so many contemporary consumers so that they can continue to move through space with the ease that they have and yet, in moving in the same way, alter their environmental future. I mean, of course, the battery-powered automobile. How much do we need to not know in order to not see from where the materials for the battery, steel, the rare minerals that will allow the computer electronics to rule the engine and operation come?

Contemporary social theory is awash in models for why and how we should walk away from the governance of the universal and its various permutations of the All and the Some and the general and specific. Certainly, Glissant provides one such model. Others include Denise Ferreira da Silva (2016), who has proposed a way of thinking about difference without separability. In the US context, Vine Deloria Jr. long ago contrasted the metaphysical conditions at stake in the struggle against settler colonialism. In his 1973 book *God Is Red*, Deloria contrasted Western understandings of revelation to Native ones. In the former case, moments of subjective revelation in specific situations are "mistaken for a truth applicable to all times and places" (Deloria 1972, 65; see also Coulthard 2014; Todd 2017; TallBear 2019). Those places that refuse this weaponized revelation machine were exterminated or forcefully converted. Zoe Todd has wondered how she can foster a relationship with her Metis dinosaur kin after they have been weaponized to serve as a toxic agent of carbon capital. She would hardly be relieved by a mere switching from their carbon remainders to her kinship with the rare minerals, ores, or waters of her lands (Todd 2017; see also Povinelli 2016, 31–56). In other words, the handkerchief does not fold itself, and the variations of the body-as-intensities that circulate across any organized body cannot be eliminated. But neither can we begin without beginning in the history of relations that continually pull all bodies toward the sheer cliff but not with the same equipment to hold fast.

In the end, all human bodies may lose their grip, but not because *the body* was not able in this instance to escape the porthole. It is because the cliff itself is the result of the ongoing parasitic gutting of the Earth and the evacuation of other bodies necessary to do so. Can we really, at this moment, still assert that it was "the scientists who threw" the situation of environmental and ecological destruction "into relief" (Galvez 2013)? What kind of Relation to crisis is the Earth in flames, and to whom will it matter—whose lands and oceans poisoned so other's lands can remain green?—this refusal to *feel* how we have been *related* to the events that began accumulating into different bodies and minds differently since the northwest passages?

NOTE

1 Janell Watson notes that the apparent sexism of Serres is leavened by his recognition, celebration, and valorization of mothers, goddesses of the hearth, and academically superior millennial women (Watson 2019).

6

WHEN WAR PERCOLATES

On Topologies of Earthly Violence in a Planetary Age

The experience of terror also dislocates time, that most abstract of all humanity's homes. —W. G. Sebald, *On the Natural History of Destruction* (2003, 154)

Since the atomic bomb, it had become urgent to rethink scientific optimism. I ask my readers to hear the explosion of this problem in every page of my books. Hiroshima remains the sole object of my philosophy. —Michel Serres, *Conversations* (Serres and Latour [1992] 1995, 15)

Topology . . . it is a state of mind. —Stephen Barr, *Experiments in Topology* (1964, 2)

"Listen: Billy Pilgrim has come unstuck in time." So begins Kurt Vonnegut his anti-war novel *Slaughterhouse-Five, or, the Children's Crusade: A Duty-Dance with Death* (2000, 19). Billy Pilgrim, the main protagonist, is a time traveler who seamlessly moves between the past, present, and future. But these time travels give Billy Pilgrim no particular pleasure. As we read time and again throughout the book, he has "no control over where he is going next, and trips aren't necessarily fun. He is in a constant state of stage fright, he says, because he never knows what part of his life he is going to have to act in next" (2000, 19). Billy

Pilgrim is a semi-autobiographical character, who, like Vonnegut, served in the US Army during the Second World War. He also was captured by the Germans and interned in Dresden, where he survived the Allied bombing that destroyed the city in 1945. And like Vonnegut, Billy Pilgrim struggles throughout the book to find a way to bear witness to the flames of total destruction that he experienced.[1] For Kurt Vonnegut and Billy Pilgrim, the result of this experience was a permanent temporal dislocation, a mode of human existence unstuck in time.[2] Several decades later, W. G. Sebald in his series of meditations on modern warfare, *On the Natural History of Destruction*, from which I draw the epigraph at the head of this chapter, made a similar point. The experiences of total bombing and devastation, which are "beyond our ability to comprehend" (2003, 25), dislocate one's sense of being in time and history and remain out of joint years, decades, and generations after the war has ended.

Vonnegut's *Slaughterhouse-Five* is a splendid book not only for its powerful anti-war message. It also offers a useful narrative form for thinking and writing about the experiences of war and mass violence. Vonnegut creates in the novel a narrative spacetime that is neither geometrical nor linear but rather *topological*. Wars, like large societal breakdowns or crises (Knight 2012, 2015), create unruly temporal topologies, and Billy Pilgrim's experience is no different. Indeed, as we learn early in the book, Billy Pilgrim comes unstuck in time only after he is kidnapped by a flying saucer and taken to the planet Tralfamadore. The Tralfamadorians introduce Billy Pilgrim to their philosophy of time that is topological. In their philosophy of time, "all moments, past, present, and future, always have existed, always will exist" simultaneously. This is in sharp contrast with Earthlings' illusion of linear and progressive time in which "one moment follows another one, like beads on a string, and that once a moment is gone it is gone forever" (2000, 22). In other words, Earthlings' conceptions of time leave, for example, the experience of war in the realm of the past and memory, separated from the present and the future. But these conceptions do not correspond with the multiple *elsewhens* toward which Billy Pilgrim orients himself (cf. Knight 2021). From the moment he encounters the Tralfamadorians, Billy Pilgrim's experience of being in time becomes more plastic, in which the past, the present, and the future are folded into each other in an indeterminate number of mutations as the story progresses. Put differently, wars rip apart not only bodies and buildings "but also families, known landscapes, daily routines, and *time itself*" (Dunn 2022, emphasis added). How time will be reassembled in the aftermaths of war remains open. *Slaughterhouse-Five*'s narrative works effectively with topological pluritemporality to describe what these experiences of temporal dislocations engendered by the war and its aftermaths might entail. In this pluritemporal perspective, an experience of war is not an event with clear temporal boundaries of beginning and end. Rather the war and its affective as well as material remainders become radically contemporaneous (Bryant 2014; Serres 2012a). They create a spatiotemporal copresence of the past violence and terror that continues *percolating* as a potentiality to strike again in the present and in the future.[3]

Over the past decade I have been researching and writing about the enduring socio-environmental aftermaths of modern warfare, with a particular focus on the long-lasting effects of lethal military waste in postwar Bosnia and Herzegovina (Henig 2012, 2019, 2020). I came to think about Vonnegut's Billy Pilgrim soon after my visit to the War Childhood Museum in Sarajevo in 2019, where I encountered *Abandoned Guitar*, one of the many objects on display in the museum. All exhibits in the museum tell a story of the childhood memories of the Bosnian War (1992–95) through intimate objects that were donated to the museum by the generation of "war children." The objects and their stories are all deeply moving. But unlike most other objects, whose stories were oriented toward the violent past, there was something contemporaneous and unfinished in the tale of *Abandoned Guitar* that caught my attention. It tells a story of Adnan—once a war child, a poet, a sport enthusiast, a father, and a deminer. The story reads as follows:

> I found this acoustic guitar in a basement apartment that had at the start of the war been a military base. This guitar was left behind after everything else had been cleared out from the room. I took it with me and started learning how to play. I kept a notebook with lyrics and chords. At prom, 1994, the names of friends with whom I celebrated the end of high school were scrawled into it in pen. I already knew then that I would be mobilized as a juvenile fighter, a member of the Reconnaissance and Sabotage Detachment. I dismantled a mine for the first time shortly before my eighteenth birthday. Today I make a living from mine removal.

After seeing the exhibition, I enquired with the museum staff whether it would be possible to meet with Adnan. I explained that I was researching the effects of military waste in Bosnia and Herzegovina, including exploded and unexploded ordnance. I provided my contact details and left. A day later I received a positive reply with Adnan's email address. I immediately arranged a meeting, and we met the following day. Our conversation lasted several hours and ranged over numerous topics from writing poetry and traveling to land mine clearance and the challenges to bear witness to the horrors of the Bosnian war from the perspective of a juvenile fighter dismantling explosive ordnance. But conversation was mostly about land mines. When we were about to depart, Adnan again contemplated the *Abandoned Guitar* and the story it tells—about his life, the war, and all the land mines still buried in the soil. Then he offered a succinct reflection on the entanglement of his life, the war, and its wastes: "My entire life is turned toward the past, removing *this garbage*, instead of being oriented toward the future" (Adnan's emphasis).[4] Although Adnan has tried to change his career several times because the job of a deminer is extremely taxing, he keeps returning to the minefields. Land mines became part of his life. "*This garbage*" keeps percolating with different intensities, forms, and effects, be it in a form of a memory or a nightmare dream or as a detected land mine he needs to remove to earn a living (Serres and Latour [1992] 1995, 139; Knight 2022). In turn, "this garbage" is an insidious remainder of the Bosnian war that is radically contemporaneous in Adnan's and other fellow Bosnians' lives, bodies, and environments. It remains in its radical potentiality able to affect, maim, or kill them in the present and the future.

I start this essay with the stories of the time traveler Billy Pilgrim and Adnan not because they are exceptional but because they are parables of temporal dislocations engendered by mass violence and war. Moreover, Adnan's reference to "*this garbage*" and Billy Pilgrim's vivid descriptions of bombing and material destructions of Dresden offer powerful reminders of large-scale devastations and altered possibilities of livability that wars leave behind. Attending to such experiences and wastes of wars remains an analytical and conceptual challenge as they defy the oft-deployed categorizations such as war: peace, critical event: closure, or past: present. While Billy Pilgrim has come unstuck in time, Adnan remains to be pulled toward the past in his day-to-day job of clearing the explosive war remains. For both of them, however, the past, the present, and the future are folding and unfolding into each other in indeterminate ways. More importantly, the two stories bring us closer to the orbit of Michel Serres and his philosophy of topological time and history as well as his ecological thinking that offer, as I shall argue in this chapter, a creative and nuanced way of attending to wars and their wastes. Indeed, there is a strong affinity between Serres's thinking on war and time and the stories of Billy Pilgrim and Adnan's *Abandoned Guitar*, namely, thinking *through* and *about* the strange topologies instigated by wars.

In his oeuvre, in which he developed a topological perspective on time, Serres rejected the modernist epistemology of historiography and historicity. His is a nonlinear-cum-capacious notion of time that is crumpled, folded, and percolating like "this garbage" to Adnan's life and Dresden's rubble to Billy

FIGURE 6.1. *Abandoned Guitar*, the War Childhood Museum in Sarajevo, 2019 (photo by David Henig)

Pilgrim (Serres [1990] 1995; Serres and Latour [1992] 1995).[5] Later, Serres further expanded this topological perspective to his arguments on the general ecology of pollution (Serres [2008] 2011, 67). While Serres's work on time has been increasingly debated in recent years across humanities and social sciences (e.g., Assad 1999, Herzogenrath 2012, Knight 2012, 2021), his thinking about war and pollution has yet to be fully appreciated (Watkin 2020, 3).[6] In what follows, I bring Serres's topological perspectives on time and pollution together to show how they can open new avenues for thinking and writing about the long-lasting socio-environmental effects of wars and their aftermaths. Drawing on my ongoing research in postwar Bosnia and Herzegovina and beyond, I retrace my encounters and resonances with Michel Serres and my *thinking with* Serres about wastes of war, their unruly temporalities, and insidious planetary effects.

Jeremiah's Cry: War and Witnessing

Michel Serres didn't shy away from tapping into the darker themes of the human condition (see Brown 2002; Shryock, this volume). Yet this side of his work is overlooked and rarely appears in the debates on war and violence. Serres is a philosopher of continuities. As a perceptive observer of relations and connectivities, Serres identifies emergent patterns of more complex continuities that can "actualize into any number of different forms in different spatial and temporal locations, patterns that are relational in a topological or non-Euclidian sense rather than relational in a conventional geometrical sense, patterns that are fluid, turbulent, non-linear and very adaptable" (Clayton 2012, 38).[7] Indeed, his topological perspective brings to light radical contemporaneities and relations that open "different spaces of enquiry, between disparate systems of knowledge" (Clayton 2012, 41; Watkin 2020, 346–47).[8] This is also evinced in his writing. In his books, we find a dynamic web of analogies, juxtapositions, and stories: Molière in conversation with Marcel Mauss, Anaximander and Thales with Poincaré, or Lucretius's atoms and those of Jean Baptiste Perrin (Serres 2012a; Serres and Latour [1992] 1995, 48). This reflects his refusal of dualistic, linear, and deterministic philosophy and historiography of modern science and epistemology rooted in Cartesianism (Serres and Latour [1992] 1995; Assad 2012; Dolphijn 2018a; Watkin 2020). Serres's philosophy has important implications for considering war and violence beyond the usually deployed dichotomies such as war: peace, winners: losers, perpetrator: victim, as I will discuss shortly. Furthermore, reading and *thinking with* Serres and his topological understanding of time, history, and pollution have pushed me beyond attending to wars and mass violence solely in terms of time-bounded "critical events," ruptures, and discontinuities (Das 2007). But these are not the only reasons why we should engage with Michel Serres. War and violence, I suggest, are generative and recurring themes in Serres's work.

My first encounter with Michel Serres was through the book *Conversations on Science, Culture, and Time* (Serres and Latour [1992] 1995). I distinctively remember two things. As for many Serresian neophytes, I was struck by the breadth and creativity of Serres's ideas that made my head spin, and I wanted to read more. And then there was Hiroshima and Nagasaki—a theme that appears early in *Conversations* as an important contextual framing for situating Serres's thought. One of the notes I scribbled in the margin while reading the book was that "Serres is giving a visceral sense of how an entire generation can be shaped by, and bear witness to, war and violence." The more I subsequently read other books by Serres, the more his ideas started percolating into my own thinking about wastes of war. Nonetheless, however inspirational it felt, I couldn't yet fully appreciate the importance of Serres's own war experiences on his thought. It was only later, when I read Christopher Watkin's excellent book *Michel Serres: Figures of Thought* that I realized that Serres's personal biography and early childhood experiences of war cannot be separated from his philosophy, which prompted me to go back to Serres and reread some more of his books (see Watkin 2020, 4).

There are at least three significant war-related biographical points of reference that are generative for Serres's thinking and percolate through his work. Serres (born in 1930) refers to this early period of his life as a "tragic atmosphere" and a "terrible era" (Serres and Latour [1992] 1995, 3–4). First, it was the experience and memory of Serres's parents with the Great War. It was "under the hail of shells at Verdun," Serres writes, when his father converted to Catholicism. He continues that "the experience of the 1914–18 war, for which he enlisted at age seventeen, brought him that religion" (Serres and Latour [1992] 1995, 19). Moreover, during the Great War, Serres's father not only had to endure the heavy shelling, but he was also gassed (Watkin 2020, 5). As Steven Connor (2003) poignantly observed, in the interwar period the idea of poison gas "took on a kind of political and phantasmal reality which has not yet diffused, and serves as a point of reference for more concerns and debates." This includes Serres himself (e.g., Serres and Latour [1992] 1995, 121) but also the work of Peter Sloterdijk (*Terror from the Air* [2009]) and Paul Virilio (*Bunker Archaeology* [1994]). And then there was the experience of his mother. As Serres mentions in *Solitude* (cited in Watkin 2020, 5), his mother was the only one of her cohort who succeeded in marrying because the great majority of prospective fiancés were massacred in the same battlefields in which Serres's father was gassed.

The second biographical point was Serres's coming of age in an era of war. As he succinctly puts it in *Conversations*:

At age six, the war of 1936 in Spain; at age nine, the blitzkrieg of 1939, defeat and debacle; at age twelve, the split between the Resistance and the collaborators, the tragedy of the concentration camps and deportations; at age of fourteen, Liberation and the settling of scores it brought with it in France; at age fifteen, Hiroshima. In short, from age nine to seventeen, when the body and sensitivity are being formed, it was the reign of hunger and rationing, death and bombings, a thousand crimes. We continued immediately with the colonial wars, in Indochina and then in Algeria. Between birth and age twenty-five (the age of military service and of war again, since then it was North Africa, followed by the Suez expedition) around me, for me—for us, around us—there was nothing but battles. War, always war. Thus, I was six for my first dead bodies, twenty-six for the last ones. . . . Violence, death, blood and tears, hunger, bombings, deportations, affected my age group and traumatized it, since these horrors took place during the time of our formation—physical and emotional. ([1992] 1995, 2)

This theme keeps reappearing in his work. For example, in *The Troubadour of Knowledge*, in the chapter on education, Serres writes in reference to his childhood:

When I hear the vibrating bell that chimes the hours in so-called institutions of learning, I know that it trembles with terror.

At home, civil and familial time had the same rhythm: bombing sirens, various alerts, the news announcing, hour after hour, after the theme song, the opening of new killing fields. Between the Spanish revolution of 1936, the Second World War and its summing up at Hiroshima, what child would have perceived the difference between these giant massacres and the merciless vendettas that brought together the cubs, the sons of wolves, and the future fathers of the same through the eternal return of the same signal marking time, the doleful law of our history writ small or large, the reflex bell of dogs? ([1991] 1997, 133)[9]

Echoing Billy Pilgrim's and Adnan's percolating war experiences, Serres sums up this generational experience as Jeremiah's cry that "comes from nowhere else but those shameful wars and the horrors of violence" (Serres and Latour [1992] 1995, 3).

And then there is the third theme—Hiroshima and Nagasaki. Hiroshima in particular is a recurring motif of total destruction. If the concept of noise is central to Serres's philosophy (Serres [1990] 1995), then the possibility of a nuclear Armageddon looms as a background noise throughout Serres's oeuvre. As the second epigraph at the head of this chapter poignantly captures, the use of atomic bombs in Hiroshima and Nagasaki "remains the sole object" of Serres's philosophy (Serres and Latour [1992] 1995, 15). Serres returned to this point in his interview with Janina Pigaht and Rick Dolphijn, where he remarked, "The most important event in my life as a scientist was the dropping of the atomic bomb on Hiroshima and Nagasaki" (2018, 170).

The motif of Hiroshima runs like a red thread through his books. For Serres's generation, Hiroshima was an important moment of reckoning with science and its relation *to* and *with* the world in the shadows of the Manhattan Project (Pigaht and Dolphijn 2018, 170).[10] Furthermore, as Rick Dolphijn writes, the experience of Hiroshima not only gave rise to new ethical considerations but also demanded a completely new form of thinking (2018b, 133). Against this backdrop, Dolphijn offers a succinct interpretive key to the figure of Hiroshima in Serres's thought: "Hiroshima showed us that all the differences to which we implicitly plead allegiance (physics versus nature, epistemology versus ethics and man versus the world) in fact point to nothing else but our short-sightedness. . . . Hiroshima . . . is the permanent goodbye to Humanism, and to all of its Cartesian and Kantian dualism that still dominate our thinking" (2018b, 136).

Indeed, as I have already suggested earlier, a move beyond Cartesianism as well as beyond ethico-epistemological demands reverberates throughout Serres's work. In *The Five Senses*, Serres persuasively articulates this demand: "Hiroshima is the foundation of contemporary science, just as the death of Socrates is the foundation of modern philosophy" ([1985] 2008, 101). And in *Statues*, he further explains, "For my generation, whole consciousness opened with Hiroshima, the same word means triumph and defeat, confidence and prudence, redoubled lucidity" ([1987] 2014, 11). The development and the use of the atomic weapon, Serres writes in *Conversations*, was a moment of reckoning with the questions of what is "the relationship between science and violence" ([1992] 1995, 16). In *Genesis*, Serres further reflected on the escalating Cold War nuclear arms race: "The stronger one presents arms and the weaker one runs away. Yes, arms are presented like monstrances. Hiroshima: the bomb ripped up the vanquished and, since then, has been getting displayed. Whoever holds it high halts the greatest violence, how many times, ere?" ([1982] 1995, 88–89). Interrogating the connections between science and violence led Serres further to the problem of scientific responsibility (Serres and Latour [1992] 1995, 17).[11] In *The Parasite*, he asks, "Science has made a deafening noise since the bombings of Hiroshima and Nagasaki. It also leaves monstrous fragments behind. Who flees at the sound of these explosions? The world? Men?" ([1980] 2007, 237).

The tone becomes even bleaker in *The Natural Contract*, where Serres writes, "Since Nagasaki we have our disappearance in our power, and the danger curve is rising exponentially" ([1990] 1995, 119). While these examples illustrate that the motif of Hiroshima and Nagasaki is a generative and recurring theme of Serres's thought—yet another Jeremiah's cry—the last reference is also significant for another reason. It marks an emergence of a new important and interlocking theme in Serres's thinking and a new ethico-epistemological demand, that of the *objective violence* (Serres [1990] 1995, 10) humans inflict on planet Earth. And this is also a point where Serres's thinking and ethics explicitly start intersecting with the concerns surrounding ecology, pollution, and wastes of war in my own ethnographic work in postwar Bosnia and Herzegovina and in the context of modern warfare more broadly.

Earthly Violence

The end of the Cold War, when the threat of nuclear Armageddon had started dissipating, intersected with another planetary crisis that became a matter of concern to Michel Serres. It was the rise of CO_2 emissions and global warming and what later became better known as climate change and the Anthropocene. In *The Natural Contract*, Serres turned his attention to the question "Who is doing violence to the worldwide world?" ([1990] 1995, 16). A decade before Paul Crutzen and Eugene Stoermer launched in 2000 the word "Anthropocene" to the orbit of scientific debates (Thomas, Williams, and Zalasiewicz 2020), Serres already anticipated,[12] like with so many other themes beforehand (see Watkins 2020, 2–4), these planetary transformations when he wrote:

> For, as of today, the Earth is quaking anew: not because it shifts and moves in its restless, wise orbit, not because it is changing, from its deep plates to its envelope of air, but because it is being transformed by our doing. [. . .] (i)t depends so much on us that it is shaking and that we too are worried by this deviation from expected equilibria. We are disturbing the Earth and making it quake! Now it has a subject once again. [. . .] This crisis of foundations is not an intellectual crisis; it does not affect our ideas or language or logic or geometry, but time and weather and our survival. (Serres [1990] 1995, 86)

Building on his earlier interests but also ethico-epistemological concerns about scientific responsibility (Dolphijn 2018a), Serres became interested in ecological issues more broadly. It was in reaction, Bruno Latour writes, to Serres's increasing concern that "the traces of our action, [are now] visible everywhere!" (2017b, 62). Serres's relationship with ecological matters and thinking is complex (Watkin 2020, 329–78). He problematized these concerns in several books in the last decades of his life. Apart from the widely acclaimed

The Natural Contract ([1990] 1995) and his writing on climate change, one ecological theme that stands out quite prominently is the general ecology of pollution (Watkin 2020, 352). And it is Serres's theory of pollution that I find particularly generative and useful for thinking about wastes of war in postwar Bosnia and Herzegovina and beyond.

Why pollution? Pollution has become a perpetual background noise of our times in which life in all its forms unfolds, from forever plastics to noise and air pollution. Anthropologists have engaged with the study of pollution since at least Mary Douglas's pioneering work (1966; 1986; 2003; Douglas and Wildavsky 1983). There has been a renaissance in recent years with ethnographic studies of waste and discard (Reno 2015) and studies of risk and pollution in late industrialism (Fortun 2012; Jovanović 2018). Yet there has been only a scant engagement with Serres's work. The theme of pollution is perhaps most succinctly articulated in his short book *Malfeasance: Appropriation through Pollution*. It integrates and further develops many of the ideas from *The Natural Contract*. The central question Serres asks in *Malfeasance* is: "What do we *really* want when we dirty the world?" ([2008] 2011, 40, emphasis in original). In other words, he asks not how but *why* we pollute. And the answer he gives is *to appropriate*. Pollution is a practice interlocked with appropriation. We can observe this logic wherever we look. Serres identifies appropriation through pollution among animals who urinate to mark and thereby appropriate their territory. This is analogous to humans erecting fences, building walls, drawing territorial borders and putting up barbed wire, or discharging toxic waste from the rich countries "in the mangroves of poor countries" as a way of "seizing and recolonizing them" (Serres [2008] 2011, 48; also Liboiron 2021). But Serres's observation-cum-critique doesn't stop at pointing out the effects of slow violence (Nixon 2013) humans inflict on each other when they pollute.[13] His critique is planetary in its ramifications (see Clark and Szerszynski 2021). "The giant garbage dumps" that encircle cities worldwide, Serres writes, "mark the collectivity's appropriation of the nature . . . at the limits of growth, pollution is the sign of the world's appropriation by the *species*" ([2008] 2011, 53, original emphasis).[14]

The same logic of human-induced earthly violence through pollution can also be found in the logic of modern warfare. As I have already suggested, Serres always tried to escape Cartesian dualistic thinking and "the Cartesian disregard of Nature" (Dolphijn 2018a, 4). Thinking *with* Serres thus also invites us to reconsider the dualistic categories with which we usually frame and think about conflict and violence such as war: peace, winners: losers, perpetrator: victim. Escaping this dualist trap, when considering wastes of wars, is important for at least three interrelated reasons. On the one hand, it is a reminder that the temporal boundaries of a conflict and its aftermath are never clear-cut but rather topologically entangled. On the other hand, the effects of wastes of war are long-lasting, and therefore the assault on the enemy's environment does not stop with the signing of peace agreements (see Henig 2019, 87). Furthermore, dualistic categories are problematic for the silences, exclusions, and omissions they generate. Serres makes this point clear in *The Natural Contract*, where he discusses the famous Goya painting *Fight with Cudgels*—a depiction of a pair of enemies brandishing sticks in the midst of a patch of quicksand. If we were to deploy these dichotomies to describe what is going on in the painting, we could ask with Serres ([1990] 1995, 1): Who will win? Who will die? Who will be remembered as a victim? Who will be a perpetrator? Or we could completely disrupt these dichotomies, as Serres does, and "identify a third position, out- side their squabble: the marsh into which the struggle is sinking" ([1990] 1995, 1). The marsh is a powerful metonym for the violence and destruction humans inflict on planet Earth.

Serres ([1990] 1995, 10) distinguishes between *subjective wars* (between adver- saries) and *objective violence* (against the worldwide world as Serres puts it). In the language of contemporary theory, we could also read the former as stand- ing for human-centric while the latter for more-than-human perspectives re- spectively (Tsing 2015). As Serres reminds us, when thinking about wars and violence, we have become myopic to the violence done to the marshes, rivers, soil, or air—the elementary components of planetary habitability for humans and nonhumans alike. To continue further with Goya's painting in mind, the great majority of studies of war and violence see the adversaries fighting "to the death in an *abstract space*, where they struggle alone" ([1990] 1995, 3, my em- phasis), where they kill, maim, or torture each other without any soil, marsh, or river or without *"this garbage,"* as Adnan put it, with which we pollute the planet when at war. In other words, when we write and think about wars and mass violence, we often don't speak about the damage and trauma inflicted on the world itself. Nor do we speak about the unruly contemporaneity of wastes of war and their life-cum-earth transforming potentialities that can strike at any point in the future.

With the unfolding ecological crises and encroaching climate breakdown, Serres writes, the once "mute world, the voiceless things once placed as a decor surrounding the usual spectacles, all those things that never interested anyone, from now on thrust themselves brutally and without warning into our schemes and manoeuvres" ([1990] 1995, 3). Or, as Nigel Clark and Kathryn Yusoff put it, "it could be said that our earth now looks disturbingly lively, and we ourselves frighteningly inert" (2017, 4–5). And it is our animated earth that also makes wastes of war lively in new ways. Indeed, the effects of climate change such as fires, floods, soil erosion, or melting ice sheets, humanitarian disarmament experts argue, can "wake up," dislodge, or expose land mines, ammunition, toxic waste, chemical weapons, and nuclear materials (Zwijnenburg 2021).

This also has been the case during my ongoing research in postwar Bosnia and Herzegovina, which is one of the most heavily contaminated countries with explosive wastes of war in the world as a result of the 1992–95 conflict, particularly with land mines (Bolton 2010). There are over 9,000 defined micro-locations affected by the contamination (directly impacting 15 percent of the population) and officially 120,000 land mines and unexploded ordnance to be found. As I have argued elsewhere (Henig 2019), most of these micro-locations are in rural areas, where the levels of contamination are particularly high for the forested (63 percent) and agricultural (26 percent) land on which the local communities are heavily reliant as the main source of livelihood. These areas are also increasingly prone to soil erosion, often due to unregulated timber extraction, that makes them more vulnerable to an increasing number of floods. In the past decade, there were several large floods that can be directly linked to climate change. The floods caused numerous landslides in the areas contaminated with explosive wastes of war. As a consequence, many minefields shifted to previously uncontaminated areas and in some cases to inhabited villages, creating dangerous new zones of military aftermath (Henig 2019; Musa, Šiljković, and Šakić 2017; Trumble 2021). To date, these shifted minefields have remained largely undocumented as the warning signs around the land mines were carried away as well.

It was after one of the devastating floods in 2014 when I became acutely aware of how the Earth can thrust itself brutally, to paraphrase Serres, into people's lives. In central Bosnia, the soil, heavy rain, and rivers became lively entangled, shifting the soil and dislodging land mines. Some land mines were carried away by a local river. It was several weeks after the floods that one of my long-term interlocutors went fishing—something he had done his entire life— and instead of a tasty catch for dinner, he "caught" a couple of antipersonnel land mines. These wastes of war were carried away from the upstream villages where the landslides occurred a few weeks earlier. In my subsequent interviews but also in my archival research, a new form of the disturbingly lively Earth has started emerging: since about 2010, there have been recurring severe rains, triggering landslides and floods that carry land mines and cluster bombs to previously safe and uncontaminated areas.

In teasing out these entanglements, Michel Serres became my crucial interlocutor. Thinking with Michel Serres, I argue, reminds us about the effects of pollution and objective violence wars inflict on the natural world. Indeed, although wars have always been ecologically destructive and European imperial warfare of the past five hundred years has been at the roots of our climate crisis, as Amitav Ghosh (2021) so poignantly documents in *The Nutmeg's Curse*, this is in particular the case for the strategies, tactics, and practices of modern warfare of the twentieth and the twenty-first centuries. Modern warfare moves away from targeting only the bodies of the adversaries toward targeting and annihilating the environmental conditions of the enemy's life (Sloterdijk 2009). Put differently, modern warfare has become "waged against the very earth itself" (Gregory 2016, 9). The full scale of this industrialized logic of annihilation was unleashed in the battlefields and trenches of Verdun, where Serres's father survived bombing of such a scale that in the four years of the Great War the soil in that area "had undergone the equivalent of 10,000 years of natural erosion" (Flyn 2021, 182). What this earthly violence leaves behind are wastes of war with their unruly temporalities of danger, violence, and uncertainty.

When War Percolates (by Way of Conclusion)

Over the years of researching wastes of war, their unruly temporalities and polluting effects, I have been conversing with Michel Serres. War is a tactic of appropriation—of territory and life, human and nonhuman alike—through waging war and unleashing violence and through seizing and/or destroying the very conditions of livability of the enemy's habitat. I am writing these lines while following the news about Russia's war waged on Ukraine and watching the very same tactic of earthly violence unfolding yet again.[15] Serres's general theory of pollution, however, brings the often omitted dimension of war pollution—hard and soft, material and discursive—into the equation. Indeed, appropriation through pollution belongs to the repertoire of tactic and logic of modern warfare whereby the entire ecosystem is appropriated by scorching and polluting it with toxic, explosive, or radioactive contaminants (see also Watkin 2020, 353). Thinking with Serres allowed me to develop *wastes of war* as an umbrella concept for tracing and thinking about these insidious war-generated pollutions.

Wastes of war refer to the forms of war-induced pollution that outlive the conflict and have a potentiality to continue contaminating and thus appropriating the environment in the future. The contaminated and toxic soil of the no-go zones of the Great War battlefields of Verdun in France (Bausinger, Bonnaire, and Preuß 2007), the ongoing consequences of nuclear bombing, testing, and fallout worldwide (Brown 2020; Masco 2021), of Agent Orange in

Vietnam (Nguyen 2016), of depleted uranium in former Yugoslavia (Nikolovska, forthcoming), of depleted uranium and phosphorus munitions in Lebanon (Khayyat 2022; Touhouliotis 2018), of land mines and cluster bombs in Bosnia and Herzegovina, Colombia, Laos, or the Korean DMZ (Henig 2012, 2019; Kim 2016; Zani 2019), and of burn pits (Logan 2018; MacLeish and Wool 2018) offer unprecedented narratives of the long-lasting socio-environmental aftermaths of wars. What these and many other examples from the twentieth and twenty-first centuries have in common is that war-induced pollution and contamination do not cease to be harmful at the moment of signing a peace treaty. Indeed, over the years of my fieldwork in the areas of Bosnia and Herzegovina contaminated with land mines and cluster bombs, I am often asked rhetorical questions by my interlocutors such as "What is the difference between land mine fields now and during the war?," which is followed by the answer "None. They can still kill you."

Wastes of war hardly ever become a matter of the past. They remain rather unruly contemporaries. Not only the deminer Adnan but the entire ecosystems continue to be impacted and dragged back to the violent past by "this garbage" as he referred to explosive war remnants. War after war, conflict after conflict, the soil, forest, trees, rivers, ice sheets, and human bodies all become containers for preserving the insidious potentialities of wastes of war. These potentialities can disappear and reemerge hours, days, years, or even decades later, when triggered, for example, by soil erosion, floods, melting ice sheets, or movements of people or animals who try to rebuild their lives in the zones of military aftermaths. The properties of wastes of war have different residence time within the local environments that can stretch over decades, centuries, and even millennia and thus intersect with the Earth's geological temporalities. The images of Hiroshima, Nagasaki, Fukushima, and Chernobyl's protective sarcophagus immediately come to mind.[16] But these are not the only examples of such unruly geological temporalities that intersect with the potentialities of wastes of war. Following Serres's concerns about scientific responsibility and the ethico-epistemological demands in the atomic age, let us in conclusion consider perhaps a less known but equally concerning case of the Iceworm project.

Between 1959 and 1966, the US Army operated a top-secret Cold War project known as "Iceworm" under Greenland's ice sheet, aimed at building nuclear missile bases (Nielsen, Nielsen, and Flegal 2021).[17] The project and its military waste were abandoned in 1967 (Colgan 2018), with the assumption that the waste would be preserved under the ice sheet for eternity. Greenland's abandoned military bases contain multiple forms of wastes of the Cold War: biological, chemical, nuclear, physical, and radioactive, with chemical toxins (polychlorinated biphenyls) being the main threat of global significance (Colgan et al. 2016). The accelerated speed of climate change and thawing of Greenland's ice sheet have turned the *potentiality* of the remobilization of these toxins, which could seep into the surface water and contaminate the marine ecosystem across the entire region, into a *possibility*, even if still distant for now (Colgan et al. 2016). This has also thrown the problem of appropriation through pollution into sharp relief. Nowadays, the question remains as to who would be liable for causing an ecological catastrophe in the making—would it be the United States or Denmark? Equally, which international treaties and agreements would apply in the case of such unforeseen and unanticipated scenarios and futures of wastes of war, entangled as they are, in climate futures? The Iceworm project is just one of countless examples of enduring earthly violence caused by (Cold) war. It also shows the relevance and productivity of conversing with Serres's arguments about pollution as a form of spatiotemporal appropriation but also about the need to sign up to a new Natural Contract.

Finally, the general ecology of pollution offers a very different angle for thinking about what modern warfare does to planet Earth. Yet to fully appreciate the unruly temporalities of wastes of war residing in local ecosystems and what makes their insidious potentialities radically contemporaneous, we need to turn to the question of *topology* once again and to Michel Serres's philosophy of temporal topology in particular. As I have already suggested, Serres rejects the idea that time is linear and develops instead the notion of complex time (Assad 2012, 91). This notion assembles "an extraordinary mixture . . . [of] stopping points, ruptures, deep wells, chimneys of thunderous acceleration, readings, gaps—all sown at random, at least in a visible disorder" (Serres and Latour [1992] 1995, 57). For Serres, this is important to acknowledge because, as he writes, "things that are very close can exist in culture, but the line makes them appear very distant from one another" or, alternatively, "there are things that seem very close that, in fact, are very distant from one another" (Serres and Latour [1992] 1995, 57; cf. Knight 2012). To understand these spatiotemporal relations between proximities and distances, we need a topological perspective—to track how things are folded and crumpled into one another and where the rifts are located. Every historical era is therefore for Serres polychronic, that is, "simultaneously drawing from the obsolete, the contemporary, and the futuristic. An object, a circumstance, is thus polychronic, multitemporal, and reveals a time that is gathered together, with multiple pleats" (Serres and Latour [1992] 1995, 60). In this topological perspective, time doesn't flow linearly but rather, as Serres puts it, time *percolates* in "a turbulent and chaotic manner."

Drawing these threads of my argument and conversations with Michel Serres together, this chapter has shown how the unruly temporalities of wastes of war are polychronic and percolating in a Serresian sense. For Adnan, Billy Pilgrim, my interlocutors in Bosnia and Herzegovina, inhabitants of Ukrainian villages and towns, or farmers in Laos (Zani 2019), none of them can be sure when "this garbage" (and in what form) will percolate once again to their present lives and with what consequences. Nonetheless, wastes of war percolate, as a potentiality, as I am often told, "to kill you."

NOTES

1   As W. G. Sebald writes, the scale of destruction was so huge that it left behind 42.8 cubic meters of rubble for every inhabitant of Dresden (2003, 3). According to the paleobiologist and stratigrapher Jan Zalasiewicz and his colleagues, in the areas of heavy bombardment from the Second World War, archaeologists nowadays excavate a distinct underlying geological strata—solely composed of war rubble that includes "concrete, brick, clinker, rock, fly ash, slag and solid chemical waste"—and this distinct strata sits between the Holocene and the Anthropocene layers (Zalasiewicz et al. 2017, 14; also Clark 2021).

2   Daniel Knight aptly describes such temporal disorientations as *the elsewhen*, that is, "the temporal confusion caused by living in different times" (2021, 46).

3   On anticipation and enduring potentialities of violence, see Hermez 2017; also Bryant and Knight 2019.

4   In the interview (conducted in Bosnian language), Adnan used the word *smeće* (literally meaning "garbage" or "waste").

5   When talking about time, Serres often gives an example of the crumpled handkerchief, which allows him to show "how time readily accommodates heterogenous objects as well as how each object can enclose within itself opposite or opposing modes" (Bennett and Connolly 2011, 159).

6   The same could also be said about Serres's thinking on violence and death, which was inspired by Serres's contemporary and friend René Girard.

7   In his *Experiments in Topology*, Stephen Barr describes it as "the study of continuity" (1964, 2).

8   This is also akin to Gregory Bateson's thinking as Arpad Szakolczai shows in his chapter in this volume.

9   On the role of (absent) church bells in the Great War's soundscape, see Morelon 2019.

10 On the Manhattan Project and beyond, see Gusterson 2004 and Masco 2021.

11 Serres explicitly mentions Simone Weil as his influence. On the ethics of responsibility in Weil's thought, see Moi 2021.

12 It is worth mentioning that Crutzen, known then mainly for his Nobel Prize-winning work on the significance of the ozone layer, is also "considered one of the foremost authorities on models that project the environmental effects of nuclear

warfare" (Grove 2019, 37) and in particular its threat to the atmosphere and the ozone layer.

In his *Facing Gaia*, Bruno Latour also illustrates how Serres anticipated many of the concerns later articulated in James Lovelock's Gaia hypothesis (Latour 2017b, 41–74).

13 While the work on slow violence inspired by Rob Nixon's groundbreaking book (2013) has become ethnographically interrogated in recent years (e.g., Ahmann 2018; Davies 2019; Vorbrugg 2022), once again Serres's work that anticipated these concerns is missing in these conversations.

14 For a productive engagement with Serres in the study of waste, see Sosna, Henig, and Figura 2022 and Sosna 2024.

15 See, for example, Zasiadko 2022.

16 See, for example, Max Planck Institute for the History of Science 2023.

17 From its outset, the project relied heavily on—and thus provided strategical support to—extensive glaciological and paleoclimatological research of the ice core. The very process of drilling and building under the ice sheet thus provided scientists with information about the past 100,000 years of Earth's climate history, and the data collected during the Iceworm project provided a key foundation for understanding and modeling past and future climate change scenarios for decades to come (Martin-Nielsen 2013).

STEVEN D. BROWN

7

FEELING SAFE IN A PANBIOTIC WORLD

Introduction: The Changing Nature of Safety

In *Malfeasance*, Michel Serres returns once again to the distinction between the hard and the soft found across the majority of his work. On this occasion, what Serres is pointing toward is a difference between two kinds of "pollution." There are the "hard" material forms of waste that are typically produced as the by-products of industrial processes and the "soft" cognitive or semiotic traces that accompany the extension of these processes into the organization of sociocultural life:

> Let us define two things and clearly distinguish them from one another: first the hard, the second the soft. By the first I mean on the one hand solid residues, liquids, and gases, emitted throughout the atmosphere by the big industrial companies or gigantic garbage dumps, the shameful signature of big cities. By the second, the tsunamis of writing, signs, images, and logos flooding rural, civic, public and natural spaces as well as landscapes with their advertising. Even though different in terms of en-

ergy, garbage and marks nevertheless result from the same soiling gesture, from the same intention to appropriate, and are of animal origin . . . in combination with hard pollution, soft pollution proceeds from the same drive . . . fundamentally it emanates from our will to appropriate, our desire to conquer and expand the space of our properties. (Serres [2008] 2011, 41–42)

This passage exemplifies some of the key characteristics of Serres's thought. It deals with contemporary problems of social organization, the ways in which we live with one another and more broadly with the generalized ecological landscape that the social weighs its increasingly unstable mass on. It offers a grand narrative sweep in which we are given to understand that the contemporary problem is not necessarily entirely novel but instead fits within the unfolding of a drive or impetus to pollute that is archaic in nature (see Watkin 2020). Then comes the twist: the roots of the "soiling gesture" are not, in themselves, human at all but are of "animal origin." We must think of pollution as a problem of a very different order from merely that of its specific forms or scale. That problem is then finally presented as one of property—the desire to "conquer and expand space." In a short series of moves, Serres shifts the problem space dramatically, with significant consequences for how we might think and act in relation to what pollution is and what to do about it.

In this chapter, I want to use this same intellectual strategy to consider how the practice and the idea of "safety" has shifted during the COVID-19 pandemic. For many of us, safety has constituted a background concern: always present, to be sure, as an issue to be managed in our workplaces, our homes, and in our sexual relations with others, but something routinized, understood to a certain degree, and for much of the time treated as something "under control." Safety is an outcome of engaging in a specific practice, of following the right guidelines (Reiman and Pietikäinen 2012). Sometimes it is already "built in" to the tools and technologies we engage with, most notably in the aviation industry, which is often held up as the gold standard for safety practices (Grote 2012). As a consequence, safety becomes most relevant at moments of sudden failure, when a series of weaknesses within a system accidentally align (the "Swiss Cheese" model of failure produced by James Reason, 1997). Whether the pandemic is best understood in this way as a terrible series of contingencies or the predictable outcome of poor preparation for emergent viruses in a globalized economy, COVID-19 has dramatically shifted both the "hard" material practices around safety and the "soft" sociocognitive processes through which we can think about enacting safety within interpersonal relations.

It would be faithful to Serres's mode of reasoning to note at this point that none of this is without precedent. During the AIDS pandemic beginning in the 1980s, the LGBTQ+ community reinvented the concept of sexual safety, far in advance of the biomedical "hard" discoveries that would ultimately enable HIV infection to become a routinized concern (Epstein 2022). The Black Lives Matter movement has more recently placed lack of safety as central to the protests around policing in the United States and elsewhere (Loader 2021). There are specific genealogies that have been traced around each case that also intersect with and inform the shifts around safety during COVID-19. But in this chapter, I want to focus specifically on the dissemination around safety guidance as a form of the kind of "soft pollution" that Serres describes. Take, for example, figure 7.1, produced by the US Food and Drug Administration as "Best Practices for Retail Food Stores, Restaurants, and Food Pick-Up/Delivery Services during the COVID-19 Pandemic."

In its most simplified form, the FDA guidance identifies four domains of concern. The first, "Be Healthy, Be Clean," concerns relations between employees. It defines the workplace as a potential locus of infection where COVID-19 can be transmitted through physical proximity or through shared contact with objects (including food). This locus is constantly at risk of being destabilized by the introduction of the virus despite the guidance for the use of personal protective equipment (PPE), social distancing, and constant cleaning and chemical scouring as recommended within the second domain of "Clean and Disinfect." It only takes one case of infection to align all the weaknesses in the system as a whole (as in the "Swiss Cheese" model of safety). As one recommendation starkly states: "If an employee is sick at work, send them home immediately. Clean and disinfect surfaces in their workspace. Others at the facility with close contact (i.e., within six feet) of the employee during this time should be considered exposed" (FDA n.d.).

The sick employee should be considered as having polluted the entire space. This requires the immediate "hard" response of eradicating potential traces of the virus in the immediate physical environment and the "soft" response of recategorizing fellow workers as now likely further sources of infection. But employees are not the only concern. Customers also constitute both a threat and a potential vulnerability to the organization, as noted in the third domain of "Social Distance." The guidance strongly emphasizes that public messaging, in the form of signs and audio recordings, be used to "educate" both customers and employees around the need to maintain social distancing at all times. But this creates a further problem. The "tsunami of writing, signs, images and logos" that sweeps through the typical retail or restaurant environment

FIGURE 7.1. US Food and Drug Administration COVID-19 sign

needs to be both increased and at the same time carefully managed (Serres [2008] 2011, 41–42). One potentially contradictory piece of guidance (FDA n.d.) recommends that the organization should both:

- Avoid displays that may result in customer gatherings; discontinue self-serve buffets and salad bars; discourage employee gatherings
- Place floor markings and signs to encourage social distancing

On the one hand, the soft "soiling gesture" of polluting the space with signs that encourage consumption should be restrained. But in its place, new signs and marks need to be added. We might reason that this redirection of semiotic pollution might result in a scattering of attention on the part of both customers and employees that is in tension with the project of maintaining an ordered physical space. A potential solution is offered in the final domain of "Pick-Up and Delivery," which effectively extends the workplace beyond the limits of the outlet itself and into the wider social networks of the surrounding community. Here the risk is transferred from the organization to the customer, who is encouraged to use "no touch deliveries" or "curbside pick-ups." This amounts to a kind of "social distancing at-a-distance" from the organization itself.

The potential complications and contradictions that arise from the FDA guidance are unsurprising. All risk management involves a trade-off that involves balancing degrees of freedom with calculations of safety margins. In fact, the FDA guidance is something of a model of clarity in comparison with

FIGURE 7.2. UK government COVID-19 sign

the safety messaging provided by the UK government during the pandemic (see figures 7.2 and 7.3).

The initial public health message was "Stay Home > Protect the NHS > Save Lives" (see figure 7.2), which had the virtue of indicating a clear lexical and grammatical referent of "home" where people should stay—most people having only one home and knowing where it is—and what the desired consequences would be of doing so (see The Independent Sage 2020). Its subsequent replacement by "Stay Alert > Control the Virus > Save Lives" was met with widespread derision given the sheer opacity of its meaning—"alert" being an imprecise cognitive state that is open to interpretation. Furthermore, the idea that this "alertness" might translate into "control" over a virus that by definition could only be perceived following infection (at which point control has been all but entirely ceded) is entirely mysterious.

Despite the lack of clarity, in both cases what is being accomplished is a deliberate shift in how safety is both understood and practiced (see Reicher et al. 2021). This shift matters considerably since it reverberates within the porous shaping of the worlds that are gradually emerging from the pandemic. The ways in which safety has been reconfigured around COVID-19 have implications not merely for future pandemics but more broadly for what "feeling safe" will mean in relation to the myriad ecological, political, and global health threats that are crystallizing (see Brown et al. 2022). When we have become accustomed to the idea that the embrace of a loved one or handling the same tools as an employee is a safety issue, our relationship to one another, other species, and the damaged planet cannot not be changed. But in what ways? And what kind of a problem is "safety" for us?

FIGURE 7.3. UK government COVID-19 sign

Hominescence and Death

In an oft-cited passage from one of his conversations with Bruno Latour, Serres reflects on how the historical circumstances of his early years formed his philosophical approach:

> Here is the vital environment of those who were born, like me, around 1930: at age six, the war of 1936 in Spain; at age nine, the blitzkrieg of 1939, defeat and debacle; at age twelve, the split between the Resistance and the collaborators, the tragedy of the concentration camps and deportations; at age fourteen, Liberation and the settling of scores it brought with it in France; at age fifteen Hiroshima. In short, for age nine to seventeen, when the body and sensitivity are being formed, it was the reign of hunger and rationing, death and bombings, a thousand crimes . . . I was six for my first dead bodies, twenty-six for the last one. Have I answered you sufficiently about what has made my contemporaries 'gunshy'? (Serres and Latour [1992] 1995, 2)

Death and violence are omnipresent features of Serres's work (see Serres [1983] 2015 in particular), as indeed they are in the work of many of his contemporary thinkers. In this passage, Serres seeks not merely to indicate his antipathy to conflict as a mode of intellectual exchange—a position that Latour (1987) elsewhere succinctly characterizes as "the Enlightenment without the critique"—but also to stress how the experience of suffering and the death of others is an embodied matter. Violence shapes the body, attunes its sensitivity to the milieus in which it dwells. Throughout his work, Serres has placed particular emphasis on how the training or physical modulation of the body is expressed in thought. The French edition of *Variations on the Body*, for example, contains the dedication "A mes professeurs de gymnastique, à mes entraîneurs, à mes guides de haute montagne, quie m'ont appris à penser" (Serres 1999, 5). To think is, before all else, to move, to engage, to taste, to perambulate, to climb, to leap, to travel. To be on the move. By contrast, that which is static, that cannot or can no longer move, is, for Serres, associated with death. In *Statues*, Serres expounds at length on how the corpse is the first object that can be properly said to no longer be a subject, to have made the transition from life to death. But in the course of becoming an object, the corpse accomplishes a stabilization of the relation between (human) subjects and objects: "The object, the subject lacking any reference, find one in and through death since the remains define the here, mark it, fix it, in space and for time. They are organized and placed, take on meaning, in relation to death; relative to that reference, they can substitute for one another. What is the object? It's the body come back, the resurrected subject, what we call a ghost—a statue" (Serres [1987] 2014, 74).

Despite the "gun-shy" nature of Serres's work, it nevertheless preserves a functional role for death within the organization of the collective. The relationship between subjects and objects—which is both extremely complex and to some extent reversible in Serres's thinking—achieves a temporary stabilization around death. The corpse and then the funereal statue that is substituted for it in some burial practices becomes the boundary marker between subject and objects, fixing them in relation to time and space. Nowhere is this clearer for Serres than in the traditional Christian formulation used on gravestones: "here lies." Because the corpse/statue does not move, it can become the point around which the spatiotemporal coordinates of the collective can be extended. Paris is exemplary in this respect, with its catacombs beneath the city: "I would have liked Eiffel to have put his tower up in the place of the lion, at Denfert-Rochereau, so as to sink the fluid foundations of the four pillars into the catacombs, the way the Abbot Suer founded the Basilica of Saint Denis over the crypt in which all our kings lie" (Serres [1987] 2014, 61).

A first attempt to define a broader conception of safety might then be *the security of the relations between subjects and objects that is provisionally secured by death*. In COVID-19 guidance, for instance, the binaries of infected/disinfected, clean/dirty, safe/unsafe are all held together by the vast unseen presence of the virus and its capacity to insinuate itself into the relations between people and the ordinary objects and tools that they handle. The threat of illness and death clarifies and orders humans and objects within the workplaces and homes where they interact. Safety is constituted through an attempted purification of space. It is tempting here to reverse the order of the UK government instructions. It is the desire to save the lives of others that commits us to enacting safety practices that help control the virus that then produce a conscious awareness or staying alert to what we are to one another and to the broader ecology that has given rise to the pandemic.

The capacity to commit to the project of saving lives is, of course, itself underwritten by the scientific and technological powers of the contemporary medical sciences. Unlike many of his contemporaries, Serres has remained live to the liberatory power of science and technology. In *Thumbelina*, for example, Serres contrasts the connective power of digital networks with the older forms of scholarly exchange of his youth:

> At the end of my studies, when I was twenty years old, I became an 'epistemologist', which is a big word to say that I studied the methods and results of the sciences, and occasionally tried to judge them. There were, at that time, very few of us, so we corresponded with each other. A half-century later every Tom Thumb on the street can make judgements about nuclear power, surrogate mothers, GMOs, chemistry, and ecology. Though I no longer claim to work in the discipline, today everyone has become an epistemologist. There is a presumption of competence. Don't laugh, says Thumbelina. When democracy gave everyone the right to vote, it did so in opposition to those who considered it a scandal to give an equal vote to both wise men and fools, ignorant and educated. The same argument applies here. (Serres [2012] 2015, 62)

It is rare to hear this kind of call for the democratization of knowledge, with all the complexities that are implied, raised with such clarity. In part, Serres's enthusiasm for modern technology is informed by the sense that it is movement that characterizes invention and transformation. To see the value in social issues like the previous ones debated "live" and on the move is then consonant with the approach that informs all of Serres's work. It is everyone's business to figure out how we should best live with one another and the Earth system rather than this being reserved for a cadre of professionals who have purified and insulated their knowledge base from the rest of the collective. (This passage is also unusual in that epistemologists are not immediately described by Serres in terms of either masturbation or defecation, as they are in Serres [2008] 2011, [1983] 2015.)

But Serres's enthusiasm for science and technology is also tempered by recognition of its unequal distribution and effects. He is well aware of vast swathes of the global population who are not, in fact, three clicks or touches away from any piece of knowledge and of the increasing poverty, conflict, and environmental destruction that are ceaselessly multiplying. In *The Incandescent*, Serres nevertheless reflects on the tension between the movement toward the global and the desire to insist on limits, to project a version of the local to simplify and map the global. He considers a range of terms beginning with the prefix *pan-*: *Pantope* (the experience of seemingly limitless space), *Panchrone* (the imagination of our place in universal time), *Pangloss* (the ability to translate between all languages), *Pangnose* (the democracy of epistemology), *Panthrope* (the possibility of living amongst all peoples), and *Panurge* (the overcoming of biological limits to human powers). In each case and in combination, Serres considers the risks that attend each of these aspects of the contemporary human condition. He concludes that it is no longer within our gift as humans to refuse the universals that are now part of everyday reality. We must instead inhabit the terrors they give rise to in order to productively seek a way to turn these powers back on themselves:

> For these questions concerning our universality are borne out today in daily practices that are numerous enough for us to now negotiate them fear and trembling, if not with prudence. Everything that we call ecology—global warming, the eradication of species or protection of the environment, ethics, prudence, sustainable development—tends to ask the world itself to put limits on our enterprises: our universal exploitation of the Universe frightens us. Our new first names with their prefix pan gives us panic. We demand of the universe to accompany, regulate and moderate with its universality the panurgy of the human. (Serres [2003] 2018, 135)

The seemingly limitless expansion of human powers has led to universal exploitation. Which in turn has led to emergent circumstances such as the capacity of an interspecies-born virus to piggyback on global travel to achieve pandemic levels of infection. In the same way that social distancing and lockdown measures tried to desperately enforce a version of the local as a way of addressing the costs of global viral transmission, Serres suggests that our common response to becoming-pan in multiple ways is, ultimately, *panic*. We insist that the world itself deliver us limits to our own powers—our panurgy—that will moderate its impact. Elsewhere, Serres puts this in a slightly more pithy formulation: "We recently went from the local to the global without any conceptual or practical mastery of this latter. These globalities have just taken on another face, one that's practical, concrete and quasi-close at hand. Everything depends on us. And through new and unexpected loops, we ourselves end up depending on the things that depend globally on us. Here, risks and chances grow as fast as our omnipotence" (Serres [2001] 2019, 10).

In this formulation, Serres reworks the classical distinction between "things that depend upon us" and "things that do not depend upon us." Panurgy has the consequence that "everything depends upon us," since we have universally exploited the Earth system. But in doing so, we have come to depend on those very same things that depend in turn on us—the soil, clean air, drinkable water. It is as though we have developed the capacity to exercise power over everything except that very power itself—"How can we dominate our own domination; how can we master our own mastery?" (Serres and Latour [1992] 1995, 172). Serres uses the term "hominescence" to describe the long journey of human evolution that has led to this point. Technology forms a crucial part of this story. For Serres, tools do not so much stand in for human powers, as they do for other philosophers of technology (see Barker 2023); rather, they are a defining part of what it means to be human. Other species acquire abilities through the gradual evolution of their bodies. This is typically confirmed through the identification of vestigial structures that have lost their archaic function over the course of evolution (e.g., wings in now-flightless birds). While vestigial structures are present in human bodies, we are also able to trace a parallel evolution in the artifacts and technologies that we use that become externalized or "set sail" from embodied actions (see Serres [1999] 2011). For instance, we began drinking water with our cupped hands, which led to the crafting of rudimentary cups. The technology of scooping then gave rise to the ability to churn the land, ultimately taking the form of the colossal technologies of digging found in the extractive industries. Yet there is something remaining of the cupped hand in the mechanical scoop that churns up rocks and minerals from beneath the soil. Serres calls this parallel technological evolutionary process "exo-Darwinism":

It took millions of years for birds to grow wings and feathers; in a few months, we build an aircraft. This gain in time defines technology fairly well. The invention of the first tools caused us to leave evolution so as to enter culture. . . . As soon as technology appears, we no longer have any need for that long patience nor for a bodily form and therefore risk disappearing less. Once the airplane is made, we embark; when making a tool is enough, the body changes little if it uses the tool. . . . Exo-Darwinism is what I call this original movement of organs towards objects that externalise the means of adaptation. (Serres [2001] 2019, 39)

The process of exo-Darwinism seems to indicate that the human body will no longer be required to change in response to evolutionary pressures, which will instead be addressed by externalized technological means. But the body is in fact fundamentally restructured through this process. Our human capacities to think, remember, feel, and perceive are taken up anew in the technologies that "cast off" from their initial locus in the body. The eye is augmented by the mirror and the magnifying glass, which set off on their own techno-evolutionary process but which return to the intimacy of our bodies as the reading glasses we can no longer do without, the smartphone picture libraries through which we curate our identities and memories, and the tiny microscopes that when inserted into our guts may help locate and cure the cancers that would otherwise destroy us. We have different bodies and vastly different lives because of the panurgic expansion of the technological. This leads us toward a second formulation of safety—*the balancing of the panic arising from universal exploitation with hope in the expansion of the limits of the human body*. The knowledge of the atom that results in the destruction of Hiroshima has a tributary that leads the MRI scanning that might save our lives.

The second formulation of safety expands the problem space beyond COVID-19 itself, toward a much broader concern with our relationship to one another and to the Earth system. But something of a blind spot remains. Where does this panic arise from? What is it about our confrontation with the universal that destabilizes safety? And, conversely, why did social distancing and lockdown measures, as a return to the local, ultimately prove insufficient in allaying our fears? To gain further traction, it will be necessary to go further upstream in Serres's account of hominescence to reflect on the changing relationship between safety and the local.

Property and Sacrifice

Understanding the origins of human collectivity and the ways in which it is constituted in a mixture of relations with nonhuman actors and the wider ecology recurs as a substantive issue across Serres's work. As he states in *The Parasite*, "What living together is. What is the collective. This question fascinates us now" ([1980] 1982, 224). In addressing this question, Serres often turns toward myth and religion, digressing into extended exegesis of biblical narratives or the founding stories of Rome. In his very last work, whose manuscript was submitted to the publisher on the day before his passing, Serres offers this pointed reflection on his use of these texts and source materials rather than works of anthropology or sociology:

> Where, then, does religion come from? Having some knowledge of astronomy and electrostatics, I am well aware that no one is hurling thunderbolts from behind the clouds with the intention of illuminating, warning, or wounding; well aware, too, from what the human sciences have taught me about the bonds between father and son, of the inanity of this supernatural creature of our fantasies. Undermined by both sides—by the sciences and by the humanities, two partial and complementary points of view—why, one may wonder has religion not entirely collapsed? . . . None of these disciplines, hard or soft, separated by analysis, inquires into the global bond, the existential synthesis through which every human relationship has its natural place. (Serres [2019] 2022, 170)

It is not then that either the hard sciences or the soft humanities have nothing to say on the matter. Quite the reverse: they speak of little else other than the nature of origins and the way they inform development and transformation, we might feel obliged to say. But, for Serres, the dominant mode in which these conversations are conducted is that of analysis, which he treats as fundamentally an act of dissociation, of cutting apart and division—"analysis comes from the Greek verb to untie or to dissolve" (Serres [2003] 2018, 151). Analysis always risks becoming a form of violence, the building up of theses through the critical destruction of competing accounts. What Serres seeks instead is a form of synthesis or federation and, more specifically, a way of grasping the way in which violence itself and our relationship to death might have a federative function that binds together rather than cleaves apart. For Serres, it is myth and religion that provide some of the most compelling narratives about what it is that might bind us together, with the challenge then being how to translate those stories into the domain of the scientific or the anthropological (and vice versa). For instance, the first volume of the Foundations trilogy, *Rome*, deals with the mythical founding of the city by Romulus and Remus. The second volume, *Statues*, argues that ritual and social technologies of exclusion remain part of the operative logic of modernity, through a comparison of the *Challenger* space shuttle disaster with Flaubert's account of the ancient sacrifice to Baal in Carthage. The final volume, *Geometry*, posits that mathematical knowledge provides the guiding thread to understand the historical relations between archaic and modern societies. From myth to science and back again.

The foundations of the human collective are not necessarily human. There must be an object that serves as the point of coordination around which the human actors can be arranged, like the points in a star figure arrayed around a center. This "first object" is a dead body: "The corpse was the first object for men. Posed before them like a problem and obstacle, lying. Any other thing, tree, stone, animal could or can enter into property, individual, collective, private, public and in this last case merchandise, stake or fetish. Before the dead body, every subject draws back: the dead body lies there, cutting out its space, larger lying down than standing, more terrifying dead than alive" (Serres [1987] 2014, 91).

The dead body cannot be exchanged, nor can it readily be divided to form a stake or a quasi-magical object of power. Its power comes instead from its completeness and from the relationship that it now maintains with those stood around it. Serres notes that the practice of lapidation—the punishment of death by stoning—recurs across religious and mythical sources. Lapidation is often associated with either the foundation or the purification of a specific site or territory. Many of the founding myths of Rome, for instance, involve stoning or burial beneath a rock. These acts construct boundary markers, with the human corpse hidden under the rock in a manner reminiscent of the great pyramids, whose immense construction concealed a labyrinth in which the body of the pharaoh was deliberately obscured—"What is a statue? A living body covered with stones" (Serres [1987] 2014, 181).

One of the most well-known instances of (attempted) lapidation comes in the Bible story in John 8:87. A woman accused of adultery is brought before Jesus, who offers the judgment "Let him who is without sin among you be the first to throw a stone at her." Serres observes that the story also has Jesus writing with his finger on the ground before he makes his judgment. Could it be, he speculates, that Jesus has written out the name of the party to the adultery and that this person is among the crowd now clamoring for blood? There is then at least one person who has a complicated relationship to what may or may not be about to happen—"What matters to me is the 'at least one.' For I know him. Not by name, but by his presence and his function—I was about to say by his usefulness" (Serres [2019] 2022, 90). Serres compares this person to that of the sole member of a firing squad who is randomly issued with a blank cartridge:

Why is this blank necessary? Why is it necessary that, amongst the executioners, this 'at least one' does not kill? Why should he be chosen, in effect, by drawing lots? For the same reason, a profound one. For in the wake of a death by stoning or by shooting, judicial review—or a palace revolt, or a popular revolution—may bring to light the innocence of the person who was executed. In that event the situation is reversed, and the people will turn on those who killed the now blameless defendant, which is to say the entire firing squad. But who among this group really killed the one who in the meantime has become a victim? (Serres [2019] 2022, 90)

What Serres describes here is an additional social function that can be added to Rene Girard's (1989) theory of "the scapegoat." For Girard, and subsequently for Serres, the collective is founded on the exclusion of one of their number, who may be subject to banishment or more likely death. These acts of exclusion serve the role of settling rivalries and conflicts within the nascent collective. The desires and contests that animate those who are collected together are resolved when they are turned on one specific individual, who becomes the "it" as though in a child's game of Hunt the Slipper or in a manner of the soccer player who has just missed a goal (see Serres [1980] 1982). The collective energizes itself through its joint efforts to expel or destroy the scapegoat, like the biblical characters who clamor to stone the adulterer. But here Serres points to "usefulness" of the word Jesus has drawn on the ground and the blank cartridge. There is one among us whose actions will differentiate the collective, who ensures that the relationship between the mob and the scapegoat may become reversible. If murder and exclusion are a prerequisite to the founding of the collective, the possibility of this future reversibility, which is known to all, provides an additional bond. Here there might be a third formulation of safety—*the presence of the "at least one" that ensures diffusion of responsibility and the possibility of reversible relations at a future date.*

If the corpse is the first object, it is certainly not the only object that may serve as a foundation. Serres notes that animals may serve as effective substitutes for humans, such that the animal sacrifice can become a means for purifying or re-collecting the bonds of the collective. In a further substitution, fetish objects such as sacred relics can replace animal sacrifice. Yet the logic of foundation remains constant throughout all the substitutions. The energy and petty rivalries of those collected together are sublimated in ritual destruction or glorification of the object that serves as the central point of focus of the collective. In the final case of fetish objects, Serres ([1980] 1982) argues that they may be accorded a kind of agency all their own that elevates them to the effective status of "quasi-objects." His example of the rugby game demonstrates this well, where it is the ball that operates to distribute the players across the pitch as they move to intercept passes. This leads to the conclusion that despite having its origins in death and expulsion, "the construction of the collective has been done with anyone and by means of anything. The furet is nothing, a ring, a button, a thing; the ball is a skin or an air bubble" (Serres [1980] 1982, 229). It is not the material composition of the quasi-object that is decisive but rather its capacity to mark out the relations between subjects: "This quasi-object that is a marker of the subject is an astonishing constructer of intersubjectivity. We know, through it, how and when we are subjects and when and how we are no longer subjects. 'We': what does that mean? We are precisely the fluctuating moving back and forth of 'I'. The 'I' in the game is a token exchanged. And this passing, this network of passes, these vicariances of subjects weave the collection" (Serres [1980] 1982, 227).

The corpse did not move, in part because it had become an object and thus needed to be hidden from sight to ensure that any remaining memory of subjectivity was occluded. But the fetish object realizes a journey in the opposite direction. It is an object that acquires something like agency—hence "quasi-object," although it might as well be called "quasi-subject"—through the way that it moves and constructs intersubjectivity by creating ripples of risk and value. If you receive the ball and transmit it on to score a point, you are the hero. If you fumble the ball and lose the match, it may well be you who symbolically finds themselves under a pile of rocks.

One outcome of the panurgic drive is the constitution of quasi-objects who possess dimensions that tend toward the universal. Serres terms these "world-objects" and defines them as "artifacts that have at least one global-scale dimension (such as time, space, speed or energy)" (Serres [1990] 1995, 15). A satellite, for example, has the global scale of speed, nuclear power the global scale of energy, and fossil fuel pollution the global scale of time. World-objects can accomplish the construction of the collective in the same manner as rugby balls or sacred artifacts, in that they become the focus of collected energies and concerns. But they also demonstrate that violence is never entirely displaced within acts of founding and re-collecting the collective. It is the threat of nuclear incineration or species extinction that is intrinsic to the function of these world-objects. In fact, Serres ([2001] 2019) notes, we have never really gotten very far from the idea of violence and exclusion as part of the organization of society. Consider how many deaths from road traffic accidents societies are able to tolerate in the name of personal mobility, the number of deaths by shooting that societies with access to firearms can tolerate in the name of individual liberty, or the multitude of slow deaths from racism, poverty, and exploitation that are balanced against uneven capital accumulation.

This tolerance for death has its roots in the appropriative character of foundation. In *Malfeasance*, Serres argues against Rousseau's famous dictum that "the first who after enclosing a piece of land thought of saying 'this is mine' and found people simple enough to believe him was the real founder of civil society" (Serres [2008] 2011, 13). This imagined act that gives rise to property rights seeks to deny the violence and destruction inherent in the appropriation of space. Someone or something must die in order to create a boundary marker. But more precisely, the burial of the corpse under the stone stains the site in a particular way that comes to define it. Serres sees this as the continuation, by other means, of an ethological strategy for marking out territory—"tigers piss on the edge of their lair. And so do lions and dogs. Like those carnivorous mammals, many animals, our cousins, *mark* their territory with their harsh, stinking urine or with their howling, while others such as finches and nightingales use sweet songs" (Serres [2008] 2011, 1). It is then necessary to rewrite Rousseau in the following way:

> Whoever spits in the soup keeps it; no one will touch the salad or the cheese polluted in this way. To make something its own, the body knows how to leave some personal stain: sweat on a garment, saliva or feet out into a dish, waste in space, aroma, perfume, or excrement, all of them rather hard things . . . but also my name, printed in black on this book cover, where my signature looks sweet and innocent, seemingly unrelated to those habits. And yet. . . . Hence the theorem of what might be called natural right. By "natural" I mean the general behaviour of living species: *appropriation takes place through dirt.* More precisely, what is properly one's own is dirt. (Serres [2008] 2011, 2–3)

Here we return to the "soiling gesture," but now in the context of an archaic ethological strategy to establish territory and "take place." The hard dimension—excrement and sweat—is ultimately transformed into the chemical and material waste of contemporary practices, whereas the soft dimension—song and signatures—becomes the semiotic tsunami of contemporary societies. Perhaps a fourth formulation of safety then follows—*the ability to tolerate and live among the excrement of others.* We feel safe when we can ignore the pain and exploitation all along the supply chains that keep us alive, when we can suppress the ways in which we collude in the destructive acts that are required to keep the collective collected together. Lockdowns and social distancing were challenging because they make this collective work more difficult to accomplish. We had to recognize that our own safety was being bought at the cost to key workers and the otherwise invisible pain of the multitudes who precariously shore up global supply chains (see Brown et al. 2022).

Panbiota

Is it possible to break with the cycle of violence and appropriation that Serres sees as intrinsic to the founding and maintenance of the collective? And if this were possible, in whatever way, how then would we be able to think of safety, particularly with regard to the range of existing and emergent dimensions in which "feeling safe" matters? In *Detachment*, Serres recounts the story of the Greek cynic Diogenes. Known for having rejected the trappings of privilege, Diogenes is living on the streets of Corinth, when he reputedly meets Alexander the Great. The king asks the philosopher what favor he might grant, what he truly desires. Enjoying the early morning sunlight, Diogenes replies, "Right now, remove yourself from my sun." For Serres, the Diogenes story illustrates the virtues of standing aside, of refusing to enter into appropriative relations with others:

> Diogenes the Cynic has forsaken this price. Diogenes has forsaken the spice of life. Appeased, in rags, alone in front of his barrel, pointing to the zero of usefulness on the nakedness of his skin, he meditates and asks: can we invent relations other than those of struggle, other than those of exchange or worship? Is it possible for me to place my hand on an object, or look at an object which is not a stake, fetish or merchandise? (Serres [1986] 1989, 69)

If we agree to not consider too closely the provenance of this story about a philosopher who is able to defy a king, the central message of the cost of inventing disinterested relations to objects and others is clear. To refuse appropriation is to stand outside the usual circuits of exchange and value. In doing so, we may be exculpated from participation in violence, but this does not necessarily mean that we ourselves may not be subject to appropriation (perhaps in the way that the story of Diogenes is here co-opted into a narrative of violence and safety). Serres is not, however, suggesting that Diogenes is somehow removing himself from all attachments rather than seeking to open himself up to them. He does not want his relationship to the sun to be mediated by Alexander and the forms of appropriation and fetishization he stands for. Diogenes wants to immerse himself in the bonds to the world that stand before the invention of property.

These bonds or cords between the world and humans have persisted within property relations, although they have become obscured considerably. A key event in the transformation of these bonds occurred roughly 10,000 years ago with the domestication of animals by humans, eventually leading to the practice of smallhold farming. In *Hominescence*, Serres describes how living together as a "common house" transformed the bonds between human and animals, since it forced them into something resembling reciprocal relations. In order to build these relations, it became necessary to construct new forms of communication and interaction, an acculturation of nature: "It's less a matter of understanding how we began to tame certain animals, therefore of giving in once again to anthropocentrism, than of seeing how the common hotel was constructed, the hotel in which the host-animals ended up living in symbiosis, at least an apparent one, since they were lodged, looked after and fed by the parasite-human; in order to prepare this common site, it is enough for the parasite to become host and hosts parasites, reciprocal domestication becoming then another name for symbiosis and this latter continuing the cultural genesis undertaken, body-to-body, next to every living thing" (Serres [2001] 2019, 87).

Farming is the decisive moment in hominescence, since it brings animals and humans into a close proximity where there is a mutual acculturation to one another. To live together requires a physical training or embodied coordination wherein species reconstruct the terms of their interaction. We humans are transformed by our close domestic contact with other species, as they are in turn by us. The biological name for this living together is "symbiosis." While the term is often used in the contemporary humanities and social sciences as a synonym for non-extractive or mutually beneficial relationships, in its technical use, it merely denotes a communal relationship where there is benefit to at least one party, with "mutualism" being the proper term for two-way exchanges of value. This is important in this case because, as Serres notes, the human remains parasitic on domesticated animals. (Parasitism is a crucial term in Serres's work—see Brown 2002, 2013.) However, through living together, the roles of host and parasite do become temporarily reversible. The necessity to make oneself a host as a condition of continuing to parasitize becomes entrenched at a bodily level among Neolithic farmers.

The value of "living together" in this way has been lost to a certain degree. In many of his books, Serres points to the massive shift away from rural farming in the Global North as marking a new phase in hominescence where we have forgotten, at an embodied level, what living together with other species feels like, what it is to feel attached to the lives of symbionts. As a consequence, we are challenged considerably when close coexistence is forced on us, such as in the case of urban rat colonies or flea infestations. Since we are no longer trained in the cultural accommodation of living together, we resort to the familiar strategies of purification and the defense of property and attempt to expel what we see as parasitic encroachment, as Serres so beautifully describes in his reading of Molière's *Tartuffe* in *The Parasite*. But to do so is to simply reenter the endless cycle of violence and appropriation that ultimately leads back to the panurgic drive for universal exploitation.

Yet our current circumstances may have curiously returned us back to a situation not entirely dissimilar to the "common site" of the Neolithic. The Pantopic globalization of movement across the globe has not only resulted in a Panthropic collecting together of all peoples (notwithstanding the violent barriers placed against migration) but also, Serres claims, a recognition that we have now entered into Panbiota—a universal living together with all other forms of life who depend on us. This clearly includes coronaviruses. To paraphrase Serres, we depend on viral and bacterial life through a strange loop that has led them to globally depend on us to be what they are. As Serres notes, Panbiota then in some sense "completes" that aspect of hominescence that began with the Neolithic. How do we then think of what living together with something like coronavirus means? It is not a question of simply ignoring its presence and "carrying on regardless," as the public health strategy in the United Kingdom has attempted and failed to do (Stokoe et al. 2022), but rather a more cautious effort at reworking the bonds between virus and human through exploring what they may be outside of relations of violence and appropriation:

> The whole of culture results from a patient, long, local and temporary management, from a comprehension and from negotiations that are as infinite as those I am recommending regarding the ineradicable accursed portion. You will suppress nature as little as you do it and suppress it as little as you do nature. However deeply your enclosure in the city and at home, in the middle of tried and proven defences, may be built, however aseptically and "culturally" your life may unfold, the violence of the wind and of suffering always returns. . . . The whole of the habitat is therefore built at once and without cease. At the same time as culture, morality occurs, and with a similar movement. Of course, one doesn't protect oneself against disease the way one does against a storm, nor against a murderer the way one does against a cold snap, but culture is born from having prepared these defences at a stroke. It only emerges from this prudence, said to be characteristic of the father of the family, whose wisdom manages nature and its constraints, life and its morbid

bacteria, humans and their violence; it is born from voluntary symbioses. (Serres [2003] 2018, 165)

This long and difficult passage contains the kernel of Serres's version of "living together." Coexistence is an infinite project that requires our full attention (hence we must, like Diogenes, ask the forces of appropriation to remove themselves from our sun). But this project will never be free from violence, from the "accursed portion" that defines life at any planetary scale. All life engages in some form of parasitism, or taking without giving, and hence must, at times, be destructive (in the same way that we must engage in analysis to arrive at federation). We have to build a home to survive, yet the fact that we must do so "at once" does not mean that we ought not to at every turn consider the necessity of welcoming others within our home. Hospitality may have to revert "at a stroke" into hostility, when we need to expel a parasite that threatens our survival, but this is an exceptional contingency rather than a principle of living. We must be hosts by default and parasites as required, with the "prudence" required to meditate on this process. *Voluntary and perpetual symbiosis, guided by wisdom, is the basis of Panbiota.* This will also serve as a final formulation of what safety means.

I want to conclude with two final reflections on the preceding passage. Serres speaks of "negotiations" with the world. This is derived in part from his argument for a *Natural Contract* between humans and the Earth system (Serres [1990] 1995). This has sometimes been derided as a well-intentioned but unworkable anthropomorphism, where that which does not speak is only accorded rights on the basis that we engage in the pretense of speaking on its behalf (Ferry 1996). But this interpretation ignores the long and careful work that Serres does throughout his writing on the dependence of legal notions of contract with notions of harmonies and laws in nature. As David Webb (2022) has argued, in order to even think of something resembling a negotiation, we must first understand the resonances and bonds that lead to federations between things that make exchange and living together possible. To negotiate with a coronavirus, for instance, is not to fool ourselves that it can be brought into language but rather to first understand the myriad cords through which we are connected and the nature of our mutual dependencies.

Finally, I read the reference to the "father of the family" as an appeal to religion rather than patriarchy (inasmuch as it is possible to distinguish the two). Serres's final work is an appeal to return again and engage in further reading of what religion and myth provide for synthetic rather than analytic thought. He speaks there of the puteal structures built in archaic times. These were walled structures built around a spot considered to be a point of passage, such as where lightning from the heavens had struck, or places where access to an underworld might be secured. These structures would typically be covered for the majority of the time, only to be revealed on festive occasions such as the *Mundus Patet* in Rome, when the "hot spots" between worlds might be open. Unlike a temple, which, by definition, encloses space, or a burial site, which transforms the human into an object as a boundary marker, the puteal is open and affords what Serres calls a "vertical binding." If we think of this binding as not between some other world and this one, but rather of an additional dimension in the relations between people and things that accomplishes a synthetic, quasi-mathematical operation of integration, we are perhaps on the right track to where Serres's thinking was going. Living together and feeling safe will ultimately require a patient, prudent, and wise exploration of what kinds of federation and symbiosis are possible.

8

MICHEL SERRES AND GREGORY BATESON

Implicit Dialogue about a Recognitive Epistemology of Nature

Sartre had edged toward envisioning a society without masters and slaves: instead of exploiting each other, men would unite to exploit the natural world. —Ronald Hayman, *Sartre: A Life* (1987, 243)

There are a number of reasons why Gregory Bateson is a choice anthropologist partner of Michel Serres in a dialogue as timely as evergreen: the nature of Nature. To start with, both were mavericks in their whole life, having almost identical, paradigmatic troubles with their PhD defense. Serres was a student at *the* top French university, the *École Normale*, with an academically most powerful philosophy supervisor (Georges Canguilhem), while Bateson was one of the first anthropology PhDs at Cambridge University, having as his professors Radcliffe-Brown and Malinowski. They were supposed to have a red carpet for their lives. Instead, Bateson never had an appointment in anthropology nor Serres in philosophy, not in spite of but *due to* the radical originality of their ideas, having nothing to do with politics. Their careers demonstrate that the intellectual history of the twentieth century must be rewritten: its center was not a presumed mainstream versus critical theory divide; the "suffering victims" were not Marxists and other political radicals but those truly original thinkers whose ideas fell outside the kind of scientistic "rationalism" that was imposed by internal academic forces.[1] Reading Serres and Bateson together is thus not an academic exercise but contribution to a genuine intellectual dialogue that can help reconstruct thinking after the damages done by "rationalism" and "critique"—as manifested through Sartre's attitude, evoked in the motto, so "self-evident" in the 1960s and so evidently absurd for us now.

Their problematization of "rationalism" is all the more astonishing as a concern with science was at the heart of their interests, already while writing their PhDs. Bateson, the son of William Bateson, one of the most distinguished biologists of his time, published an article with his father in 1925, up to his last writings integrated ideas from biology into his thinking, and considered himself a real scientist—actually, *more* than those who simply followed or imitated the mechanical rationalism of Descartes and Newton. Serres similarly studied science, mathematics, and the history of science instead of following fashionable philosophical currents and considered the failure of philosophers to keep up with these developments a most problematic aspect of modern intellectual life. Yet both were unreserved critics of standard rationalism and the unlimited technological application of scientific results to life, identifying in a presumed "mastery of nature" and not in capitalism, class struggle, or bureaucracy the main culprit behind our current condition.

This is because they were among the first to perceive, beyond any dualism of nature and culture, or hard and soft sciences, the fundamental unity of our world as "Nature." It is this recognition that made them among the first modern investigators of the unity of Nature and diagnosts of the ecological crisis.

My encounter with the work of Serres, just as Bateson, was similar, rather flat as a story, and yet might have its method-logical relevance.[2] I encountered them, like most of the other thinkers who exerted an impact on my work—Weber, Foucault, Elias, Mumford, Voegelin, Girard, van Gennep, Turnbull, to name a few—in a library or in a bookstore. My background education was economics, especially econometrics, and then survey research methods, so my teachers could not orient me toward historical sociologists or anthropologists. However, as I realized that imitating the "natural" sciences and pretending to understand social and human life through statistical methods is meaningless, a dangerous dead-end street (at the age of twenty-two, I wrote computer programs on Texas Instrument machines to analyze survey data and started to write my notes on the back of computer punch cards), I had to reeducate myself on my own—receiving important help from colleagues and later even students rather than teachers. Concerning Serres and Bateson, while I already encountered their works in the early 1990s, for a series of reasons, I only *recognized* their importance much later—which has the further method-logical significance of introducing the difference between mere cognition and *re*-cognition.

In the following, I start by reconstructing the core argument of the *Natural Contract* and then map, on this basis, some convergences between the ideas and projects of Serres and Bateson. This will be a sketch, as bringing out the full significance of these parallels would require a monographic treatment.

The Core Argument of *The Natural Contract*

Beyond any debate concerning "authorial intentions," in the case of *The Natural Contract*, it is particularly easy to identify the core argument, as it is contained in the section "Natural Contract" of the chapter "Natural Contract" and its direct context.

A "natural contract" starts from "social contract" as a model, claiming that there is nothing wrong with social contract; it must only be complemented (Serres 1992a, 67). The term is taken from Rousseau and Hobbes, and its meaning is elaborated extensively in the book; in fact, much of its first chapter is devoted to what a "*social* contract" means, while the last chapter explores the meaning of "contract."

The social contract answers to a state of war, a civil war that has become all but permanent. The book cover shows a painting by Francisco Goya (1746–1828), *Fight with Cudgels* (1820–23), a stunning image with all kinds of archetypical and anthropological references. The image was painted on the wall of his house by the artist, then about seventy-five, as part of a cycle with deep personal and philosophical significance and should be read together with the 1791–97 home-wall-paintings of Giandomenico Tiepolo (1727-1804), of similarly testament-like and philosophical-anthropological significance, having at their center Pulcinella, the Venetian trickster figure of commedia dell'arte as metaphor of the times, the French Revolution, and postrevolutionary warfare. The background of Goya's image is the Napoleonic Wars in Spain, recently analyzed by René Girard, a lifelong friend of Michel Serres and a major anthropologically inspired social thinker, as direct precursor of the total wars and totalitarian regimes of the twentieth century that produced the "militarization of civil life," being "the jolt that caused this change in European societies" (Girard 2010, 10).[3] The painting also evokes two extremely important images, important not just for art history but for a whole series of socio-politico-anthropological reasons: the incision *Battle of the Nudes* by Antonio Pollaiuolo

(1470) and Albrecht Dürer's 1494 drawing *Death of Orpheus*, centrally analyzed by Aby Warburg in his anthropologically inspired art history. Both images represent the same stunning violence, with men not simply fighting each other but bent on demolishing the other, without any attention to their own life and security, as if inspired by madness. Those images, depicted at the end of the Renaissance, had a tremendous impact on the entire history of European culture, magnified by the fact that Pollaiuolo's image was one of the very first incisions made and thus was proliferated by the then-invented printing press—though printed images are strikingly and almost systematically ignored by the standard reading of the "Gutenberg effect."

All this helps clarify the meaning of a "social contract," as for Hobbes or Girard the "state of nature" is a permanent war, and in the case of Girard is specified as a mimetic rivalry that in absence of an explicit legal framework is always threatening to break out. Such a meaning attributed to nature is quite problematic but plays an important part in the way modern science and rationalism justifies its own effort to conquer and subdue nature—the reason why we now need, according to Serres, a "natural contract." Serres, so close to Girard, especially in his books written in the decade before this book, seems to accept Girard's position concerning "violent origins," which is problematic, but this cannot be pursued here.

Turning to the meaning of "contract," as so often, Serres offers illuminating etymological insights, also illustrating the hyphenation singled out for attention in the introduction to this volume, calling thus attention to aspects that otherwise would have gone unnoticed. Thus, "con-tract" means to draw together, a game of ropes or strings, emphasizing connectedness, a kind of pre-reflexive or existential basis for the legal regulation of relations.

The crucial significance of this analysis is that, staying with anthropological parallels, Serres liberates the idea of contract from its strict technical legalism, showing the genuinely social and not technical foundations of the law, just as Mauss did in his essay on the gift, in tracing the origins of legal terminology to gift relations. The two are even closely connected, as taking, in contrast to giving, has an inherently violent aspect (when taking is not accepting a gift), while the direct opposite of pulling together is either the violence of pushing away or the anti-sociality of an act performed alone.

Returning to the central section, it is necessary to add a natural contract to the social contract because of our constitutive parasitism on Nature—an untenable and most dangerous situation, as "the parasite—our actual status—condemns to death what it despoils while living inside it, without realizing that in this way it condemns itself to disappearance as well" (Serres 1992a, 67).

Such parasitic relationship must be turned into reciprocity and symbiosis; instead of just taking from nature, our usual way, we must *give* to it.

This is the core of Serres's message, but in order to understand it properly, we need to go into further details, almost concerning every word, as in his often seemingly straightforward, even trivial accounts, every word has its weight, which he indeed underlines by frequent semantic and etymological discussions of his terminology.

So "we" and "now" are parasites, but who are "we," and what is "now"? And what actually is the meaning of a parasite—especially of us and now being parasites?

Starting with "us"—does this mean us moderns or us humans? This is a non-trivial question and should be posed to much of social and political theorizing that all too often generalizes a contemporary reflection into a presumed universalistic truth. Even Serres was close in *The Parasite* to considering parasitism as some kind of original condition, coming even before gift relations, but this is certainly unacceptable, ultimately eliminating the difference between a parasite and any living being. Our parasitism can certainly be traced to agriculture—calling human existence in the Paleolithic parasitical makes no sense, but even in agriculture we need to make further distinctions. Serres had an experience with Chinese agriculture, told at the start of *Detachment*, where the utmost, exploitive "rationality" of Chinese agriculture is characterized as parasitical, in contrast to historical European agriculture, which he encountered still in his childhood, and where something was always left apart (*écart*), for wild animals and plants, and so exactly *not* all resources were used to the limit, as dictated by modern economic "rationalism" (Serres 1986, 8–17). Here, in this central section, he adds the word "beauty": we need to *give* now to nature, just as "the cultivator, in times now passed, returned in beauty, by his care, what is owed to the land, from which through his work took some fruits" (Serres 1992a, 68). This implies recognizing a specific rhythm in the interaction between Man and Nature: every increase of human knowledge or every technological innovation, due to the specific character of our humanity, its capacity for inventiveness, takes something away from Nature, disturbs and abuses it, so we must be conscious of the need to compensate for this by our care.

Thus, starting with the term "us," we have already discussed the "now," and it could not have been otherwise, as things are always interconnected; it is not possible to break down an analysis into strictly separate parts—this is already the heart of the problem Serres, and also Bateson, attempts to discuss. But we must now return to a more detailed discussion of the "now," all the more so because Serres calls attention to it, starting his central section by the paradoxical expression "since then [*dès lors*]"—certainly "modern rational" textbooks would severely warn their readers against starting any section by such a reference (Serres 1992a; [1990] 1995, 38). But Serres "must have had" a good method-logical reason for doing so, and indeed there clearly is one, certainly outside modern rationalism, a central discovery of political anthropology (through liminality) and philosophical hermeneutics (through Dilthey): that we always and inevitably start in the middle.[4] The gesture of the modern rationalist philosopher, from Descartes through Kant and Hegel up to Husserl and beyond, is meaningless: nobody can pretend to lay down *the* foundations by offering the "first word"—among others, it appropriates the role of God, according to the Gospel of John, so is an instance of absurd self-divinization.

The expression "since then" thus alludes to the argument of the previous sections, indicating that central as this section is, one cannot just start from there but most go back to the previous ones, as part of a "circular" and not just linear reading.[5] Indeed, the previous sections directly cast forward the argument of this, discussing the Declaration of the rights of man (Serres 1992a, 63-64), the parasite (64-66), and the question of equilibrium or balance (66-67). "Since then" thus evokes the moment in which *us*, modern humans, became parasites without realizing it, the moment in which *we*—again—lost our balance with Nature. Again, because it was not the first time in world history—a much earlier and major crisis, it does not matter whether first or not, was the so-called crisis of the Neolithic, when the discovery of agriculture stretched the limits of land abuse.[6]

So, and again, the expression "since then" returns us to the previous section, or the question of balance or equilibrium, in fact of a meta-equilibrium, as it implies an equilibrium between natural and social equilibria—and to the way in which such balance, in so far as it existed before, became lost and due to which we ended up living in a system of "reciprocal modern abuse" (Serres 1992a, 66), which poses the need to think of a new balance between these systems of balances (natural and cultural). Here Serres offers another etymological excursus, now on thinking (*penser*), through its links to "compensate," another hyphenated "con"-word, as "thinking," *penser*, and "weighing," *peser*, have identical origins.[7]

This has its own importance, as "equilibrium" and "balance" are identical words—"equilibrium" simply means equal weight or measure, so a balance. Thus, thinking implies weighting and measuring—certainly not in the sense of exact quantification.[8] Here two connected etymologies can be added—English-German "thinking" and "thanking" have a common root, which takes us close to thinking as recognition (see French *reconnaissance*, "gratitude," thus directly bringing in grace), while *gondolkodik*, "thinking," and *gondoskodik*, "caring," are identical in Hungarian. Knowledge as recognition, thinking as careful balancing, establish a semantic universe in which the task of humans is to carefully balance their acts among themselves and inside Nature. Constitutive of modernity as parasitism is a serious imbalance in this delicate process—whether it is toward other humans as capitalism or communism or whether it is toward Nature, this is a secondary issue; the primary issue is to promote thinking as a return to balance.

A second important "linguistic" consideration concerns the terms "world" and "nature." Intriguingly Serres, otherwise so concerned with words, does not see an issue here; thus, in this crucial section on "natural contract," he uses *monde*, "world," and *nature* interchangeably. But this won't do, as the meaning of "world" cannot be taken for granted. It has a most interesting etymology, *wer*, "man," and *auld*, "age," or literally "the age of man," thus with evident connections to rites of passage and liminality. Moving to semantics, the Greeks had two words for "world," *cosmos* and *chaos*. The difference is evident: *cosmos* in ancient Greek means an ordered universe, while *chaos* a disordered one. This has its theoretical, even theological significance, as God or the gods certainly could not have created chaos. They created order not out of nothingness (*nihil*) but out of chaos. Creation out of nothingness seriously makes no sense; it is a theological lapse from reason, just as magic is a lapse from religion, and not the other way around, a central idea of Bateson (Bateson and Bateson 1988, 56; Bateson [1979] 2002, 197–98).

Even further, our contemporary term "world" has three very different meanings. At one level, it can mean anything: everything ever existing anywhere in the whole universe. This is fine, but this does not mean much; in fact, it is almost irrelevant, for *us* and *now*, in the *concrete*, into which we were "grown" together and that after is all our concern—which *should be* our concern. So, second, world for us really means "Nature," or life and everything on planet Earth, as it exists, as it was *given* to us—by whom, it does not really matter, but for which we first of all must be *grateful*, or "reconnaissant," acknowledging its unique beauty, in a circle of wisdom/knowledge and grace/recognition instead of doing such an insane alchemic operation as trying to make money out of it, as that is only bound to destroy it for us and certainly for our children's children. But third, "world" also has a cultural as opposed to "natural" meaning, the evangelical sense of a social construct (see especially the Gospel of John), the "public sphere," in the sense of Dickens's "society" (see especially *Our Mutual Friend* but also *Bleak House* or *Little Dorrit*): not the genuine social world of gift relations, *charis*, participation, presence, love, friendship, and sociability, but the empty public arena in which everybody is fighting with everybody else, whether with words, budgets, or arms, moving back to the image of Goya, and recalling the hostility of Serres (1992a, 24–25) to "debate."

This has its own, not negligible importance for science, as "science" correspondingly also has three different meanings. We usually understand "science" in the sense of "natural science," which supposedly offers us an insight into the world as it really is, beyond ideologies, subjectivism, and so on, and so the modern "social sciences" are the sciences dealing with human life, imitating the "natural sciences." But, again, what "world" are these sciences dealing with? Here the first trouble is that the "world" as everything in the entire universe and the "world" as Nature in our Earth are by no means identical.

Let me explain this through a trivial but not irrelevant example. "Science" tells us that the Earth rotates around the Sun and not the other way around. Anybody questioning this here and now would be considered an ignorant fool. Well, it might well be, but there remain two, not negligible problems. First, most of us, not being physicists or astronomers, would have real trouble convincing someone who would argue, "Look here, I'm no fool. Here is the Sun, rising every day in the East, setting in the West. I can see this with my own eyes, so do not try to take me for a ride." How would *you* convince our fellow? But the second issue is even more delicate. It is that, for anybody working with plants and animals, in agriculture or anywhere, it is vital to know where exactly the Sun will rise and set and when, and if those would be convinced that this does not matter, as it is the Earth that actually rotates around the Sun, then we would all die in hunger.[9]

Thus, the knowledge of "the world" and the knowledge of Nature are not identical. What we misconceive as "natural sciences" are actually "*un*natural sciences." This does not mean that they are wrong, just that they have a specific, universalistic perspective that is not the same as *our* perspective, living on Earth, in the here and now.

And that is not enough, as the issue is that the unnaturality of the universalistic sciences is a genuine plague for us, in that the objects they produce, by their knowledge, might be *incommensurable* with our life and reality.

Incommensurability as a concept was developed by Agnes Horvath, in her analysis of liminality and the void. It has fundamental affinities with the works of Francis Yates, one of her sources, concerning the alchemic origins of science. It can also be translated into the terminology offered by Mary Douglas, one of the most important anthropologists of the past century and—it is important in *this* particular context, trying to overcome Western technologized rationalism that indeed is significantly "male"—another woman, through her definition of dirt as "matter out of place" (1966, 44), as indeed everything produced by technology based on "universalistic" science is "matter out of place," or technically dirt: garbage, waste, decay, desolation, destruction, devastation, once it stopped being "useful" (which, thanks to modern advertisement, happens ever more quickly). Technically and literally so.

And the situation, here and now, is actually even worse, as these "sciences of the world," meaning universe, are also the ways to progress in the "world," meaning power, fame, success and money, promotions, and ERC grants: if you are in the social sciences, you must imitate as much as possible the natural sciences; set up experiments, if not directly, then by simulation; get data for computer modeling, *loads* of data; and then the "sciences of the world" will bring you success in "the world."

Except that these "worlds" are not the same as our world, as "Nature," incommensurable even with culture, will be only further unbalanced and destroyed. So we need to return to Serres and then Bateson and see what they mean by world, nature, science, and the mind. What are the ways these two great maverick thinkers offer to escape the dreadful trap into which the "rationalistic" sciences of "world" are pushing all of us?

The Ship Metaphor

A new natural contract, beyond the social contract of Hobbes and Rousseau, requires a new "eco-politics" and first of all the invention of a new political man. At this point, Serres turns to an example as classical as possible, the ship metaphor in Plato's *Republic*. The interpretation, however, is new and illuminating, as it starts from an aspect ignored by Plato, perhaps because of its self-evidence: a ship navigates on sea and not on land. Such neglect is not a minor issue, says Serres, as the example assumes an *exceptional* state: those on a boat are in a particularly delicate situation, distant from the stability of the land, and so cannot ignore the fragility of their condition and must compensate for the solidity of the land by their solidarity, which excludes any possibility of violent conflict, as that would directly lead to their destruction.[10] Thus, their social contract comes directly from nature (Serres 1992a, 70).

Paying attention to the *exceptionality* of a ship on sea brings in the state of emergency, or liminality. Thus politics, according to Plato and Serres, turns to the problem of *limits*. Fragility can be thought of through the question of limits: through the force of our power/knowledge, we have "reached the limits of our global habitat"; we live "as if" we had all boarded a ship; we have all become "a compact and united group that reached the very limits of objective forces"

(71–72). So Plato's classic metaphor turns out to be particularly timely, as we now live in a continuous state of emergency or permanent liminality—in flux.

This poses the problem of government, beyond politics, the title of the next section (*Du gouvernement*). Serres does not evoke but certainly implies that the word "government" is derived from the Greek *kybernáō*, which in ancient Greek only applied to navigation. Serres here first elaborates on the exact meaning of this art, which implies a continuous attention to circumstances and conditions: currents and waves, as many external constraints that the wheelman must take it into consideration, producing a series of "curbed interactions" among the will, the ship, and the obstacles. The argument closely evokes Bateson's *Mind and Nature*, and in fact at the start of the very next paragraph, Serres alludes to cybernetics in the modern sense, part of the same semantic horizon, and with a focus on the interdependence created by being *on* the limit.[11] Serres illustrates his argument by variations on the term "exchange": our current situation is produced by our reciprocating the gifts (*dons*) received from nature by inflicting damages on it, which then for us become new facts (*données*) with which we must deal. As, if we don't, then, being out in the high sea, we risk being lost there (*Désemparé*, title of the last section), or shipwrecking.[12]

This metaphor also brings in the connection with porosity, the title word of this book, another approach to liminality. Porosity was discussed explicitly by Serres in *The Incandescent* ([2003] 2018) in connection with certain frontiers or even cultures but is present in his works by the broader associations of the term (Hermes, parasite, angels, and so on—see the introduction to the volume). As Serres was certainly well aware, Poros was father of Eros in Plato's *Symposium*, the word meaning "passage" and the figure standing for plenty—in contrast to the mother, Penia, standing for poverty—with the etymology going back to the PIE root *per*, one of the most important such roots, meaning a dangerous travel or passage and analyzed extensively by Victor Turner (1985). Porosity is thus closely connected to "becoming," the other title word of this book, which also captures the passage between two states of being, or liminality. Here the Hungarian term for becoming, *vál(ik)*, is of particular importance, as this root is at the origin of an entire range of words connected to liminal situations and rites of passage, like separation, election, selection, choice, divorce, crisis, or excellence, to name just a few, establishing an intimate linking between such liminal moments in Hungarian, with the proviso that the "real" outcome of such liminal situations is as if directed by natural endowments or character essences—as the root meaning in Hungarian implies the passage of a boy or a girl to adulthood or of a butterfly to an insect and not just any arbitrary whim; in the language of Heraclitus, a testing of one's character.[13]

The focus on interdependence enables Serres to problematize the main idol of contemporary social theory, the term "public," arguing that "publicity" in fact reveals the essence of the term, the exclusive focus on the social collective as "world," and to argue that the new natural contract must conform to the original meaning of "nature": the conditions into which we all are born—or must be reborn (1992a, 74–75).

Returning from politics and government to history in the next section, Serres offers a more precise definition of natural contract as based on a *recognition*—and of a *metaphysical* kind—that we all live in the same global world (meaning planet Earth as an entity) and confirms the use of the adjective by emphasizing that it implies going beyond the ordinary limitations of specialized knowledges, in particular physics (Serres 1992a, 78). As a further clarification on the term "recognition," Serres argues that, just as politics and society are based on recognition of other states and other men, the natural contract must also be based on recognizing the world (or "Nature") as our equal partner (Serres 1992a, 78–79). And, just as any contract, this new natural contract must also be based on links, or relations, now unifying the Earth (*la Terre entière*) (Serres 1992a, 79).

Such focus on links or relations directly introduces the next sections, the last two sections of the chapter, which immediately qualify the term.

Concerning religion, as Serres would make it explicit on the next page, the link with linking is direct, as the Latin etymology of "religion" is *religare*, or bind again, according to some accounts, while the other account only reinforces the point, as it is "collect" or "read together," *relegere*. However, as always, Serres takes the point further, in an uncharted direction, pointing out that the opposite term is "negligence," *ne-legere*, "not-linking" or indifference. Not surprisingly, this is what the "neutral spectator" of Adam Smith and Kant or the "neutral observer" of the scientific method prescribes as a model. Thus, Serres draws the conclusion that "modernity neglects, absolutely speaking" (Serres 1992a, 81; also Serres [2019] 2022).

Yes, but it does not work—probably never worked—and it certainly does not work now. We can no longer neglect the ties that connect us with the world/nature. And here we can return to the starting paragraph of the section where—as a rare reference to ethnographic material—Serres evokes priests as they perform acts of devotion, not to some idols but to the world (meaning Nature) (Serres 1992a, 79–80). Here Serres moves very far from Girard and his exclusive (though in itself important) focus on the Bible, recovering the underlying wisdom and not just lie of tradition.

This further underlines the unacceptable uniqueness of modernity in *neglecting* literally everything—neglecting the world, or Nature. No human community or culture ever could neglect nature. The cycles of nature, life and death, growth and decay, the rising and setting of the sun were at the center of any religion. This is why time and again these processes could be abused, developing into bloody rituals of sacrifice by which certain *Magi* could maintain their power. The point is that until modernity, *neglect* never developed, could not develop, into a civilizational principle. And even our current situation only demonstrates how extremely stupid it is to worship neglect—indifference, exteriority, nonparticipation, and the like. The science of neglect, of ignoring the links, the "pattern that connects" (Bateson [1972] 2000), might give short-term benefits, following the alchemical logic of universalizing science, but eventually—in fact, already in our present—the foolishness of this perspective is bound to become evident.

The return not simply to nature but to religion or the sacred by both Serres and Bateson, two of the most perspicuous prophets of the ecological crisis, has something paradigmatic about it. This is because they were born into the two cultural traditions that were most uncompromisingly hostile to "religion" and the "sacred." Bateson was not simply British, from the homeland of capitalism and the industrial revolution, modern empirical and experimental science, journalism and the public sphere and political parties, but was born into a family of scientists who were atheists going back five generations (1991, 301). Serres was French, from the homeland of the Revolution and republicanism, cult of the Supreme Being, and while his immediate upbringing was very different from Bateson's—the countryside, having at home the Bible as the only book—in his broader family background, the situation was similar, as his father was born into a radically republican atheist family and only the experiences of World War I turned *him* away from mainstream modernity and back to the Church. The significance of their case was not that they had a personal conversion experience—not that such cases are insignificant, and anyone sticking to a straight atheistic position should look into, with an open eye, the *real* historical case of Bernadette Soubirous—but that through their experiences as *scholars*, with a specific emphasis on *science*, they came to realize that the deep *anthropological* wisdom connected to the sacredness of nature can be neglected only at our greatest peril.[14]

The foolishness of ignoring such connectedness is all the greater, and more absurd, as beyond "relationality" as another abstract—and thus necessarily neglectful—category, among others ignoring the *character* and *quality* of such relations, it encompasses two of the most important and valuable aspects of human existence, love and beauty.

Love is the theme of the last section of the chapter "Natural Contract," rhyming with the ultimate sections of chapter 1 on beauty and peace. Hard-core moderns would certainly raise an eyebrow or two seeing the section title, as they—"we"—were certainly educated to realize that humans are not guided by love but by "rational" self-interest and that love is at best a code word for sex. Undisturbed, Serres starts by simply stating that "without love there are no links of alliance" (1992a, 82). And, again, without such alliance, we will all be lost, as a ship on the high seas.[15] But Serres does not bring in love just as a trick to produce the "necessary" connectedness, to "ease" the fact that our interests must be harmonized with the interests of others, but simply as a fact (rather a "given," in the sense of a gift) of life. We are all linked, humans and Nature (even the "and" is problematic), by profound solidarity and love.

"Nothing is as real as love, and there is no other law than its"—last sentence of the chapter (Serres 1992a, 84).

Moving to Bateson

Of the manifold connections between the thinking of Serres and Bateson, I will focus on connectedness and its links with the sacred and with beauty—easily the most important aspect of Bateson's lifework.

Already the central theme of *Steps to an Ecology of the Mind*, Bateson's most famous book—though it was a collection of articles—was the indissoluble unity of man and nature, which Bateson eventually came to thematize through the unity of the mind. It is this connectedness that Bateson wanted to explore in an accessible book form in *Mind and Nature*, but the book, as it turned out and as he made evident in its last pages, the mock conversation with his daughter titled "So What?" (Bateson [1979] 2002, 191–200), ended up as only a necessary preliminary to the question he really wanted to explore, the connection between consciousness, aesthetics (or beauty), and the sacred, themes "that were implicit in his work for a very long period, [but] again and again pushed back" (Bateson and Bateson 1988, 8), suggesting that that theme would be explored in his next book, provisionally titled *Where Angels Fear to Tread*. At this stage, however, cancer intervened, and the book *Angels Fear* was eventually published only after his death, based on notes and drafts, by Mary Catherine Bateson.

If Serres was a philosopher whose work has vital anthropological interests, then Bateson was an anthropologist whose work shows similarly vital philosophical interests, way beyond the kind of issues professional philosophers are concerned with in our days.[16] The central issue is to overcome the dualistic break or scission of experience, the split between the body and the mind.

While much of modern philosophy is a philosophy of cognitive consciousness, from Hegel through Husserl to Sartre and beyond, Bateson offers the outline of a philosophy apart from or against such consciousness. Consciousness is part of the problem and not the answer; this is visible especially clearly when contrasted with aesthetics and the sacred: "To be conscious of the nature of the sacred or of the nature of beauty is the folly of reductionism" (Bateson [1979] 2002, 200). The problem is not with consciousness per se, he comments later (1991, 299–300), but its selectivity: it always tends to focus on something specific, thus necessarily failing to grasp the broader picture, while "notions like the sacred and the beautiful tend to be always looking for the larger, the whole."

Problems with consciousness are broader and lead us to the epistemological problems of modern science. Problematic terms include "function," which necessarily ignores the whole, as only parts have function (Bateson 1991, 303–4); "purpose," which can easily imply a narrow, preconceived end, making one ignorant of the broader picture (Bateson and Bateson 1988, 56); and the connected prior "separation" and "dichotomization" that, furthermore, are technically alchemic (Yates, Horvath) and are bound to be destructive: the dualistic mode of thinking, from Descartes through Newton and Locke onward, and its main attributes "have simply torn the concept of the universe in which we live into rags" (Bateson 1991, 305).

A main error with such focus on consciousness and purpose is that this simply misconceives the nature of human perception, which is *not* conscious. A recurrent example Bateson offers is image formation. We orient our conduct by perceiving things, this perception proceeds by forming images in the mind, but we are simply not aware of how this is being done: image formation is still mysterious for us and is unconscious (Bateson and Bateson 1988, 96). Gaining awareness would only disorient us, rendering our lives unlivable: after undergoing some experiments about visual perception, Bateson's faith in his own image formation was shaken so much that he could hardly cross the street (Bateson [1979] 2002, 34). So we are *lucky* that we cannot be aware of how we create our mental images—even further, this fact can be taken as almost a proof not simply for the existence but the goodness of God (Bateson and Bateson 1988, 96; Bateson [1979] 2002, 35)!

Bateson's recognition is particularly important in the contemporary context, where the evocation and manipulation of images and the attempt to control human behavior, even the mind, through them are increasingly an explicit concern of communication technology, through artificial intelligence, forecasting the real danger of mind control.

There is one term that captures as if in a nutshell Bateson's epistemological interests, though he hardly used it, at least until his last publications—and even there it is rather used by his daughter—and this is *recognition*. Bateson's interest is not how we consciously come to know things, or knowledge and cognition, but how to *recognize* them—recognize their particularities, but especially meaning and significance. This can be seen particularly clearly in his reference to some lines of T. S. Eliot, at the start of his "Last Lecture" (Bateson 1991, 307), which was also selected as the motto to the foreword of *Mind and Nature*, about arriving at the end where we started and *thus* knowing it for the first time, recalling Serres's cycle, as discussed previously.

This can also be seen in his famous definition of information, which is not the standard one about bits of knowledge but rather the perception of a *difference*, which always implies a background or context in which such difference appears or can be *recognized* as a *difference*. This can help one understand and recalibrate his interest in cybernetics—and even the related interest of Serres (1992a, 73–74). He indeed might not have understood all that "others" were meaning by the term (Dupuy 2009, 88), but this is because cybernetics was a prelude to contemporary *cognitive* science, and Bateson's interest was rather in a would-be *recognitive* science.[17] Such recognition can be directly tied to the philosophy of Plato, so central for Bateson's interest in beauty and Plato's concern with knowledge as recollection, so much tied with memory and thus with the formation of images.

Bateson's epistemological position concerning science and religion, or the sacred, is not easy to map, as its exposition belonged to the heart of his unfinished project and its reconstruction, an all but hopeless task anyway, would require at least a full paper on its own. Here only a few allusions can be offered, insofar as they establish a dialogue with the ideas of Serres.

I start from a formulation contained in one of the introduction drafts to the *Where Angels Fear to Tread* book project, published as chapter 5 of the joint-authored book version. Such drafts have the advantage, or disadvantage from another perspective, of being particularly blunt, thus bringing out Bateson's concerns with great clarity—and as his daughter judged it publishable, it cannot be considered as excessive. At the end of this chapter, Bateson situates his project in between two extremes, "crude materialism" and "romantic supernaturalism." This, in itself, would be of little interest. What makes it intriguing is that "crude materialism" is characterized through "quantitative thinking, applied science, and 'controlled' experiments," or what actually is considered as the heart of "scientific methodology," and that this position is characterized further as being of the same category as "romantic supernaturalism": a "nightmare of nonsense" and, shortly before, by implication, as one "of the more ludicrous and dangerous epistemological fallacies fashionable in our civilization today" (Bateson and Bateson 1988, 64–65).[18] Given that this is in a draft written by a seventy-five-year-old man with a mortal disease, one could dismiss these claims with a patronizing smile, except that they are perfectly compatible with Bateson's position in his published writings, especially *Mind and Nature*.[19] An epistemological position that approaches the complex phenomena of our world, tearing them out of their context, through mere quantification, submitting them to mechanical experiments in which different segments are forced under certain predetermined conditions, would strike any normal person in any decent culture as being stone mad—and we now, having been pushed by applied science or technology to the brink of an unprecedented "environmental" disaster (note that the term "environment" is itself glaring nonsense and fallacy, as it simply means our Earth, or practically everything, redefined, in one of the most absurd and dangerous revaluation of values, as what only "surrounds" the scientist and his "experiments"), should carefully reconsider which position is more "true" or "real." On a slightly different perspective, while "research ethics" in our days has become a major slogan, often for the wrong reasons, the problem of "controlled experiments" is never posed, though for a combined set of epistemological and ethical reasons, this technique, the heart of contemporary "scientific methodology," should actually be prohibited as being intrusive and parasitic on the unity of any living being and especially on the unity of Nature.[20] Without entering a long discussion, this can be seen in the incompatibility between any knowledge gained through "controlled experiments" and "wisdom." One could perform as many "controlled experiments" as one wanted yet would not move an inch closer to "wisdom," as "wisdom" implies knowledge and awareness about complex life situations that can only be gained through life experiences and not "controlled experiments." This is fundamental for teaching, education, and (mis)education, central for the recent work of Tim Ingold, just as for Bateson: the appendix to *Mind and Nature*, with the weighty title "Time Is Out of Joint," ends with a simple question: "As *teachers*, are we wise?" ([1979] 2002, 210)—as if offering an echo to the title question of his daughter's contribution to his Festschrift (Bateson 1978). So what kind of teacher will be the person who in his whole life only performed "controlled experiments"? Bateson, arguably, has an answer and at another "last word" place, the end of the introduction to *Angels Fear*, where he claims that the "impatient enthusiasm for action" shown by so many applied scientists "covers deep epistemological panic" (Bateson and Bateson 1988, 15). The point is well taken, as panic is the radical opposite of wisdom, a wise man can never be in panic, and furthermore a panic is a typical imitative mass phenomenon, a key feature of biblical "world" and Dickensian "society," and is bound to overtake people who lack inner security.

This brings us directly to the heart of Bateson's central idea, the "sacred unity" of mind and nature, indeed of the entire biosphere (Bateson 1991, xiii) or the sacredness of the world, the living world of which we are part, a perspective that returns to religion and the sacred as an epistemological question, as "the *sacred* [is] related to a knowledge of the whole" (Bateson and Bateson 1988, 86–87). This is also a question of faith and belief, to the extent that epistemology itself is based on faith or belief, as any epistemology starts from perception and perception implies faith in the image-forming capacity of our senses; it means to approach a field in which, again programmatically and paradigmatically, "angels fear to tread."

The main sign of the sacredness of Nature, helping us recognize it, is beauty: "the beautiful" that "persists" in spite of everything (Bateson [1979] 2002, 5), the "'invisible and unchanging beauty which pervades all things,'" or "'the ultimate unifying beauty,'" as Bateson keeps quoting from Plotinus the neo-Platonist and from St. Augustine (Bateson [1979] 2002, 2, 13, 16–17). In that sense, the last work of Bateson is based on a recognition, the gradual awareness that by the unity of nature argued in *Mind and Nature*, "he was approaching that integrative dimension of experience he called the *sacred*" (Bateson and Bateson 1988, 2). Beauty helps us recognize the unity, and therefore the sacredness, of the world, as this unity cannot be disturbed without destroying or at least endangering its beauty, and for this reason aesthetics (or beauty) is also directly a matter of epistemology: "It's not a new idea that living things have immanent beauty, but it is revolutionary to assert, as a scientist, that matters of beauty are really highly formal, very real, and crucial to the entire political and ethical system in which we live" (1991, 311).[21] This is why *Angels Fear* has as its subtitle *Towards an Epistemology of the Sacred*, a position argued already in *Mind and Nature* and furthermore as an explicitly "Platonic thesis": "Epistemology is an indivisible, integrated metascience whose subject matter is . . . the science of the mind in the widest sense of the word," which is "always and inevitably *personal*" and yet at the same time is also "unconscious of the greater part of the processes" in which one is involved, in contrast to the Promethean hubris of mainstream modern science and rationalism, and any epistemology, while recognizing sacred unity through beauty, is also based on *giving oneself up* to the ultimate reality of the world/Nature as a *gift*: "I surrender to the belief that my knowing is a small part of a wider integrated knowing that knits the entire biosphere or creation" (Bateson [1979] 2002, 81–82).

Here Bateson again and conclusively encounters Serres, as "nothing is as beautiful as the world" (Serres 1992a, 45).

In Conclusion

Contemporary intellectual slogans flatter us with claims that we live in a knowledge or information society, producing unprecedented well-being, now even in the Anthropocene, while constitutionally ignoring the damage inflicted by the faulty methods and means of science and technology on the world.

Instead, Serres and Bateson help us realize that the primary activity of the mind is not purposeful consciousness or cognitive construction but *recognition*. As the prefix in the word indicates, this cannot be a foundational act, as any act of recognition is preceded, whether at the level of an individual, a group, or an entire culture, by countless acts of perception and cognition. But a recognition is exactly an act, in the sense of an event that suddenly happens: it is when we come to appreciate what we only took for granted before.

Concerning the heart of human understanding, the most important such recognition, the recognition to which Serres and Bateson lead us, is to recognize that before any of us start to perceive and think about the world, *our* world, the Earth, first of all, exists, and not just exists somewhere outside, as an "environment," but *we* only exist as part of it, so we *participate* in it, are *present* in its *presence*; and it not simply exists, but is primarily beautiful—and so its given existence is indeed something like a *gift*; thus, we are bound to reflect on what kind of gift this can be and how is it that our very existence is a gift inside a gift.[22] Through such recognition, we no longer simply "know" things but can become "wise"—which implies the immediate realization that we can no longer act as parasites of our global home, imagining that we can safely despoil it without paying for the consequences.

Such realization of course is not that new, as it can be traced back to a thinker evoked by Serres and Bateson in prominent places, Plato—the *real* Plato, and not the presumed father of "idealism," the position of modern rationalism, misrecognizing him as its predecessor. The recognition of beauty is offered by Plato in crucial junctures in his work as central and furthermore directly implying love: realizing the beauty of the world implies that we must love it as well, that our life must be governed by love, and that this is the proper frame of mind for the most important human endeavor, the engendering of beautiful children and thus moving this beautiful world further. This is expressed best in the *Symposium* first, then in *Phaedrus*, and in a way finally in the *Timaeus*, offering Plato's cosmogony, a set of dialogues defining what his own undertaking is, even naming it as *philo-sophia*, or love of wisdom. Philosophy does not mean a set of verifiable propositions, mere exercises in logic, or a university subject matter, but rather the love of beauty as wisdom: wisdom as love, love as wisdom.

This is fundamental to the way Serres helps reinterpret such central terms as "balance" and "equilibrium," away from the trivial, static, and positivistic meaning they gained in Newtonian science and economic theory, reducible to the way a ball stops moving at the bottom of a bowl. This is not a proper metaphor for the way a balance can be reached in human, social, or cultural life, which rather requires, way beyond the mechanical stabilization of opposing forces, a delicate adjustment of one's *own* inner states and emotions, anticipating, accommodating, and incorporating those of the surrounding others.

Thus, through Serres and Bateson, and also others like Tim Ingold and Alfred Gell, the originally anthropological *and* also "ecological" or "ecopolitical" philosophy encounters the heart of an anthropology that has become philosophical.

NOTES

1 See Szakolczai and Thomassen (2019) for details.
2 See also Szakolczai (2022), chapter 4.
3 In case this requires pointing out, the current times don't represent a move toward ecological understanding but rather the extension of such civil militarization to epidemic disease, following Nixon's "war against cancer," quite in line with the arguments of Michel Foucault and Giorgio Agamben concerning Panopticism and bio-politics.
4 By "method-logic," I mean almost the same thing as "methodology," or the reasoned manner (*logos*) in which the "way" of a research (*met'hodos*) must proceed, except that I found it necessary, fully in the spirit of Serres and Bateson, to indicate my strong dissent from the way "methodology" is made into an object of almost cultic veneration by various powerful (neo)positivist and (neo)Kantian currents in social thinking, fully confirmed and proliferated by funding bodies, the ERC leading the way, necessary for success in the "world"—at the expense of "nature" and "culture."
5 This type of circularity is also central for Bateson ([1979] 2002, 96–102, 181–82).
6 The arid, rocky landscape around Göbekli Tepe was the area were some of the first plants cultivated by agriculture were experimented with—until total land erosion destroyed their fertility. Note that agriculture developed around a major religious center and not the other way around, as rationalist materialist evolutionism long assumed.

7   Further playing on the same motif, "concrete" means *con-crescere*, or "growing together"; in a given "con-text," *con-tessere*, or "woven together," fully in line with Serres's argument.

8   Note that this term will play a central part in Bateson's attack on conventional quantitative epistemology; see Bateson and Bateson (1988, 62–63).

9   My other favorite example about the meaninglessness of science is: What can science say about the smile of my three-year-old granddaughter? Leading to the Nietzschean question: How important is then "science" for life?

10 Solidity and solidarity are etymologically linked.

11 Note that Kantian rationalism is unable to discuss this issue. For Kant, a boundary is only a mental construct; one cannot be "on" the limit. But in real life, this happens; we now *are* on the limit, and our condition can only be understood if we resolutely *abandon* thinking along Kantian lines. This is one of the main reasons why anthropology is so fundamental today for serious thinking.

12 The metaphor of shipwrecking is also centrally used for the modern condition by Roberto Calasso (1983, 48, back cover), another contemporary master thinker with an anthropological orientation, in his case departing from Mauss through Dumont and against Durkheim (Calasso 2010, 410–45). See also Szakolczai (2022), chapter 3.

13 A proper comparative study of Hungarian, with other agglutinative languages, would require the attention of a Sapir or a Whorf, unfortunately not likely to happen soon. Official Hungarian linguistics is hopeless for any in-depth understanding.

    On Heraclitus, see the famous fragment *Ethos anthropoi daimon* ("Man's essence is character") (Heraclitus, 247, in Moreiras 2020, 165).

14 See also Milton's ([1667] 2013) *Paradise Lost*, IX.99-119, a genuine hymn to the Earth. Note that, paradoxically, such recognition is offered by Satan before he starts acting to destroy it.

15 Again, to avoid misunderstanding, this does not justify WEF-Davos style world government.

16 For a close contemporary parallel with Bateson, see Tim Ingold (2022), who recognized that his recent work is drifting toward philosophy and even evokes Michel Serres.

17 Central to this are the works of Alessandro Pizzorno (2000, 2007, 2008), another key Maussian thinker.

18 Note Serres's (1992a, 45) term "epistemodicy," particularly significant here, given that Serres wrote his thesis on Leibniz.

19 Mary Catherine Bateson (Bateson and Bateson 1988, 5–7) argues in her preface that Bateson in his last years in Esalen *lived*, in his "essential alienation," this tension between scientists following faulty epistemological positions and devoted followers misconstruing his approach.

20 Mentioning only one wrong reason, declaring oneself "doing research" certainly makes any genuine "participatory observation" in modern life impossible, not to mention that according to the method-logical principles of "total participation," developed by Colin Turnbull when he encountered Turner's liminality (see Szakolczai and Thomassen 2019, 190–95), really important research is based on a participation that started even before one had gained the idea to conduct research.

21 For Bateson, unity and whole are always connected to recognition, not construction. Constructing wholes for others to live in is even worse than the reductionism of parts, though it is a necessary consequence of alchemic thinking: "synthesis" always must follow "analysis."

Bateson's position is very close to the position of Agnes Horvath (2021) about the centrality of the complementary term *charis* for the very structure of reality. Note that the journal *IPA* was launched by a lead-off article of Agnes Horvath and Bjørn Thomassen (2008) on Bateson.

22 Note that "present" is a synonym of "gift."

PART III.

KNOWLEDGE

QUESTS

TOM BOYLSTON

9

ANGELOLOGY

At any given moment of the day, the breeze plays on your cheek, and since it carries codes from everywhere,
it's telling you about the state of the body of the world —Michel Serres, *Angels* (1993, 29)

This chapter considers angels in Ethiopian Orthodox Christian practice and in the work of Michel Serres, particularly in *La Légende des Anges*, published in English as *Angels: A Modern Myth* (1993). This is a disorientating task because of the simultaneous proximity and distance of Serres and Ethiopia. French philosophy has not much to do with Northern Ethiopia, for the most part. But Serres's immersion in Catholic iconography, especially in this text, gives a common language and common reference points, most notably in the figures of Gabriel and Raphael, who feature prominently in Serres's text and in the lives of the Christians I have worked with in northwest Ethiopia. If we think of Christianity as a giant message-bearing system spanning a couple of thousand years and large parts of the globe, it is a system in which both Ethiopian Orthodox praxis and Serres's thought participate. Ethiopia also reminds us that neither the origin of the Christian system nor its contemporary centers of greatest activity are intrinsically European.

In the long debates about anthropology's problems of subjecting mostly non-European ethnographic material to mostly European-derived conceptual analysis, part of the problem is in the installation of a secularized master discourse of theory that closes off the possibility of describing or even noticing certain connections. Recent work on anthropology and theology has explored this in some detail (Robbins 2020; Lemons 2018; Bialecki 2018). But Serres is not a theologian, certainly not in any way my Orthodox friends would recognize. Still, he speaks in an angelic register that suggests dialogue is possible.

I am certain that many of my Ethiopian Orthodox friends would consider Serres's angelology heretical (as, indeed, do many French Catholics), and how often does ethnographic writing take on even the possibility of speaking heretically? To take on this risk implies writing in a way that is subject to the other's critique, that risks entering other spheres of judgment than those of the academy. The hope of this chapter is not to "apply" French philosophy to the interpretation of Ethiopian data; it is to explore the figure of the angel as a being that connects diffuse realms and that makes communication possible. In the material I consider, angelology is a deeply practical endeavor concerned with asking how messages get where they need to and how the world is organized such that meaning is possible.

What Angels Are

Before the Orthodox new year, at the end of rainy season, comes the five-day intercalary month of *P'agumen*—a time when spirits good and bad are known to be active. The third day of *P'agumen* is the annual feast of Raphael. On this day, Raphael turns the rain and the entirety of Lake Tana, on which Zege sits, into holy water. Children dance naked in the rain; people descend to the lake to be healed and to wash their things; angelic presence is palpable while the angel remains invisible. There is no clearly translatable message in this mediation, just a subtle transformation in the environment and climate and our orientation toward them, that alert us to the benevolent presence of an intermediary figure. This kind of mediation is sensory, felt on the body in relation to the environment one lives in, and it is this sense that I have been trying to capture in years of arguing that mediation matters in Orthodox Christianity, as practiced in Zege, in a way that modernist secular thought tends to miss. Angels carry messages between God and humankind, but this is only the foreground part of their activity; it is the background alignment between bodies and environment—the angels, the rain, and the lake—that tends to go unnoticed. As John Durham Peters (2015) reminds us, mediation is not just about connection but about the elements—earth, air, fire, and water—in which we dwell.

The Amharic and Ge'ez word for "angel," *mäl'ak*, is cognate with the verb *mälak*, "to send." Angels, as one priest explained to me, are God's messengers. They go both ways, carrying our prayers to God and announcing key messages to humanity. Icons of the archangels adorn the gates of the inner sanctum of any church—winged human figures, often in warrior pose, slaying monsters and saving the innocent. Minor angels, usually arranged around the edges of any other icons or the church joists themselves, are simply faces with wings. In practical terms, the archangels have a similar role to saints; they are warriors as much as messengers, with the Archangel Michael in particular being a major figure in the fight against demons. Archangels are figured as male, human-animal hybrids, but people tell me that in truth they are beyond species and gender (and in the Bible, they are more alien still). They are often known through miraculous acts, especially the discovery of holy spring or in relieving people of demonic possession. Angels, then, are fearsome, thought of as having hybrid or strange bodies, warrior messengers who are often known through ambient, environmental effects.

Angels for Serres, writing in the mid-1990s, are the figure of all messengers and hence of the information age, "a new labour process, of message bearing" (Serres 1993, 293). The angel is a "skeleton key" in a world construed as a general message-bearing system, from the tides and winds through to contemporary communications media. Angels do not simply carry messages from A to B; they mediate between different scales (heavenly and earthly, singular and multiple, artistic and technical, hierarchies and meshworks).

The point about hierarchies is important. Angelic messengers have universalizing, equalizing potential, and yet the trajectory of the information age has manifestly been one of inequality and domination. At the heart of this problem is the relationship between production and transmission: Serres (1993, 87, 136) accurately diagnoses the shift in power from producers to transmitters (think of social media companies but also of logistics; note that Facebook and Amazon do not build much except the channels of transmission). Creative power is divine, but the risk is ever present that the messenger will misappropriate the power of the message they bear; this is the fall of angels. It is inscribed clearly in the repeating cycle of new media: a promise of a new era of human connection, followed almost instantaneously by the emergence of new monopolies.

It is not that angels are a metaphor that helps us grasp new media; it is rather that the symbolic, technical, and material worlds are linked through angelic means. Angels can act as "interchange agents" that mediate not just between sender A and receiver B but between systems. The capacity to mediate between systems (as my computer screen now mediates between code and language and between electrical and optical signals), crucially, is located in bodies—Serres (1993, 168) mentions a plug adapter that allows us to shave in different countries.

In the deepest sense, we can describe angels as agents of negentropy; in transmitting messages, they ensure that information is preserved. Their function is algorithmic, a key term for Serres: "a procedural operator that generates local instances of order without any necessary sense of a pre-existing grand unity" (Watkin 2020, 78; see Serres [1993] 2017). This is not a denial of universalism but an observation that the universal can be constructed from the concrete operations of particular algorithmic agents. In *Angels*, Serres (1993, 86–87) describes this process in terms of angels constantly building a ladder or tower to the heavens, building hierarchies from the bottom up, sometimes tumbling back to the bottom. At other times, he speaks of sunbathing: Are we lying in the sun's rays, or are we feeling the sun itself on us? Are angels emissaries of God or tiny perspectival parts of divine unity (1993, 107)? This linkage between the particular and universal is heretical both from a poststructural orthodoxy that would deny any universality (Watkin 2020, 87) but equally from an Ethiopian Orthodox theology in which God is the universal that precedes all particulars.

So angelic mediation can connect heterogeneous systems to one another (and thus is universalizing—interchange agents are the means by which the world can operate as a single message-bearing system of systems), and mediation is embodied, perhaps in bodies that are understood as intermediary or hybrid. To this we can add two more angelic functions: angels can be guardians, which maintain a boundary, and they are prepositional; they bring things into relation: "Angels are God's prepositions" (Serres 1993, 146). Prepositions "are active elements of messages, which are message-bearing elements in themselves, and which are pre-posed agents and subsequently nodes in the network of message-bearing systems" (Serres 1993, 294). Prepositions are good mediators because they vanish—we do not notice them, but they enable everything else to come into relation. If angels fall when they appropriate the message for themselves, perfect mediators vanish or die in the process of transmission. The central example is Christ—which makes Serres interesting in dialogue with Christian ethnographic settings.

The Stakes of Mediation

In a decade or so writing about a small Ethiopian Orthodox Christian community, I have spent most of my time looking for a framework to express the importance of mediation and mediators, as I understand it, in Ethiopian Orthodox life (Boylston 2018). This is clearest in the proliferation of religious intermediaries, human, saintly, and angelic, who permeate the Orthodox world. But it also connects to a pervasive materiality, a sense that divine patterns and rhythms are carried in the body and the lived environment. There is an ecological aspect to Orthodox Christianity in Zege, a sense of systemic interdependence evident in rhythms of fasting, labor, and ritual activity. Something I have particularly struggled with is how to describe this ecological, angelic dynamic, its ethical tones and the importance that inhabitants place on it, without romanticizing or utopianizing and while accounting for the social authority of the Orthodox Church and its historical and contemporary role in political domination, enslavement, and the extraction of wealth. This is a common observation among anthropologists of heavily institutionalized religion: a devotion among the faithful to their church as a repository of truth and value along with a highly critical attitude toward the actual behavior of both the church as an institution and its officeholders on an individual basis (Mayblin and Malara 2018; Bandak and Boylston 2014). The dynamics are comparable to the contemporary decolonizing the university movement: a deep attachment to higher education and to the university as a *potentiality* exacerbates the deep frustration with, and even hatred of, the university as an *actual* institution acting in the world.

Both the university and the Orthodox Church are in this sense angelic institutions: bearers of messages of great beauty and attraction, with an equal tendency toward self-aggrandizement and monopoly based on their status as authorized messengers. Serres shows that there is a fundamental duality to these organizations, a tendency toward political-administrative power that is inseparable from what we might call their sacred message-bearing duties.

A substantial literature on religion and media has grown around the insight that religious revelation is itself a form of mediation; that revelation and transmission are inseparable (Vries 2001; Debray 2004). A major contribution of this work has been to reveal a modernist fascination with immediacy and transparency, to show how this fantasy of immediacy is built on ever more complex networks of mediation, and to connect the ideology of immediacy to Protestant and reformist religious movements (Keane 2007; Meyer 2011). I have suggested that the Orthodox emphasis on mediation in Zege is precisely a riposte to what is seen as a Protestant-modernist project of severing all the crucial ties of religious mediation that link body, church, and environment into an interdependent system (Boylston 2018).

But a problem with this line of thought has been that, as soon as you focus on the material, mediatic aspects of communication, everything starts to look like mediation. And if everything is mediation, what does mediation actually mean? Is it not just a banal truism that things are connected to other things? Here Serres's angelology provides a framework for a viable philosophy of mediation—or, at least, for describing the stakes of mediation more clearly. Mediation involves interchange between scales, which is always embodied. Mediation is prepositional—in that it arranges relationships among things but also in that it "pre-poses" or prepares bodies for communication. And mediation is always subject to the fall of angels and the theft of the message—in a different Serresian idiom, it is parasitical. Focusing on these distinctive properties—embodied, prepositional, parasitical interchange between systems—will help elaborate the stakes of mediation beyond simply saying that things are connected. Bodies need to be prepared very carefully if they are to serve as agents of systemic interchange (Serres [1999] 2012).

Pre-Posing Bodies

In one sense, angels on the Zege Peninsula are known by their representations—principally, the archangels Michael, Gabriel, and Raphael. Seven monastery churches across the peninsula contain painted icons of the angels, usually guarding the doors of the inner sanctum that houses the *tabot*, the seat of consecration. Michael, in particular, always holds a sword. Angels are warriors as well as messengers.

And then angels are known by the practices that address them. People pray to Michael, particularly for deliverance from demonic affliction, and to Gabriel, particularly for help with fertility issues. Farmers observe the monthly holidays of Michael and Gabriel and do not work the fields on those days; in this sense, their presence is imbued into the land and the productive cycle.

There is, then, an obvious sense in which angels are intermediaries and messengers. As Orthodox Christians in Zege explain it, angels carry your prayers to God, as do Mary and the saints. God is omnipresent but so superior to us that it would be improper to approach him directly; angels do so on our behalf. Protestants who deny the role of mediators therefore seem to deny the superiority of God as well as the deep connections between land, labor, bodies, and religion. Note that many Ethiopian Orthodox Christians deny that Christ is a mediator—because Christ is God and God cannot mediate with God. For this reason, other mediators are necessary, and one way that people might suspect a person of having Protestant leanings would be in invoking Christ too frequently or with too much familiarity. But there is also a less obvious aspect of angelic mediation. Angels are not just carriers of messages; as negentropic agents, they play a crucial role in arranging the world so that message-bearing is possible. We can think of this as prepositional work.

It is hard to pray directly to God, so angels do so on our behalf. This "on our behalf" is crucial, baked into the Amharic prepositional infix *le*, "for, per, to, on behalf of." Amharic is a deeply prepositional language, in which verbs are easily and frequently modified to occur by, with, from, for, or because of something else. It is there in the standard greeting *t'éna yist'illiñ*, "May He give you health on my behalf," where *yisit'* would be "may he give," *yisitiñ* "may he give me," and *yist'illiñ*, "may he give on my behalf" or "may he give *for* me." Perhaps the greatest lesson I learn from Diego Malara's (2018) ethnography of Orthodox practice in Addis Ababa is that "doing on behalf" is a central logic of Orthodox morality. We pray on behalf of others, to mediators who bring those prayers to God on our behalf. A central virtue of holy water is that we not only can consume it ourselves but can fetch it on behalf of those who are sick or working or otherwise unable to get to church. Beggars call for alms on behalf of Mary, the saints, and the archangels, and in giving to them, we enlist the mediation of the beggars' prayers of thanks.

This brings up an important and difficult point in Serres: the idea that the destitute have an angelic, annunciatory function, that they bear a message of bare humanity, and that in their humility they embody the disappearing mediator (Serres 1993, 17–19). In one of many moments of Serreso-Ethiopian resonance, an Ethiopian friend describes beggars to me as *sämayawi säwocch*, "heavenly people" and intermediaries of prayer. The idea that destitution could have meaning in itself is offensive to a modernist sensibility, in which the only purpose regarding poverty and suffering is to strive hopelessly for their eradication. This has been one of the central controversies around charitable work in Ethiopia: foreign and non-Orthodox parties take exception to the notion that the function of charitable work in poor areas is to build churches rather than produce some kind of development or direct humanitarian aid. The modernist settlement can accept angelic intermediaries only in the form of charitable administrators.

The importance of almsgiving is based on a religious ethic of doing on behalf of others. Secular charity is unidirectional—you help the poor. In the Orthodox ethic, you may help the poor person, but they simultaneously help you by carrying messages on your behalf. Suddenly no relationship is dyadic; other mediatory agents are always involved. The transmission of messages is not more important than the arrangement of relationships that goes with it. Serres's work on prepositions allows us to dig beneath the obvious function of angels as messengers and go-betweens to a more diffuse angelic presence in the background, a more ambient and climatic mediation, prior to the formation of signs or symbols.

Serres's pun is between *preposition* and *préposé*, meaning an official or post-man. It is not coincidental, he says, that the deliverer of letters is like the hyphen, which both connects and separates (Serres 1993, 142–45; Bandak and Knight, this volume). He adds to this the idea of *pre-poser*, to "pre-pose" or set a body in position—for example, to begin a dance or children's dolls arrayed to enact a play scene. Dancing, tumbling, surfing, and skydiving are all arrangements of bodies with regard to the message systems of bodies, winds, and tides. Serres talks about the activation of a writer's body before an idea has taken shape, perhaps posed at a desk, becoming sensible of some kind of awareness or curiosity. Bodies, we might say, are channels of messages at various scales and often need to be appropriately prepared to act as such. I want to suggest that Orthodox fasting is like dancing; a collective preparation of bodies for the kinds of message-bearing that come before meaning.

Fasting occupies a substantial portion of the Orthodox year. In previous work, I have described fasting as lacking in propositional content—you can fast without knowing why or attaching any particular message to it. With Andreas Bandak (Bandak and Boylston 2014), I have described fasting as a form of deference and deferral, in which meaning is offloaded onto experts to be determined at some indefinite point in the future. I was influenced by Bloch's (1974, 2005) famous analysis of ritual, in which formalized expression, emptied of propositional content, becomes a general mechanism in the maintenance of hierarchy. The problem, as we began to explore in that piece, was to do justice to those aspects of fasting that seemed to us to be creative, imbued with world-making potential, and that practitioners described as both ethically desirable and even enjoyable.

Dancing can be expressive without being symbolic. What if we were to think of fasting in the same way, as both an expressive practice and a technique for orienting bodies to the world and to each other? In a Serresian idiom, fasting "pre-poses" bodies toward receiving and communicating particular messages, sensations, and affects. Ethiopian Orthodox fasting is not hugely stringent—often it may simply involve abstaining from consuming animal products, perhaps taking no food until noon. But fasts are regular and frequent: every

Wednesday and Friday, fifty-four days of Lent, the eves of major feasts, and a scattering of other calendrical events. Some, such as Wednesdays, Fridays, and Lent, are very widely observed; others are mostly maintained only by the clergy. In addition, individuals may conduct their own fasts as penance on advice of their soul-father priest, and nobody may consume food or water before entering church, especially if they are to take the Eucharist.

As I have explored throughout my work, people give varying explanations of why fasting is important or what particular fasts mean. Crucially, you do not need to know doctrine to take part in the fast; you just need to do it. But it is worth highlighting what I believe to be a widespread understanding of fasting as a kind of spiritual technique, a preparation (or "pre-posing") of the body for certain kinds of moral experience. Metacommunicatively, to fast is also to express commitment to the shared practice of fasting itself. The focus is on what fasting does rather than what it means. And what it does is to make you hungry—and, as practitioners explain it, to weaken the body and dampen emotions such as anger and lust. Fasting makes us more conscious of our bodies and so less apt to be distracted by impulses (see also Serres [1999] 2012). It therefore enables spiritual reflection and meaningful penance and makes it easier to pay attention to saints and angels. Fasting is not a message but the foundational condition for religious message-bearing to take place. Serres writes, "The dancer, being inarticulate, precedes articles. Adept at all positions, he expresses pre-positions" (Serres 1993, 127). The same goes for those who observe the fasts.

The upshot of such regular fasting is that every day of the year is marked, at a felt, pre-symbolic level, as fasting or non-fasting. After you have spent days hungry, sufficiency gains a new meaning by comparison. The bodily level at which this takes place is crucial; according to Serres, "in the bodies of angels and cherubim, and of all intermediary professions, one might say that there exists something like a swinging pendulum, a reversible metronome . . . Far from making a system fragile, this is what stabilizes it" (Serres 1993, 171). Fasting-feasting is metronomic in this way, oscillating between reflective, spiritual abstemiousness and exuberant, fleshly plenitude. This indicates that a process of interchange between different systems is taking place. We can think of this as interchange between flesh and spirit so long as we do not take this as a hard dualism, but precisely one of systemic interface and message-bearing: fasting demonstrates more than anything that spiritual life is achieved through practices of the flesh. Interchange, as Serres (1993, 101) tells us, takes place through oscillating, hybrid, or pre-posed bodies, and so translation between incommensurable systems is possible (although, *traduttore traditore*, "the translator may be a traitor").

When I talk to people in Britain and the United States about Ethiopian fasting, I often sense that people find it disturbing and see fasting as a kind of self-harm. Countless scientific articles have been attempted on the effects of fasting on pregnancy health (though most priests hold that one should not fast if there are compelling physical or practical reasons that it would cause harm). Modernist ontology can only see fasting as an irrational premodern hangover, unless it is for dieting or fitness purposes consistent with regnant ideals of self-fashioning and subsumed into capital in that nonconsumption becomes a form of consumption through protein shakes, diet books, and gym memberships. The idea that consumption itself might have a dimmer switch, and at a collective level, is increasingly hard to comprehend. Vegetarianism would seem to be different and to represent a form of active nonconsumption that takes the form of embodied, ethical message-bearing. But even then, the recent rise of yuppy "plant-based" foods for the upper-middle classes indicates that even this form of nonconsumption can become consumption. Nonetheless, I would suggest there is value in thinking about vegetarianism through a Serresian lens as a form of pre-symbolic mediation. This is even more so for ethical vegetarians who eat meat when it is offered in hospitality, a very nicely Serresian example of bodily interchange between systems. Fasting is expressive, but it also alters the conditions (ethical, affective, environmental) under which communication takes place.

Dead Mediators

The condition of the flesh is a key concern. Angels, Serres (1993, 102) tells us, must be unsexed because "If the messenger gets pleasure, the transmission becomes obstructed"—and anyway, angelic life is a different kind of ecstasy in the fusion of worlds (Mayblin 2014). This is true to Ethiopian practice—where in addition to fasting, strict purity rules keep sexuality separate from the sacred. It is also perhaps a limitation of Serres's angelology. Could we not imagine a cosmology in which the pleasure of the mediator was celebrated, whose heroes were Lucifer and Prometheus? Serres toys with the idea, but only briefly. Perhaps this is only the realm of cult leaders and corrupt televangelists, or perhaps these corruptions are precisely perverse outcomes of the general repression of the mediator's pleasure.

In any case, in Ethiopian Orthodoxy, the austerity of the mediator is crucial. I would go so far as to say that a religious mediator's trustworthiness is proportional to the degree in which they are already dead: from the sacrifice of Christ; to angels, who were never quite "alive"; to saints, who long ago suffered violent death in defense of the faith; to monks, who are strongly coded as dead beings; to priests, who marry and have children and who therefore, regardless of how knowledgeable they may or may not be, never quite carry the same authority.

The social authority of monks is a case in point. While it is known that individual monks may occasionally become corrupt, it is axiomatic in Zege, as elsewhere in rural Christian Ethiopia, that deference to the teachings and advice of monks is essential, a moral cornerstone of social existence. This authority stems not just from monks' ordination but from the bodily status they maintain: monks must be virgin, live in humble accommodation within the monasteries, and strictly observe the routine of daily prayers and all of the 250-odd fasting days of the Orthodox year. Van der Weyer (2009) describes admission to monastic orders as follows: "Traditionally the candidate is wrapped in palm leaves, the funeral dress of the poor, and the monks sing the funeral requiem to signify his death to the world." Friends in Zege, too, have told me that monks' funerals are different because "they are already dead." As such, as vanishing mediators, they can be trusted.

Structurally, monks are dead to the world but also in constant interaction with it. Its ritual activities and prayers are on behalf of (again, note the preposition) the community at large. In an area like Zege, where the monasteries are within the community, they function also as people's regular churches of worship; because of their authority, monks are frequently called on to help resolve disputes or generally provide spiritual advice. Monasteries are also economically entangled with the community, formerly in selling the surplus produce from their landholdings and today as the focus of historic tourism. What do tourists perceive in the monasteries? What draws them from thousands of miles away to visit? Not, apparently, the hope of blessing, cure from ailments, access to holy water, or divine assistance with fertility. More of a mediation through time: the churches are old and beautiful, and their icons are old and unusual. Tourist literature and journalism consistently describe Orthodox Ethiopia as a sort of living medieval relic, a place where you can peer into the past, an idea that is strangely compelling to the visitor. Tourists do not think they come for the angelic presence in the churches (though, if they come to see the icons, is this not in a sense what they are doing?). But without the angelic presence, the historic density and seeming continuity that attract the tourists would not exist.

Monks, being partially dead, become able to carry messages of eternity. Being partly of the flesh, they can go between worlds as angels do. There is a neat structural logic to this, which I am convinced is correct as far as it goes. But the life of monks and all clergy and the church as a whole is not only determined by their structural position. As Serres shows, we need to attend to the fundamental duality of institutions not just as message-bearers but as competitors for the authority to determine which messages are correct.

Actual Institutions

I have met few Orthodox Christians in Zege who expressed any opposition to the Orthodox Church as a general entity. Even for some Orthodox scholars who wish to see reforms in the Church, the prospect of schism is concerning, and such views are expressed indirectly if at all. One man in Zege who openly criticized the Church was widely considered mad; others convert to Protestantism, though this carries a risk of being disowned by one's family. For most, however, I have no sense that loyalty toward the church, deference to the monks, and commitment to fasting are in any way feigned or halfhearted.

This being said, even the most loyal Christians frequently tell stories of the wrongdoings of members or parts of the Church. I hear that a monk has been caught in a brothel, that a priest was abusing his confessional position to try to persuade a parishioner to marry him, that the Church Synod is politically compromised. There is widespread concern about fake monks stealing alms.

We may add to this critiques leveled by non-Orthodox scholars. Ahmad Abdussamad (1997, 1999) has detailed the deep involvement of Zege's priesthood and nobility in the slave trade between the nineteenth century and the 1930s. Tihut Yirgu Asfaw (2009) argues that the Orthodox Church in Zege has impeded the distribution of aid and other resources to Zege's poorest. At the time of writing, amid a chain of civil conflicts, factions of extreme ethnonationalist preaching have become prominent in Orthodox circles, some of whose rhetoric is outright genocidal. This rhetoric draws on narratives of Ethiopian Orthodox Divine election—to the exclusion or marginalization of all others—that the mainstream Church has long promoted.

And what of gender? The Orthodox Church represents an almost complete capture of the angelic role by men, with Mary as the high exception who guarantees the rule. There are exceptions in the traditions of nunneries, of important female religious scholars, and of new movements of women taking part in important ritual activities. To say that these remain marginal is not to deny the huge role that women play in Orthodox ritual life—in the organization of festivals, in the support of priests, and as the majority of those who actually attend church—but precisely to point out how this contribution is marginalized.

None of this is particularly surprising—to a large extent, this is what institutional churches do, especially when they become closely involved in the activities of the state. What is less recognized is that loyal members of the church know very well that every part of their church is prone to fall. The daily realities of church life include priests grumbling about being posted to the hinterlands and their low pay, local power struggles of all kinds, and fierce disagreements at every level.

Yet despite (or perhaps because of) this, still they are loyal to that Church that exists beyond the actions of any of its members. This is the double nature of the Church: as the unity of the faithful in the body of Christ and as an operating institution. This duality figures heavily in the anthropology of Catholicism (Mayblin et. al. 2017; Pina-Cabral 1986; Badone 1989), not least in work in the wake of the Church sexual abuse scandals (Orsi 2017).

Here I do not know how to read Serres. On the one hand, the fall of angels and the emergence of "powers, thrones, and dominions" appear to be inevitable in a universe of communications. The message is always vulnerable to appropriation by the messenger, and the one who controls the channels holds the power. Yet Serres seems to hold open an answer in the incarnation; the egalitarianism of fleshy immanence, in the straw, with the animals. The institutional church itself seems to vanish as the word becomes flesh: "Angels will still be able to continue expressing their language, writing and singing, transporting and coding messages, distinguishing the symbol and the devil . . . but henceforth their role will be subaltern, their age will have come to an end, and both their role and their age will have been fulfilled, because the message is here, in this place, in this living space, in this stable with animals, in this cradle surrounded by quadrupeds, in this smelly immanence and real and tangible actuality, at the crib" (Serres 1993, 285).

Yet, implicitly, this is not the same as the critique and negation of institutional churches mounted by various Protestantisms in the past five hundred years. It is a fulfillment, not a negation, of the angelic function. It is Catholic without the Church. There is, it must be said, little indication of how this vanishing act is to be achieved. I must also note that my Orthodox friends would find the notion of dissolving the Church appalling. As I have said, we must entertain the possibility of heresy.

Conclusion

Serres's turn to incarnation leads us back to what I have found to be the most profound and productive aspect of his mediology: that of preposition and preposing, of the preparation of bodies for message-bearing, and of bodies that ride the winds and currents of the world.

"The true messages are human flesh itself. Meaning is the Body."

"Or the World."

"Love is fleshly" (Krämer 2015, 60).

Superficially, Ethiopian Orthodox practice may seem to despise the flesh. Flesh is widely invoked as the source of sin; holy figures who transcend the need to eat are venerated. But as we have seen, fasting is a technique that operates on the flesh and one that practitioners frequently describe as both joyous and morally necessary. While fasting enables non-fleshly reflection, it does so by amplifying our consciousness of our bodies and their desires, not by repressing it. It is also something that can be done on behalf of others; monks' observance of the full cycle of fasting is for the sake of the very community that they withdraw from. In fasting, all the focus of life is directed toward the humble neediness of the human body.

If a focus of the anthropology of religious mediation has been on the intractable materiality of symbols (Keane 2007), Serres's angelology posits that the material world is message-bearing all the way down. What matters here is the way we position ourselves toward message-bearing systems, whether in hierarchical ladders or overlapping circles, whether bodies are attuned to the points of interface between different systems. Religious messaging—a paradigmatic example of communication between different systems and scales—exists not just in discrete, explicit utterances and exchanges but in an ambient prepositional labor that is easily missed.

I am left with the feeling that Serres's angelology has brought me closer to something important and profound in Ethiopian Orthodox practice, but in ways sufficiently divergent from doctrine as to raise the question of heresy. In Josephson's account of Euro-Japanese cultural contact, heresy was a term deployed by both sides to mark the other as deviant, dangerous, and misleading. Heresy emerges prior to the concept of religion per se (which presumes the existence of multiple religions that are somehow equivalent and so downgrades both Buddhism and Christianity from ontological primacy). It marks the other as doing something recognizable but distorted.

Let me propose that secular speech, the anthropological norm, is not heretical because it operates in a sphere in which "religion" has already been bracketed into its own domain. In Orthodox Ethiopia, heresy (*mänafik'*) is very much a live term, mostly applied to Protestants and those suspected of smuggling reformist distortions into Orthodoxy. When an atheist is labeled heretic, it is because their atheism is assumed to be basically devil worship, since nobody could reject God without satanic interference (Malara 2022). Secular anthropology is rarely so bold as to explicitly occupy this position; religion is "taken seriously" in the sense that it is not ridiculed, but care is taken to keep this, by and large, within the realm of cultural difference. Presumably, if I were to really take Orthodoxy seriously, the first thing I would do would be to fall to my knees and beg for forgiveness.

From an Orthodox perspective, much secular anthropology is simply not relevant, since it has foreclosed on all the questions that really matter. Heresy is that which speaks within the orthodox domain, into the same questions, while also deviating from regnant institutional norms. To risk heresy is thus to risk being wrong rather than just irrelevant. It raises the possibility that I have a stake in understanding not just the internal logic of what people believe but the world within which those beliefs are anchored, in belief's conditions of possibility.

10

FORMS OF PROXIMITY

. . . irrespective of porous cultures, absurdities still exist as outmoded as borders between nations. —Michel Serres, *The Incandescent* ([2003] 2018, 71)

The river of Love and Hatred mixed together, productive and destructive of the global whole and the tiny facts of knowledge, shapes the evolution of species, the collective history of the human race and the behaviours of individuals. —Michel Serres, *The Incandescent* ([2003] 2018, 151)

I arrived in South Italy in search of difference. My personal journey had brought me late to social anthropology, via a schooling in sports science with methods based in biology and the medical sciences, accompanied by a heavy dose of Greek nationalism that preached the importance of sameness—all Greeks sharing collective belonging, underpinned by an uninterrupted genealogical line to their glorious ancestors. In biology, I was taught to identify the essential criteria of *being human* through shared DNA, muscle movements, and the physical and chemical capacities of the body. A nationalist curriculum preached the biological roots and routes of all Greeks, which ensured cultural and intellectual continuity and a sense of security in numbers. Both—fundamentally evolutionary—pathways placed "Greek humans" at the very pinnacle of "natural" and "cultural" worlds.

In stark contrast, my graduate years in social anthropology in the United Kingdom had been driven by the deconstruction of sameness, reflexive self-questioning of everything I had known about identity politics, spurred on by my proudly postmodern teachers. Cultural relativity and individual becoming were the core concepts of this new discipline. My challenge in my doctoral research in South Italy was thus to break down the assumptions of my formative years in the natural sciences. I was tasked with understanding difference among a linguistic minority in Calabria fighting for recognition in a state preaching homogeneity.

And so, as I entered the field of the Grecanici linguistic minority of Calabria, located on the very toe of the Italian boot, my pursuit of difference was twofold. On the one hand, the small population of around twenty thousand sought to be recognized among the hegemonic majority; they claimed difference in a variety of political and cultural domains. Although Italian citizens and Catholic, the Grecanici distinguish themselves as "the brothers of Greece," speaking Grecanico—a language composed of archaic Doric, Hellenistic, and Byzantine as well as local Romanic and Italian linguistic elements—alongside official Italian and the Calabrian dialect. They trace their roots back to the foundation of the first cities of Magna Graecia in the eighth century BCE (Pipyrou 2016). On the other, I also wanted to deconstruct the myth of their Hellenic sameness, as being part of the evolutionary pinnacle that is "Greek humans" so strongly propagated in my own education, and to dive into the murky waters of claims to Homeric origins and ancient ancestry, in a similar way to how I was learning to critique myself in this newfound life as an anthropologist.

Eighteen years after my initial visit, I reflect here on how I have come to understand the Grecanici in a world of proximity—to each other, their families, various state and suprastate bodies, categories of belonging, and a foundational humanity—as they mingle connections that provide power and representation. The work of Michel Serres has, retrospectively and introspectively, provided an avenue to think about the messy scalar proximity of humanity in its many guises as the Grecanici zoom in and out, play with categories of sameness and difference, tie together disparate historical events and futural orientations, all to facilitate comprehension and then performance of their position in the world. Serres has helped me locate both the Grecanici and my own background on the accordion scale too readily polarized by cultural relativity (the arts, individual diversity) on one end and human nature (biology, universalisms) on the other.

Among the Grecanici, I encountered similar concerns with social life as my supervisors whose research was on vastly different topics in Greece and Pakistan—family feuds and love stories, the politics of resource sharing, notions of limited good, and power games aimed at local and national government. I found myself connecting their analogous points of interaction with the natural environment and equivalent struggles to place the human in the vast network of relations that make up worlds of politics, technology, and global systems of power and knowledge.[1] Questions proliferated concerning the relationship between individual and collective identity, national belonging, and existential quandaries on human control of eventedness and crisis. Thus, like so much of Serres's oeuvre from *Genesis* ([1982] 1995) to *Variations on the Body* ([1999] 2012), I struggle with placing human beings—here, my Grecanici interlocutors—within a nexus of scalar proximity where individual and societal truths can be combined in a movement toward more universal concepts of humanity-in-cosmos, what Serres alludes to in the second chapter-opening epitaph as global wholes and tiny facts, infused with the spectrum of love and loathing, that make up species and individuals. As presented in the introduction to this volume, for Serres, this is an exercise in algorithmic thinking, or preserving everyday individual and collective truths constructed in particular socio-historical environments as building blocks toward universal understandings of the human condition (cf. Watkin 2020; Serres [2003] 2018). Or, in other words, an interconnected method of biology and sociology, of the natural and social sciences, working together rather than in arbitrarily separated categories. Critiquing Plato's version of one single truth, Serres poses that a "thousand nocturnal sparkles" of knowledge are navigated by "free people" (2015a, 18). This shimmering of knowledge with its "billions of glorious and timid colored suns" holds "innumerable truths linked together by a thousand related networks" that are all required to construct a complete interpretation of the world (2015a, 22). The appreciation of anthropology's situated truths builds toward shared biological and physical realities of human-in-the-cosmos. This baseline sameness does not necessarily spell harmony; violence and power struggle, as per Serres, are readily present in the formation and negotiation of natural and social contracts. As we will see later in the chapter through putting Serres in conversation with Sigmund Freud, even in the most commensurable relations, microfissures of difference are rife with violence and subjugation, where porosity is replaced by a brick-hard membrane.

Indeed, it could be said that Serres's worldview rides the squeeze box of proximity and distance and of topological connectedness. It is in many ways his method, moving in and out of ratio, mingling and mixing dimension, folding and assembling the very large and very small, universal and relative. It is a method of constant flux that calls for symbiosis while not shying away (indeed, embracing) the violence at the heart of forming connection. Serres is concerned with the breaking of boundaries of category, advocating a focus on the seepage, overflow, and porous borders of concepts often glossed as identity politics (kinship, nation, culture, Other). This is perhaps most readily accessible through his personifications of spatiotemporal complexity, including Hermes (Serres 1980), who rodeo-rides timelines of past and future to produce proximity between disparate events. Thinking beyond bounded categories of space and time, at the seepages and overspills of concepts, allows for new connections between relativist and universalist knowledge, the soft and the hard, difference and sameness. The connection, for Serres, is not one between a set of polemics or dichotomies, but rather he likens proximity by way of connection to "veil, canvas, tissue, chiffon, fabric, goatskin and sheepskin . . . all the forms of planes or twists in space, bodily envelopes or writing supports, able to flutter like a curtain, neither liquid nor solid, to be sure, but participating in both conditions. Pliable, tearable, stretchable . . . topological" (Serres [1982] 1995, 45).

Quasi-abstract and unapologetically metaphorical, and often crumpled up in plain sight, connections fade in and out, take on different angles and appendages, get mixed and lost in the wash: for Serres, "the world is a mass of laundry" ([1985] 2015, 100–101), after putting connections in the washing machine, one might end up novelly wearing odd socks to work or a pink-tinted shirt where the dye has run. This might incite ridicule from workmates and cause particular tension with the boss, but Serres asks us to own the novel connection, navigate the violence, and perhaps even encourage others toward audaciously embodying their own polychromic connections.

One such category breach is between Self and Other, as I was finding in my personal challenge to transition from the natural to social sciences while entering what for me were surprisingly liberal political spheres in the UK university [what Serres has discussed as the general truths of natural sciences and the individual truths of the social and later the transcendent power of religion ([2019] 2022)]. Among the Grecanici, sameness and difference, I concluded, were not dichotomic categories to be neatly boxed but were to be explored in their kaleidoscopic complexity, their seepages, connections, and the search for power and representation that a playing with these categories could provide. It is Serres's approach to the world as composed of topological proximities and distances that help simultaneously hold relative and universal truths that forms the basis of the current chapter. Proximity to relations drawn from numerous boxes into assemblage of a novel truth that represents the individual or small-group identity was precisely how my Grecanici interlocutors experienced the world; not through a pandering to hegemonic top-down truth about what it is to occupy the category "minority" but rather a fearless navigation of multiple sources of knowledge to forge themselves a novel branch of existence.

However, I also intend to play devil's advocate as the chapter progresses by asking what happens when there is too much proximity, when connection is unwanted or threatening, or when institutions—such as the state—operate through categories as they search for security and systems of governance. There is often a grind between Serresian fluidity and ethnographic reality: categories of ordering and sorting, such as "minority," "refugee," and "ethnicity," carry indexes of power and serve disciplinary functions. The state requires bifurcation, separate branches connected according to security policy, law and order, and the rigid maintenance of the myth of superiority (such as the bio-social evolutionary pinnacle of the "Greek human"). In both top-down and bottom-up manners, categories of inclusion and exclusion, sameness and difference *are* enforced—as Serres notes in the first epigraph—whether based on premises of scientific enquiry or political governmentality. Crossing the Ionian Sea, I present the case of another minority population—the Pontians of northern Greece—to argue that proximity is always a recursive game, the branches always well fertilized with power, and strategically played in relation to wider boxes of identity politics.[2] When encountering in the field situations where sameness and proximity become problems in need of division and distancing, I propose that Serres can be read in conversation with Sigmund Freud's "narcissism of minor differences" ([1930] 2010; Pipyrou 2021). Too much proximity is an ethnographic reality and a problem for the state that wishes to maintain difference on the basis of security and categorical purity. An analogy of magnetic repulsion serves to illustrate the need to keep subjects, objects, and histories apart. For Freud, it is often not the completely unfamiliar or new but rather the "secretly familiar" that is frightening and threatening and must be sorted (Freud [1919] 1973, 245).

Whereas the Grecanici navigate hegemonic power and representation, of often violent categories of belonging and deservingness, by way of fearlessness and brash (re)appropriation, Pontians have tended to subtly play the game of residing on the seepages or blurriness of boundaries; they phase in and out of commensurable relations. In the customary style of Serres himself, bringing another figure of thought into the conversation helps graft a branch in the connection between numerous bodies of knowledge—here, anthropological observation on the search for sameness/difference, Serres's pondering on hard and soft categorization, and Freud's focus on uncanny narcissism. I conclude that Serres's proximity and Freud's narcissism are alternate angles on the same problem—how much connection is desirable, and where does categorization become an inevitable tool of governance for sorting and sanitizing difference?

Variations of Proximity in Serres

My Grecanici interlocutors move between nexuses of relations, playing with the categories of sameness and difference to create proximity between themselves and ancient Greece, the European Union, UNESCO and state minority recognition programs, the Greek Ministry of Culture, and international tourists. At times, they promote or conceal their sameness/difference to other linguistic minority groups on the international stage, Catholic religious beliefs, Italian citizenship, and civil society associationism. This map of connection and proximity is fluctuating and mobile, navigated as a series of potentialities toward power and self-governance. Through multiple forms of relatedness, Grecanici make their way through rebounding, intersecting, and overlapping channels of political representation and power (Pipyrou 2016, 22). The minority fearlessly seek political representation for their cause through crossing categories in "ensemble movements" through "possible roads" (Serres and Latour [1992] 1995, 105). Their claims to minority status recognition are based in the folds that hold sameness and difference in mutual embrace. As a minority group, they emphasize the "shared diversity of humanity" to access resources provided through networks that demand the accentuation and silencing of difference. The morphing in and out of different proximities to categories of identity recognition provide the Grecanici with pathways to diverse sources (and scales) of power. At once, they entangle a global category of minority, play to nationalist idioms of lost brothers of Greece, are religious patrons of both Catholic and Orthodox churches, and make claim to being archaic and modern, embodiments of an illustrious past worth preserving and emblematic of modern multicultural European futures. They twist and turn, shoot across categories, and forge new branches of association by animating and numbing aspects of their minority status. They are simultaneously proximate to multiple nodes of knowledge and appropriation.

In *Conversations on Science, Culture, and Time* ([1992] 1995), Michel Serres speaks with Bruno Latour about simultaneity and connection. For Serres, disparate peoples, places, and eras exist in networks of proximate relations. Topologically rather than geometrically connected, networks twist, turn, and contort in a state of constant movement, what Serres compares to "the dance of flames in a brazier," fanned by mobilities and the unexpected (Serres and Latour [1992] 1995, 58). In *Hermes V*, Serres likens the movement of the flame to tectonic plates, not being linear and Cartesian: it flickers, changes directions, surges, and fritters out (Serres 1980; also Watkin 2020). Add some extra gas, a hand pump, or the right chemical concoction and the flame gains energy, burns brighter, sets fire to adjacent furnishings. The connection, if you like, becomes stronger if the circumstances are right or dies back to a slow underlying burn at other times. The fuel for the connection can be environmental, social, provided by individual agents or planetary events; there is vast scalar potentiality in how connections are formed. For Serres, the intermingling of dancing flames can be encouraged by forces operating beyond the normal plane of existence: the metaphorical angel or messenger (see Boylston, this volume; Serres 1980, 1993).

Through the analogy of the flame, Serres makes clear how, at times, connections become closer, are proximate, and then dissipate in obscurity. Connections that might once appear distant become simultaneous, famously illustrated in Serres's example of the folded handkerchief (Serres and Latour [1992] 1995, 60; cf. Bennett and Connolly 2011, 159; Knight 2012). The similarity in concerns about the social and natural world expressed by my Grecanici interlocutors fade in and out and are made visible, felt, and expressed in tones and textures depending on the dance of the flame, the porosity of connection between time and place. Thinking with Serresian analogies, this appears to be what Serres is referencing when discussing the constructive and destructive margins of tectonic plates where new material (knowledge) is created, being forced into the light of day, while at other points knowledge is crunched under the fold and back into the bowls of the foreboding Earth ([1982] 1995). As political actors, the tectonic movement of Grecanici between categories of creation and destruction—highlighting certain aspects of their stories—lays bare and conceals sameness and difference depending on the intended audience. But I may caution, perhaps this is not so much a story of exception but rather a constant presence in the dance of human lives as we enact our relations daily. There is a general familiarity in the dance of the flame as it makes strategic connection, whether located in a minority civic association or on a university campus. Perhaps it is only anthropology's search for difference that makes such an observation interesting at all.

Proximity runs throughout Serres's oeuvre to explain a cosmos of spatial and temporal flux. The pendulum of proximity, surging and lulling in connection to others, objects, and places, is a fundamental aspect of being human—and being in time. For one, Serres talks about the proximity of figures of thought—Plato, Leibniz, Jules Verne, Voltaire, Venus—with whom he converses without citation or index of spacetime. Further, his messengers in angelic and classic form tie together events as disparate as wartime hunger, the *Challenger* space rocket disaster, and biblical revelations. In *Troubadour* ([1991] 1997), the impish harlequin has absorbed all the colors of possible existence. Stripping back the harlequin's layers one by one, we are led to the conclusion that the universe is white, composed of all possible hues of the spectrum of experience. Infinitely colorful but also uniform: the Humanities and the Sciences. White is thus the ultimate node in the connection of everything, at once bland and unsurprising but eternally complex in its composition. Serres is unrelenting in his holistic pursuit of proximity.

Yet, in the classic mold of anthropological debate on structure versus agency, Serres also acknowledges that this fluidity in connection cannot be *lived* by human beings without some idea of stability. This is where categorization provides order to the unorderly and contains the violence of unrestricted becoming. For the purity of truth to be maintained in every connection (that each connection be taken on its own merit) while building toward a universal truth of human inhabitation of the planet and the cosmos, Serres offers hyphenation, or the bridging of concepts with a physical "line" to hold seemingly opposing ideologies in close proximity, not melting one into the other or conflating meaning but rather maintaining both uniqueness and difference (Serres 1993, 142–45). The point here is not to bring together two homogenous boxes of material but rather the work of the hyphen itself in fashioning out a shared space that preserves both difference and sameness. This is perhaps why Serres is so interested in prepositions (Boylston, this volume), as linkages that bring subject matter together. According to his obituary in *Philosophy Now*, "Serres was, to say the least, unusual: on a quest to minimize violence . . . a polymath who sought always to build bridges and linkages. Two prefixes mark out the intellectual landscape fostered by Serres: 'with' (*syn* in Greek and *con* in Latin), and 'between' (*inter*). He articulated what he called a philosophy of prepositions, not one of nouns and verbs, in which 'with' and 'between' played central roles" (Boisvert 2019, 56).

The two-way bridge of the hyphen allows for simultaneous sameness and difference. Rather than thinking about a world operating by way of dichotomy, as tempting though that may be in both philosophy and anthropology, Serres proposes bringing structural difference together in a third way where structure is provided by the bridge. It is on the bridge where the fine-grained detail of relations pass by, intermingle, and exchange the atoms that both structure connections and provide them with vibrant dynamism. Back to the obituary, "Philosophers have traditionally favored metaphors involving solids with fixed, rigid barriers, kind of like statues distributed in Euclidean space. Serres believed it was time for a shift to liquid metaphors, with their emphasis on flux, temporality, interplay and, importantly, *porous boundaries*" (Boisvert 2019, 56, my emphasis).

Porous boundaries providing movement, flux, and interplay between the biological and the social—hence the title of this very collection—establish the essential link for moving between relative to universal truths while maintaining atoms of knowledge. Serres's philosophy is not of static objects, eras, events, and personas but of movement through the (idealized) porous membrane of the universe, forming symbiotic connection across scale and meaning as we go.

Arguably his most important text on the metamorphoses of the biological body in the social world, *Variations on the Body* tells of the "fair harshness" of the physical world that

> teaches the truth of things . . . without pretense. . . . Written or spoken, repeated without danger, language, conversely, causes the proliferation of parrots who, immobile, fidget and reproduce. The other is reduced to nothing in it, by dead messages, and the thing is reduced to its recording media—wax, screen, paper—lastly oneself is reduced to its neurons, to the I, to thought. The risk of truth disappears, whereas the world inimitable, extracts movements and actions whose pertinence it immediately sanctions, and the group destroys itself there in proportion to its lies. But, in fair exchange, this world shows its phenomena, in all obviousness, and gives its data for free. ([1999] 2012, 7)

Culture takes away the "risk of truth" by wrapping up the physical world in layers of language and recording media; the "proliferation of parrots" who reproduce themselves to escape the naked danger of the natural world; the "proliferation of parrots" who call for difference to be arbitrarily boxed by categories of minority, refugee, gender, or ethnicity—culture erects boundaries to porous becoming (cf. Serres [2003] 2018). For Serres, there is an original shared state of variations of bodies in the world, which are drawn toward sociocultural clustering. The key word here is "variations." Serres is not claiming each body is identical or that the world sanctions bodily movement in the same way for each individual. Or that each individual necessarily has the same capacity for movement in the world and between relations, as would be the premise of Kantian cosmopolitanism. Rather, there is a structuring set of potentialities in how the biological body could and does interact with the social and material world (see Povinelli, this volume). From this point of encounter, variations or branches of symbiotic metamorphosis between nature and culture surge forth. Biology and sociology have an almost predictable metronomic rhythm that produces sets of shared yet different routes.

With Serres, I argue that humans reside in proximity to biological truths with variations on comparable interactions with the natural world. The branches of sociocultural potentiality operate across the bridges of hyphenated categories of nature, culture, and sociality. This multidirectional flow, present in all relations and things, is what Jane Bennett (2010) terms "vibrant matter," or the active relationship between humans and nonhuman forces. Vital materiality, for Bennett, runs across the Serresian bridge, pumping lifeblood into our veins through the connection of biology and society. The mixing of hard (universal, scientific fact, the physical) and soft (social meaning and conceptual constructions) is indistinguishable in everything we do, not forming "separate subject and object" ([1985] 2015, 26). But this innate connection is often obscured by the assumption that body and society are different entities, bounded categories that do not intermingle, and by the proliferation of parrots who populate each disciplinary box calling for categories to "cut the network" (Strathern 1996). In *The Five Senses*, Serres provides ample examples of "mingling" of biological and social—sound waves being converted to sentences with meaning, the driver feels the freeway through their fingers and sees the hazards that are converted into sensations and moral judgements ([1985] 2015, 138). With a background in Greek traditional dance, I experienced the body as the primary receptacle through which the world (material, immaterial, social, natural) is filtered. When dancing, the biological organism that is the body facilitates a performance that communicates (and can potentially traverse) the categories of the social. Coming from a background in the biological sciences toward anthropology, by way of an interest in dance and performance, for me the discovery of ethnomusicology was revolutionary as it provided the bridge connecting bodily to social understandings of music and dance.

The world is experienced through forms of codification and categorization—technological and social—that at once provide differentiation and integration (Serres [1985] 2015, 156). Now, I wish to return to a statement from the beginning of this section, where I cautioned against seeing the Grecanici as an exception in how they move across categories of proximity, sameness, and difference, both purporting connection with disparate times, places, and peoples while simultaneously striving for distinction. The dance of the flame in making connection, I suggest, is as much linked to the biological sensibilities of the human as it is to social peculiarity: That is to say, we all do it, and anthropology's desire to find difference glosses shared human attributes expressed in their sociocultural form. I turn here to another of Serres's favorite terms, "audacity," to explain why the Grecanici are able to cross categories and form connection in a manner not universally observable. I propose that the Grecanici embody an attribute that allows them to pursue connection not readily found in all ethnographic fields, namely, fearlessness.

Fearlessness

Serres asks his readers to courageously embrace the dangerous, the conflicting, and the possibly fatal in the search for alternative truths (2015a, 22). Almost paraphrasing Serres's observations in *The Five Senses* ([1985] 2015), the Grecanici talk of the need for "sharp senses" of seeing, hearing, feeling, and imagining to anticipate and embrace possible hazards of those wielding the axe of political hegemony (Pipyrou 2016, 7). Their senses, located and absorbed in the body by way of social immersion, direct them toward new relations; and form bridges between networks of kin, civil society, quasi-legal organizations, and various programs offered by the European Union and Greek and Italian states. In doing so, the minority invert preconceived notions of a subjugated and powerless people and provide a prime example of *porous becoming*—the ability to absorb, transform, and connect natural and social stimulants, piecing together lifeworlds and novel versions of truth. And the minority are *very* confident in their version of truth, garnered from various pools of knowledge. They audaciously navigate the topology of human perception not to negate violence, fear, failure, and stigmatization but to capture their sublime qualities for a new genesis. From fearlessness, "murky" coordinates codify "shifting geopolitical, moral, and legal axes" as well as doubt and risk to produce new assemblages with generative potential (Ben-Yehoyada 2012, 113). The minority incorporate hegemonic power and knowledge by way of hyphenation, rather than resistance, as might be expected or endorsed by common perceptions of *"what it is to be a minority in Western Europe"* (subjugated, powerless, fighting the system). They fearlessly connect domains of power and representation by employing figures such as Spanish knights, St. Michael the Archangel and the Madonna, and ancient Greek ancestors as messengers who bind the history and politics of minority identity. In Serresian eyes, "the miracle of tolerance" (Serres [1991] 1997, 80)—that is, the hybrid coexistence of good and evil, of parasite and host, of minority and national government—is a process of mutual learning that instead of producing a vicious circle of repetition provides momentum toward innovative solutions. This joy amid awareness of death— UNESCO lists the linguistic minority as "severely endangered"—is down to an openness to incorporate different versions of truth from across hegemonic categories of ascription.

Fearlessness, then, is what distinguishes Grecanici in their pursuit of proximity while maintaining distinction. They cross relations and tap into bodies of truth in a relentless pursuit of knowledge of what it is to be a minority in modern Europe. Their "sharp senses" allow them to artfully switch direction, to traverse a new passage between categories of representation, just in time to avoid death. Fearlessness is what defines their method of connection; not to flinch, to embrace brave new worlds, to add a symbiotic mishmash of sources as pedagogies to the category "minority." Moving between and across diverse scales and disciplinary genealogies provides the porosity key for members of a minority population to captain their own ship on the choppy waters of contemporary politics that sometimes pose an existential threat. The confident and daring quality of intrepid boldness, their audacity signals not return or recovery or doubling down on small-scale definitions of what it is to be part of a minority group but rather the challenging of mainstream understandings of the very concept of "minority" and piercing the rigidity of that accepted knowledge (see Serres [2004] 2020).

Grecanici pursuit of novel connection and their risk-taking in the face of existential threats can be read in parallel to Serres's discussion of an organism in a critical state, presented in *Times of Crisis* ([2009] 2014). When an organism is endangered owing to growing infection, the body automatically makes a decision on how best to adapt; it either cleanses itself of the offending material or symbiotically incorporates the new condition. This decision is fraught with danger since both rejection of the known and the incorporation of new material may result in death. If the crisis is survived, the body learns and takes an entirely different path when faced with the same scenario in the future. Instead of returning to its earlier state, which would "imply a loop-like return to the original course leading to crisis" (Serres [2009] 2014, xii), the organism remodels itself and finds a new route through a new connection or relation. A critical event "propels the body either toward death or to something new it is forced to invent" (Serres [2009] 2014, xii). A return backward is simply no good—the problem must be fearlessly challenged head-on since a recovery to the previous status quo would lead straight back to crisis. Porous movement across categories and relations, then, is the only way to survive, assembling as we go.

Fearlessness for Grecanici is a practice that cuts and forms new relations to help redefine widely accepted assumptions of minority life. Threats are part of everyday existence and must be navigated but can also be potentially harnessed and exploited to form a more desirable assemblage of relations. Bringing numerous sources of knowledge, different truths, and resources into close proximity—tapped as they are from the EU, civic associations, kin groups, state organizations, and the dark economy—allows Grecanici an existence outside the "minority" category as classically defined. Into their self-shaped category of sameness are incorporated a variety of truths, with hegemonic top-down versions of minority status often twisted, folded, and turned on their heads. Power and representation of the minority category as a top-down diffusion of ideas are thus reappropriated. But this is a dangerous assemblage in need of careful management since mixing together such diverse versions of truth ultimately makes a Molotov cocktail of powerful and often conflicting relations, always on the threshold of detonation.

Inviting Freud to the Conversation

The approach of constantly morphing proximity aids a more general reassessment of current Western political systems modeled on diffusionism with a center from which powerful global influences are sourced. Whether neoliberalism, democracy, the definition of minority status, or the very concept of civilization, diffusion from the West as blueprint to global power and influence places exploitation of social and natural resources at its heart. This single-source umbilical approach to a singular truth that the Grecanici so successfully undermine inevitably leads to a world of gasping aspiration, extraction, and, as Serres has put it, one-way parasitism (Serres [1980] 2007). As demonstrated by the Grecanici, bringing disparate sources of knowledge into proximity to be remolded as assemblages that serve local purposes seems to move toward addressing systemic vulnerability caused by a center-periphery diffusionist approach. Rather than one central universal truth to which all inhabitants of a category relate in the same way, with audacious pathfinding, knowledge can be twisted, molded, reappropriated. But in practice, proximity can also harbor resentment, swell competition, and provide too much similarity when resources are scarce.

Here, I would like to turn on proximity as a generally positive mode of connection, the untainted corridor to understanding and critiquing human similarity and difference. I earlier posed that despite perceived differences, all humans are proximate in their very humanity, connection is everywhere, and the hard/soft sciences would do well to build two-way bridges rather than exceptionalist categories. But what happens when proximity and sameness become thorns in the crown of everyday social relations and governing apparatuses? Too much proximity can be a bad thing, and, as Serres ([2019] 2022) says, sacrifices and scapegoats are often required before people potentially find peace with their own paths. I bring to the conversation Sigmund Freud and his concept of "the narcissism of minor differences" to consider alongside Serres's proximity and connection to form a basis to further pick at the sameness-difference binary. For Freud, alterity relies on the degree to which Otherness is or is not tolerated, which is of particular concern in minority studies. In the same vein, traits of minority identity, evoked from the top down (state to grassroots) or bottom up, may promote certain levels of (in)security for minority populations where fearlessness does not abound. Ethnographically following when (and when not) difference is invoked by governments, international policy groups, and local people helps unpack fundamental questions of proximity as premised on exclusive notions of belonging and reflexive engagement with the Other; Serres is "inclined to think" that "every evil in the world" comes "from belongingness" ([2003] 2018, 75). Unpacking these containers of truths helps further understanding of how the proximity of people that are "the same but different" can provoke "anxiety" rather than fearlessness as "the key emotional response to danger or threat" (Freud 1919, 236 in Murer 2009, 123).

Through numerous publications, Freud developed a concept to understand claims to difference arising between groups that share similar categories of identity, be it religion, ethnicity, or class. He noted that there is a strong tendency among neighboring states and closely related peoples to exaggerate their distinctiveness from each other in what he called the narcissism of minor differences (Freud [1930] 2010). The bottom line is that contrary to what might be commonly believed, similarities and not differences perpetually threaten each group's sense of identity; thus, each one clings to some small distinguishing marks, investing them with disproportionate significance. It is the commonalities between them that drive people to seek and create differentiation from one another, often through violence, in an attempt to manage "the endogenous unease in human society" (Figlio 2012, 8; Pipyrou 2021).

At this point, I would like to turn to my work among another minority group, Pontian communities in northern Greece, whose families were relocated from the Black Sea region of Anatolia to Greece as part of the forced population exchange implemented in the 1923 Lausanne Treaty. Religion being the agreed basis for relocation, at this time, Greece received almost 1.5 million Christian refugees and Turkey around 500,000 Muslims (Pentzopoulos 1962). Despite sharing a common religion with inhabitants of their new lands, the relocated populations spoke different languages and dialects and held their own varied customs, performances, and rituals. They were at once proclaimed to be the same ("Greek," "Christian," "brothers") and different, causing a sliding-scale effect between patriotism and antagonism. Not only was the newfound spatial proximity between local Greeks and the Pontian migrants problematic and they competed for limited natural and political resources, but also they experienced the same-yet-different proximity of an array of identity categories.

Although the national government recognized Christian refugees from Asia Minor as "Greeks" on the basis of religion, the reception of displaced people on the local level was often negative. The displaced peoples were often locally labeled as "Turks," "Turkseeds," or "leftists" (Kirtsoglou 2003), echoing Serres's lament that cultural exceptionalism is "bound to territories, to places, to climates" ([2003] 2018, 87). Subjected to the nationalistic processes of the Greek nation-state, Pontians felt the pressing need to belong to the national cultural corpus, and to this end, they engaged in selective remembrance and a reshaping of their identities as simultaneously privileged and disadvantaged members of Greek society. Over the following decades, Pontians reconstructed their collective system of representation by shaping categories of identity relating to national Greek history, language, and dance. In so doing, they engaged in historical constructivism from below, appropriating and engaging with Greek nationalistic history in order to claim proximity to a Hellenic narrative of belonging, otherwise reserved for "true Greeks" of perceived linear bio-cultural descendance (Zografou and Pipyrou 2011). Pontian civic groups simultaneously promoted Pontian identity traits and cultivated commonalities between Pontians and Greeks, thus critically influencing the ways in which proximity was historically and politically imagined.

On the one hand, there was the need to perform their proximity to the Greek nation and their new lands of residence, yet, for local people, the resemblance of these new inhabitants raised suspicion—they *were* and *were not* the same, and their proximity (in language, custom, and, suddenly, geography) was uncomfortable and was the source of a great deal of resentment. From a Pontian perspective, difference was at first carefully downplayed in order to fit and secure their place in reconstituted categories of belonging (Zografou and Pipyrou 2011; Pipyrou 2021). Then Greece came face-to-face with European modernity in the 1980s, when the coming to power of the socialist government initiated an era of political and cultural awareness. The then socialist prime minister, the late Andreas Papandreou, even went as far as visiting the monastery of Panagia Soumela (the par-excellence Madonna of the Pontians) as an act of publicly acknowledging the support that his party received in the general elections of 1981. This public proclamation of sameness, which included promoting the importance of the Pontian Madonna in the Greek national conscience, connected history, religion, and politics under a collective banner of shared struggle. Papandreou embraced the proximity of the resettled populations as important members of the electorate and as modern European citizens of Greece.

For decades after the displacement of the minority, the Greek state was opportunistically selective in recognizing the proximity of aspects of Pontian identity that were secure enough to be openly adopted. The Pontians were still a threat because they were "not really" Greek but closely related enough to be a political resource. At the heart of Greek nationalism is this opportunistic narcissism—the assumption that minority traits that share, in Freud's ([1919] 1973) words, an "uncanny" resemblance but that do not pose a threat to the nation can be adopted on the premise that the Greek nation-state will reap the political or economic reward. Narcissistic opportunism involves calculated strategies of branding minority traits as pan-national, but this only pertains to traits that are deemed secure enough not to pose a threat to categories of national homogeneity. An ontological principle of sameness—a fundamental feature of narcissism that allows the state to see the "Self" in the "Other"—acts as the driver behind nationalism, and so minorities unavoidably represent a disturbing challenge to the claim of a homogenous national truth. To contour this threat, nations such as Greece have developed their agendas through an operation of opportunistic narcissism: the process of underscoring minority differences, territorializing, and finally nationalizing them, sanitizing categories of minority truths, cleansing perceived threats, before incorporation. "Evil prowls," Serres tells us, along boundaries of belongingness, "ensuing from comparisons and rivalries they incite, it being roused by libido's heat" ([2003] 2018, 75–76).

This is where Freud and Serres converge. Since states see difference and diversity of relations as threatening, they require sanitation. The process of incorporating minorities into the national corpus is not one of symbiotic assemblage or porosity since difference is approached as pollution. Serres employs the term "malfeasance" ([2008] 2011) to describe the social and psychological pollution of the norms of the modern world. The narcissism of minor differences ejects a form of psychosocial pollution, an intolerance of difference, that taints the universal category of the human, of biological sameness. To follow the line of malfeasance, this cleansing is a morally dirty business, a way of claiming humanistic proximity between different groups by washing away unseemly cultural difference. This desire for cleanliness, for flattening at the expense of individual truths, is the real dirt of the modern world, says Serres, and constitutes a form of soft pollution (Henig, this volume). Minority difference is sanitized not to strike a boundary between the state and Otherness, but rather to create secure spaces within which the state can continually control difference while persisting to fantasize sameness and homogeneity. The narcissism exemplified by the Greek state in selectively claiming sameness/difference, proximity/distance, therefore involves the appropriation of threatening "minor differences" that could otherwise be magnified and turned into conflict between categories of belonging.

Conclusion

Thus we always confuse belonging and identity. Who are you? On hearing this question, you state your first and last name, and you sometimes add your place and date of birth. Better yet, you claim to be French, Spanish, Japanese; no, you aren't, identically, such-and-such, but, once again, you belong to one or the other of these groups, of these nations, of these languages, of these cultures. Likewise, you say that you are Shintoist, Catholic, Democrat or Republican; no and no, once again, you merely belong to this religion, to some political party, to some sect full of obstinate people. . . . So say your identity. The only truthful answer: yourself and only yourself. —Michel Serres, *The Incandescent* ([2003] 2018, 72)

For Serres, topology is the order of the day, and nowhere is this more apparent than in forms of proximity. Proximity can mean shared biology of the species, similar social concerns in diverse ethnographic contexts, assembling and connecting relations from disparate categories, and two or more labeled groups learning (or not) to live together. Proximity also needs porosity, for what is it to be proximate if not open to absorb and exchange knowledge?

Minorities often find themselves boxed by national and supranational governments and assigned and denied identities (and pathways to knowledge) according to ideals of the national corpus. Broadly speaking, both the Grecanici and Pontians have had to un-make knowledge of "minority" before rebuilding, foregrounding, and concealing knowledge selectively. This has taken two different trajectories. The Grecanici have fearlessly opposed hegemonic labeling and pursued overlapping and entangled routes to power from bodies as diverse as the EU, Greek/Italian states, kin groups, and illegal organizations. They have assembled an often-perilous cocktail of relations that provide representation by allowing Grecanici to be simultaneously proximate to seemingly competing truths. The dance of the Pontian flame has been to flicker along the very hyphen of connection, between preservation of difference rooted in a geographically and increasingly temporally distant past and gradual acceptance into a highly nationalist state trying to selectively prove its own belonging in Western modernity. The position of the century-long Pontian shuffle along the hyphen depends on geopolitical trends, existing in the seepages, cracks, and overlaps of categories. The folds and wrinkles in this "mass of laundry" are sometimes crisply ironed out, indicating seemingly flawless connection between minority group and hegemonic Other—a sameness based on shared humanity—while at other times the basket of clothes is thrown under the bed, out of sight, out of mind, and quickly gathering unseemly polluting dirt.

A reading of the turbulence of minority politics and fluctuating forms of proximity to human condition and social relations can benefit from a partnership between Serres and Freud, I suggest. Freud's narcissism of minor differences seems apt in identifying where and why proximity might become too much or undesirable. Operating as Freud does on an assumption of shared bio-social sameness (the trunk of humanity), narcissism of minor differences is located where the branches and twigs start to diverge from the corpus—at the points of bifurcation, one might pose. Such a conversation between Serres and Freud helps in simultaneously holding relative and universal truths in hand since the sameness of proximate humanity explodes into social categorization that is an inevitable tool of governance, sorting, and sanitization. The narcissism of minor differences is precisely the "proliferation of parrots" Serres refers to where universal truths are destroyed by social vanity and egotism; it is also where porosity starts to rot. It is in this fold that categories of minority, refugee, gender, or ethnicity gain traction. I find myself, the scientist from a staunchly nationalist nation retrained as anthropologist, in this fold where holistic human truth gazes across the gorge at narcissistic opportunism, both ready to race down the valley sides, Braveheart-style with swords and shields in hand, to meet in a messy swell of dangerous competing knowledges and just maybe some symbiotic novelty. For Serres ([2003] 2018, 43), behind the relativity of cultures, universality of corporal relations is revealed; "we become rediferentiated in myths and technologies, fashions and cosmetics, become species again by means of knowledge."

NOTES

1   Serres notes that almost at the very moment of discovery of DNA that gave hope of defining the human in terms of exactitude and generality, the natural started losing ground on the cultural to the point where "human nature" became negated. He rather accusatively allots the cultural description of acquired behaviors to anthropologists and ethnologists ([2003] 2018, 87).

2   Pontians, or Pontic Greeks, are not a minority per se, not officially recognized by Greek law and most readily classed as ethnically Greek. However, their refugee status when relocated in 1923, linguistic differences, and cultural variations have led to classification as a distinct subgroup. De facto, Pontians can be considered a minority in the vernacular sense, and it is this working definition I employ here.

11

COMEDIC TRANSUBSTANTIATION

The Hermesian Paradox of Being Funny among Stand-Up Comics in New York City

Premise: Serresian Comedy

In his essay on Molière's Don Juan (Serres 1982a, 3–14), Serres speculates that if Nietzsche takes Dionysus to be the father of tragedy, we might consider Hermes as *the father of comedy*. "Is he the god of the crossroads of thieves and of secrets," Serres wonders, "this god sculpted on milestones and adorned with such conspicuous virile organs who, like Psychopomp, accompanies Don Juan to Hell?" (Serres 1982a, 14). In *The Parasite* (Serres [1980] 1982, 211), we are told that "comedy is first of all a feast. One eats, speaks, speaks of eating, stops eating to speak, all amid the noise. Thus the passage from the material to the logical occurs." Indeed, "laughter is the human phenomenon of communication" (Serres 1982a, 14) where comedy captures the circulation of all things. Bound to inhabit the porous no-man's-land between different realms of codified structuration, Hermes consequently makes comedic use of all the tricks in the book to aesthetically organize and transactionally weave together otherwise incommensurable and tattered registers of information.

A Serresian exploration of comedy and humor begins with Hermes. In a social universe, Hermes and Don Juan, his "worthy heir" (Harari and Bell 1982, xxxv), operate as parasitical tricksters who disturb the frictionless circulation of information. Serres considers these disruptive operations as *variations on the tobacco theme* (1982a, 5,10).[1] In the first scene of Don Juan, the assistant, Sganarelle, declares that "there's nothing so fine as snuff. All the best people are devoted to it, and anyone who lives without snuff doesn't deserve to live. Not only does it purge and stimulate the brain, it also schools the soul in goodness, and one learns in using it how to be a true gentleman" (Molière 2021, Act 1, Scene 1). Although appearing as merely an object of pleasure that is casually enjoyed during feasts, tobacco introduces an ethics of exchange in social life. Sganarelle continues: "You've noticed, I'm sure, how whenever a man takes a pinch of snuff, he becomes gracious and benevolent toward everybody, and delights in offering his snuff-box right and left, wherever he happens to be. He doesn't wait to be asked, but anticipates the unspoken desires of others—so great is the generosity which snuff inspires in all who take it" (Molière 2021, Act 1, Scene 1). It is this moral script that is disturbed from within by Don Juan, Hermes's "worthy heir." Giving and receiving create ties of reciprocal obligation that obviously restrict individual independence. Don Juan cannot accept any limitations to his "theatrical sovereignty" (Braverman 1987, 76) that is so crucial for an aristocrat of his prominence, and he ends up taking without giving and giving in words instead of kind. Rather than confirming the laws of exchange that otherwise guide social life, Don Juan thereby introduces an alternative transgressive economy of desire and power that properly fits his "sovereign ego" (Braverman 1987, 76). He refuses to settle the account and is de facto out of the transactional "game" (Serres 1982a, 7). As a parasitical disruption, then, Don Juan is responsible for transforming a festive occasion into "comedy as a theatricalization of manners" (Braverman 1987, 73). Serres considers this process as "a passage from the material to the logical" ([1980] 1982, 211), and here I follow Braverman, who coins the transformation from table to theater, that is, "from manners to mode," as "comic transubstantiation" (1987, 73).[2]

In order for the exposition of giving and counter-giving to be a comedy, it has to be a feast (Serres 1982a, 13). As Serres ponders, "Who does not know that such feasts are only dramatic representations of gifts and remittances, only dramatizations of the law of exchange? Are we at the very birth of comedy?" (Serres 1982a, 5).

According to Serres, Molière's Don Juan can be considered as a dramatization of "the law of exchange," which consolidates a relational economy. In the end, things do add up, and the relational economy is preserved even through

Don Juan's parasitic disruptions, which accentuate the entrenched hierarchical relations that his sovereign ego feeds on. In Peter Hertz-Ohmes's illuminating discussion of Serres's and Deleuze's approaches to comedy and humor, the dramatization of match and counter-match is therefore portrayed as a conservative "form of cost-efficient circulation" (1987, 244) where the score is eventually settled and shared conditions of truth are confirmed. With Deleuze, Hertz-Ohmes distinguishes comedy from humor, which does not require consistency and truth in order to operate. Rather than significance, humor speaks from nonsense—not as an absence of sense but as "sense zero grade" (Hertz-Ohmes 1987, 245). It can be characterized as an emergent actualization that does not offer any directionality but exists through the sheer aesthetic enjoyment of its own "thereness." "Nothing unfolds," Deleuze claims, "but things happen sooner or later, everything at its own speed" (Hertz-Ohmes 1987, 112).

Considering this presumed tension between structural consistency (Serres) and emergent actualization (Deleuze), it may appear that Serres leaves us with a notion of comedy where a mischievous and parasitical playfulness potentializes a transactional system that can best be described as circulatory conservatism. By focusing on the mercantile ethics of exchange, as it were, Serres seems to suggest that comedy reflects a global structural aesthetics, whose dynamism emerges from the sequential repetition of a form that is continuously being disrupted from within. In her discussion of Serres's double affirmation of the global and the local, Hayles alludes to a consistent push in his writings toward a globalizing theory while equally emphasizing the fragmentary and turbulent singularities that resist such encompassment (1988, 4). Put somewhat differently, we could say that while Serres insists on the primary importance of the singular, he is also trying to develop a universal theory that "holds true regardless of local circumstances" (Hayles 1988, 4). This intricate scalar tension between the local and the global has been commented on in a number of works on Serres (e.g., Mercier 2019; Serres and Hallward 2003; Serres and Latour [1992] 1995; Watkin 2020). To many commentators, the question is obviously whether Serres succeeds in maintaining a universal that is not an immediate synthesis of the fragmentary and singular and, equally important, what the purchase of this double affirmation might be (cf. Watkin 2020, 86–87). In his seminal discussions with Latour, Serres insists that he wants to "describe global relations that are as fluctuating as those in turbulence" and, later, that he seeks to "compose, to promote . . . a syrrhèse. A confluence not a system, a mobile confluence of fluxes. Turbulences, overlapping cyclones and anticyclones, like on the weather map" (Serres and Latour [1992] 1995, 116, 122). To Serres, then, the relationship between the singular and the universal is not stable. While a globalizing concept may reflect totalizing tendencies, Serres indicates that its realization may amount to a congenitally failing operation. Let me therefore hypothesize that comedy is Serres's analogical vehicle for exploring the conceptual workings of the turbulent relationship between the local and the global, the singular and the universal. If global systems and universalizing theories operate not through the assimilation of local particularities but, rather, by way of a mobile juxtaposition or confluence of differential relations, could it not be that comedy and humor constitute an optimal conceptual space for charting their internal dynamics? And may it even be so that comedy and humor extend Serres's synthetic project by isolating the Hermetic inflection points where the relationship between the local and the global is perpetually disrupted? If so, comedy also functions as a topological space for capturing the scalar moves that allow the global and the local, the universal and the singular, to coexist as fluctuating forces that reject a complete and permanent synthesis.

In the following, I want to explore the relationship between (global) form and (local) content in relation to one unique comedic modality, which is stand-up comedy as it is being performed by up-and-coming comics in New York City. Even in its experimental forms, stand-up comedy is first and foremost a genre of entertainment that aims to get laughs from an audience through verbal performances (Mintz 1998). Its origins can be traced to vaudeville, burlesque, and variety theater, but it was during the 1950s that stand-up comedy was defined as a separate performance genre (Nesteroff 2015). Since then, it has gone through several cycles of massive popularity followed by periods of widespread public indifference and even scorn. The most recent "comedy boom" began around 2008 when the observational and somewhat absurdist comedy of that era was challenged by the return to a more personal—"confessional"—and anecdotal style embodied most prominently by comics such as Louis CK, Maria Bamford, and Patton Oswalt. Boosted by the expansion of global streaming services (Comedy Central, Netflix, HBO) and social media platforms (Twitter, Facebook, YouTube, and TikTok), stand-up comedy has since become a pervasive part of public life in the United States and several other parts of the world and is considered by many as an important medium for reflecting on and understanding the conditions of contemporary (and mostly Western) societies (Holm 2017). Today, many New York–based comics develop their comedic material through what is widely known as "confessional comedy," that is, subjective reflections about the hardship and challenges of their personal lives. At the same time as they labor to honestly convey their inward struggles, however, comics are also acutely aware of the importance of making their audiences laugh. As I will shortly show, in New York and elsewhere, contemporary stand-up comedy is therefore caught by a critical tension between form and content, that is, between the need for laughter and the need for addressing the dilemmas of social life. At the outset, we might say that the need for laughter invests the comedic form with a conservative progressive aesthetic, which advances from premise and setup to the eventual punch line through the expected surprise that occurs when it is revealed that the initial target assumptions have an endpoint that is (ideally) unanticipated by the audience. With Serres, I will consider the myth-like opposition (form::content) as a creative driver of stand-up comedy. Through the narrative aesthetics of the joke, a Hermetic passage emerges that allows for what might best be described as "comedic transubstantiation" of content to form. By way of the passage that is created at particular inflection points, the form comes to operate on the content as a vehicle of transformation. The outcome is, as I suggest toward the end of this essay, an exhaustion of the comedic form, which unsettles and may even liberate the humor from the confines of the joke.[3]

Serres famously insisted on the importance of allowing for confluences of symbolic meaning and scientific knowledge (Serres 1982b, [1991] 1997). Across his varied body of work, Serres would often use myths, fictions, and fables as narrative vehicles for his investigations of societal-cum-moral dilemmas and philosophical conundrums. "In some respects," he explained to Latour, "a well-told story seems to me to contain at least as much philosophy as a philosophy expressed with all this technical voluptuousness" (Serres and Latour [1992] 1995, 24). In *Genesis* ([1982] 1995), Serres introduces a form of inventive reasoning, which builds from the turbulent intermediary state between the solids and the fluids, the unitary and the multiple, the universal and the singular. "Chaos appears there, spontaneously in the order, order appears there in the midst of disorder" ([1982] 1995, 109). Returning to our point of departure, we could say that in Serres's writing, it is Hermes, the God of communication, who mythologically finds new passages across an intermediate turbulent space and weaves together otherwise disparate realms of knowledge. My ethnographic exploration of stand-up comedy is guided by Serres's suggestion that myths may contain fluctuating and complex forms of knowledge that go beyond the exact sciences (cf. Harari and Bell 1982, xxx). In the following, I will therefore analogically consider stand-up comedy "bits" in relation to mythic accounts, which share "with the sciences and philosophy the passages on the road to what it means to know" (Assad 2012, 94).

Setup: Confessional Comedy in New York City

The audience was still laughing at his last joke, but the host of the night's show continued unabated to introduce the next stand-up comic: "Put your hands together for Ian Fidaaaance!" The song "Under Pressure" by David Bowie played through the loudspeakers while Ian Fidance jumped onto the narrow stage, where he shook the host's hand and proceeded to take the microphone from the stand, which he moved to the right. As he walked a couple of steps back to the center of the stage, Fidance began his five-minute performance by addressing the audience:

> What's up? Hi everybodyyyy! (paces the stage) Hello! I look like drunk Richard Dreyfuss. Hello! Oh yes. Good to be here! Hello! (points at audience members in the front) Good evening! Hi . . . yes! Hello! Good to be here . . . Who here is depressed? Oh yeah? (makes karate moves) That's me fighting my depression! Hayahh! Get outta here, sadness! Yeah! I'm manic depressive, dude. I'm up, I'm down, I'm swinging around. What's up? I don't know . . . (smiles) Yeahh! I'm not outrageous. Sometimes you feel like a nut. Sometimes . . . you don't get out of bed in June, you know. Yeah, it's hard to admit, man. Who here likes cheese? Cheese? (high-pitched cheers from the audience) Yeah? (points out over the audience several times) Mozzarella? Mozzarella? Yeah, you fucker! You ever take a ball of mozzarella and just bite into it like an apple, huh? (imitates holding a huge ball in his right hand close to his mouth) Yeah? (smiles broadly and nods; points out over the audience) That's a fun way of letting people know . . . (points out over the audience) . . . you're depressed. Yeah, we figured it out! We're all fucked up, man! (makes a quick forward toss with his head) I ain't the only one! Now we figured out what we do to fix it. What I like to do . . . I like to do little stuff like . . . my phone depresses me so what I do is I turn that off and . . . this is crazy: What I like to do is I like to go outside and (makes quick forward toss with his head) talk to other people! Can you imagine? (points at an audience member in the front and smiles broadly) It's nuts, dude! (shouts) Oh my god! I love it! Eye contact and head nods! (imitates sticking a syringe into his arm) What's up! Oh yeah! Holding the door for a stranger (imitates snorting a line of coke) WOW! (shouts) Yes! It gets my nipples hard, man! (pauses; smiles and moves hand across mouth) But . . . (holds up open hand toward audience) I'm gonna be honest . . . If I hold the door for you and you don't feign just like (lowers his voice and holds the tip of

his thumb and index fingers on right hand a few centimeters apart) a little bit of gratitude . . . (pauses and laughs; continues to smile but the smile looks fake) . . . well, I'm gonna murder you! (nods and smiles broadly) Yeah, that's right! The other day I held the door for this dude. He made eye contact (points two fingers at his eyes and makes quick sweeping gesture with his arm to the side) with me and did nothing else! I lost my shit, Jason! (looks at the pianist sitting to the right side of the stage) I lost it (shouts) and . . . that's right, I confronted him. Yup, me! Ian, your old pal, the hero! That's right! Except . . . the thing is . . . when I confront people, I kind of panic and my voice raises five octaves. (lifts his right hand over his head) So, like, in my head (points to his forehead) I wanna be like (points toward the audience) "Hey man, that was pretty rude!" But what came out of my face was (shouts in high-pitched voice): "You're not the only one in this world!" Yeah . . . Why . . . why did that happen? Yeah . . . He thanked me and someone was like (points to the side) "Yeah, that lady is right!" "Yeah, that Italian woman with the mous- tache, . . . she's correct!" And I was like (with Italian accent) "ababadibub grazie" (laughs and paces the stage). (long pause) I love my mom. . . .

This excerpt is from one of Ian Fidance's regular performances at the leg- endary Comedy Cellar stand-up comedy club in New York City in the fall of 2017. At the time, I had known Fidance for several months without ever having considered his physical resemblance with the American actor Richard Dreyfuss. But as soon as Fidance made the analogy, there was no going back. With his over-the-top positive energy, somewhat erratic and seemingly unco- ordinated physical movements, curly dark hair, thick moustache, and heavy glasses, Fidance did indeed project a drunk Richard Dreyfuss.

There was an eerie honesty to the performances of Ian Fidance and many other comics who I met during the seven months that I carried out ethno- graphic fieldwork in New York City in 2017–18. Depending on the format of the show, a stand-up performance—a set—gave the comic anything between five and fifteen minutes to make the audience comfortable with their stage persona and produce a sufficient number of laughs. Still, despite the clearly defined parameters of the comedic performances, I was often left with a strange feel- ing of having experienced something genuinely truthful and almost intimately revealing about the comics. By way of often silly and even absurd anecdotes and one-liner jokes ("I just broke up with this guy a little while ago. He was a lot older than me, but I didn't really notice the age difference until he started bathing me in the sink"), many comics seemed to give the audience access to an intimate realm of their personal life, which exposed with brutal clarity their weaknesses and fears, insecurities, and social awkwardness.[4] At the same time, comics working in stand-up clubs were driven by a supreme objective of having to make their audience laugh. New York comics are widely known for their economy of words, where they get to the punch line as fast and efficiently as possible in order to maximize the number of jokes in a set.[5] Watching the comics perform at the city's comedy clubs, bars, and restaurants hosting comedy shows or at "open mic" venues, it was difficult and often impossible to determine whether it was their uncompromising willingness to expose seemingly intimate and personal truths or the capacity to make an audience laugh that was the secret to their unique art form.[6] While eating a late-night bowl of dumplings in a small Chinese restaurant on the Lower East Side, Kate Bigalow, who is now based in Los Angeles, described her ideas of truthful comedy:

> The whole goal of comedy is to be as truthful as possible. . . . And the only way to do that is to find truth in your own personal life. To be as open and honest as possible. It's fucking hard. Looking in the mirror and being like "Oh, shit, I'm a bad friend. I'm falling short in my daily duties of being a function in human society. Why? Why is that? Where can I find a humor in my shortcomings?" If you think people should be sterilized at the age of eleven because we're in a fucked-up world, why is that? What experiences led you to believe this wild thing?

To many comics, this credo of honest and truthful comedy requires complete transparency. There are no personal flaws or shortcomings that could not potentially be used on stage if that offers the audience a deeper, funnier, and more meaningful experience, which they might carry with them long after the show has ended. With Serres, who suggests that "transubstantiation, the sharing of bread become flesh" is the optimal theory on the collective bond ([1982] 1995, 92), we could argue that the transformation—or transubstantiation—of honesty into humor constitutes the making of the comedic collective that momentarily connects comic with audience.

"Act-Out": The Hermesian Paradox of Being Funny

With Hermes as our mythological "weaver of spaces" (Serres 1982b, 46–50), in this section, I try to pinpoint the precise inflection points at which the form of a comedic narrative or joke begins to operate on the content as a vehicle of internal transformation. In an early essay, Serres considers a structure (or form) as "a set of undefined significations . . . grouping elements in any number (elements the content of which is not specified) and relations, in finite number, the nature of which remains undefined, except for their function regarding the elements" (Serres 1968, 22 in Mercier 2019, 4).

In contrast to the post-Saussurian structuralism of Lévi-Strauss (1955, 1970), which operates from the differential magnitude (content) of related sets of social and cultural meaning, Serres's methodology highlights the formal relations between elements. According to Watkin (2020, 45), in Serres's structuralism, an analysis is structural only insofar as "content is prevented from taking the role of a structure."

In order to allow for an ethnographic engagement with Serres's ideas about mythic meaning-making and thereby investigate the relationship between form (global) and content (local), I propose a transformational structural analysis that straddles the divide between Serres's algebraic methodology and central influential post-Saussurian approaches. As also indicated by Clayton (2012, 43), there are interesting overlaps between the two: although Lévi-Strauss focuses on the unpredictable transformations that occur within a myth, ". . . in its actuality, there lurks beneath this surface appearance of a self-similarity of (virtual) structure." As I will shortly suggest, then, based on a generalized structural grid, we may come to detect the precise inflection points at which form begins to operate on content and thereby disrupts the scalar relationship between the universal and the particular, the global and the local. While this (admittedly tentative) approach is deeply indebted to Serres, it is operationalized through a reworking of the transformational structuralism promoted by Lévi-Strauss and other key scholars. In previous writings, I have suggested that such structural forms can be detected in numerous stand-up comedy performances (Nielsen 2018, 2019a, 2019b). For this ethnographic investigation, I will use Ian Fidance's opening bit from his Comedy Cellar set that was previously introduced.

Schematically, the sequence of the bit can be considered as two consecutive movements (see figure 11.1) that are preceded by an initial introduction ("I look like drunk Richard Dreyfuss . . .") and followed by a finale that transitions to the next bit (pause and then "I love my mom . . ."). The first movement focuses on Fidance's struggles with manic depression ("I'm manic depressive, dude. I'm up, I'm down, I'm swinging around. What's up? I don't know . . ."), which he resolves by engaging in random social relationships ("What I like to do is I like to go outside and talk to other people!"). The second movement is structured around Fidance's response to being ignored when trying to reach out and help other people ("If I hold the door for you and you don't feign . . . a little bit of gratitude . . . well, I'm gonna murder you!") and the eventual ridicule that is prompted by his over-the-top reaction ("He thanked me and someone was like

'Yeah, that lady is right!'"). The main target assumptions of the bit are established, problematized, and resolved in the first movement. Fidance describes the hardship of living with manic depression, which involves periods of almost catatonic paralysis and a recurrent need to do seemingly absurd things simply to feel alive. And he then suggests that connecting to other people might be a viable way of dealing with these serious challenges. So while the first movement revolves around Fidance's continuous struggles with manic depression, it is also a collective story about the power of social relationality to relieve personal pain. In this sense, the bit finds a logical conclusion with Fidance's expression of excitement about reaching out to other people ("Yes! It gets my nipples hard, man!"). Instead of transitioning to a new bit, however, Fidance moves from an *extrospective* and collective discussion of how to deal with manic depression to an *introspective* and individual examination of his moral qualms about not being recognized for trying to relate to other people. We could say that the second movement becomes a commentary to (or even a shadow version of) the first movement, which inevitably ends up with Fidance as the butt of the joke embodying the female character that was suggested to him/her by some spectator having witnessed the unfortunate incidence.

What holds these consecutive movements together as a coherent comedic bit, I will argue, are two inflection points at which the internal logic of the performance is parasitically disrupted and eventually exhausted by the form operating on the content. Both of these moments seem to be woven into the narrative cloth, but in their consequence, they come to work against the comedic performance almost as if having instigated an aesthetic incongruity between form and content (cf. Oring 2016). The first inflection point happens after Fidance has described the paralyzing effect of manic depression ("Sometimes . . . you don't get out of bed in June, you know"). Apparently, this insight is taken forward into a new set of reflections ("Yeah, it's hard to admit, man") but only to be disrupted by an awkward question posed to the audience ("Who here likes cheese?"), which immediately destabilizes the sensitive mood that Fidance had otherwise established. The question is posed without setting up a clear premise or outlining the contextual parameters, and so the listeners are invariably caught by surprise almost as if they were being presented with an absurd punch line to the previous joke. The second inflection point happens when Fidance's story of confronting other people turns into a series of one-liner jokes about his "hidden" persona, an Italian woman with a moustache ("He thanked me and someone was like 'Yeah, that lady is right!'"). Similar to the first inflection point, a sudden shift of tonality occurs, which does not fit the introspective examination of moral self-righteousness, this time by the bystander's reaction

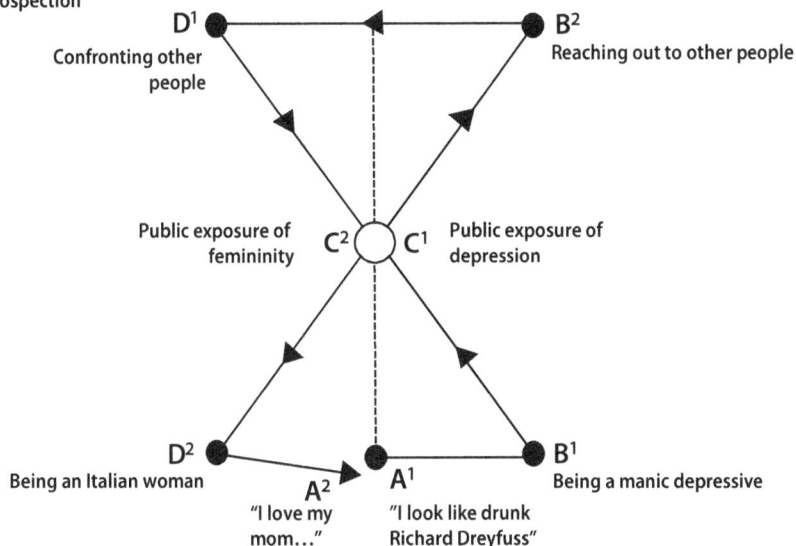

A: Point of departure
B: Extrospection
C: Inflection point
D: Introspection

D¹
Confronting other
people

B²
Reaching out to other people

Public exposure of   C² C¹   Public exposure of
femininity   depression

D²
Being an Italian woman

A²
"I love my
mom..."

A¹
"I look like drunk
Richard Dreyfuss"

B¹
Being a manic depressive

FIGURE 11.1. Comedic transubstantiation in Ian Fidance's stand-up performance at the Comedy Cellar (by Morten Nielsen)

to Fidance's burst of anger. In terms of the narrative's overall arc, this change of focus that leads toward Fidance embodying the suggested persona ("And I was like 'ababadibub grazie'") offers no resolution to the moral dilemma that was introduced at the beginning of the second movement. Consequently, the audience is momentarily stranded with Fidance's unresolved personal qualms until he transitions into a new bit ("I love my mom . . .").

Carefully shielded by the structure of the joke, we detect a tropic modality actualized at two critical inflection points that does not operate in terms of the conventional progression of setup and punch line. It seems to me that this tropic modality somehow introduces a transformative—transubstantive, we could say—logic that runs counter to our expectations even of how a surprise should be orchestrated within a (conservative) comedic format. Serres reminds us that Hermes, the trickster God of communication, is located "in the most profitable positions, at the intersection of relations" ([1980] 1982, 43), from where he reconnects different systems of meaning. Could it be, then, that it is at the critical inflection points that we first detect the coming into being of a new albeit fluctuating and inherently unstable swerving of significance? In

Genesis (Serres [1982] 1995, 67), Serres coins such a minimal form of differentiation a "surge."[7] We might imagine, Serres tells us, a high-pitched sound that barely lifts itself out off the background noise of the wind. It is not a singular sound in and of itself, but it has already established the conditions for its own transformation into a "tiny little meaning, a local concept" (Serres [1982] 1995, 67). In order to expand on the Serresian idea of a "surge" emerging at critical inflection points within the comedic structure, I need to briefly leave the discussion of Fidance's comedy bit to introduce relevant myth analysts, who have investigated the processual-cum-differential play of tropes.

Tagging the Joke: Tropic Differentiation

In *Structural Anthropology*, Claude Lévi-Strauss tells us that "mythical thought always progresses from the awareness of oppositions toward their resolution" (1963, 224). As Lévi-Strauss shows in his detailed exegesis of American myths, however, mythic narratives make promises that they cannot keep. Transposed into a narrative form that figures the principal dilemmas of society through a dramaturgy of counteracting cosmological forces, sequences of internal contradictions continue to be generated until the desire for resolution is definitively eliminated. The myths that Lévi-Strauss discusses involve various forms of asymmetric pairs and pairings, which are in a continuous state of dynamic disequilibrium. In *The Story of Lynx* (Lévi-Strauss 1995), for instance, the workings of semiotic contradictions are reflected in the analysis of impossible twins in the myths of the Salish-speaking peoples of the Pacific Northwest.

Based on his ethnographic research among the Daribi people of Papua New Guinea, Roy Wagner has developed a particular understanding of such mythic paradoxes that emphasizes the "processual form" of tropic constructions (Wagner 1979, 32; 1981; 1986). In Daribi mythic narratives, series of consecutive symbols act upon each other as innovative elaborations on a conventional background. Previous semiotic relations are supplanted—or *obviated*, to use Wagner's term—by new and self-contained relations in a continuous sequence of transformations that only closes when the narrative returns to its point of departure. In *Symbols That Stand for Themselves*, Wagner concludes that "a myth, then, is an expansion of trope, and obviation, as process, is paradoxical because the meanings elicited in its successive tropes are realized only in the process of their exhaustion, and exhausted in that of their realization" (1986, xi).

At the moment of tropic exhaustion, that is, when a conventional semiotic relation has collapsed, the new symbol emerges as a holographic and self-contained image that "stands for itself" (Wagner 1986, 2001; see also Dulley 2019). It recursively encompasses the thing that it symbolizes and thereby effectively produces its own tropic grounding before eventually sedimenting as a conventional semiotic relation (and the transformational cycle repeats itself).

Returning again to the analysis of Fidance's performance at the Comedy Cellar, it is interesting to compare the two critical inflection points that hold the consecutive movements of the comedic bit together with the sequential transformation of tropes that we find in some mythic narratives. As I have suggested, in stand-up comedy, we find an irresolvable tension between form and content, which seems not to be articulated through the progression from setup to punch line. To recall: At the critical moments when Fidance exposes his severe manic depression and, later, his dormant femininity, an internal contradiction is intensified of the form (the surprising and discontinuous shifts of tonality) working against the content (the honesty of his public confessions). Already at the first inflection point (C^1), this disruptive contradiction invests the bit with a progressive force that allows Fidance to move beyond what would otherwise be the logical conclusion of the bit, namely the moment when he has "resolved" the dilemma that structures the extrospective (first) movement (B^2). As we move further inwardly toward the "single master metaphor" (Josephides 1991, 156) of the performance, the second inflection point (C^2) exaggerates the contradiction that has already been established by irrevocably undermining the reliability of the narrative. Similar to the initial inflection point ("Yeah, it's hard to admit, man"), we are first led to believe that Fidance will offer some reflections on the implications of his own actions ("Yeah . . . Why . . . why did that happen?") but that narrative trajectory is immediately impeded by the intervention from some bystander ("Yeah, that lady is right"). In fact, although the narrative at this point is infused by the turbulent and disruptive energy released at the first inflection point (Serres [1982] 1995, 24), Fidance could probably have produced a viable resolution to the dilemma of the second movement (his problematic moral self-righteousness) by finishing the joke then and there. But through his acceptance to embody the suggested persona ("ababa-dibub grazie"), the narrative logic is recursively disrupted at the moment when his femininity is publicly exposed (C^2).

We could therefore tentatively argue that comedic transubstantiation offers a processual form of tropic construction that analogically resembles some types of mythic narratives. The comedic bit engages in a "never-to-be-resolved dialogical whirl" (Wiseman 2007, 173), where problems are being introduced without providing definitive solutions. As embryonic "tiny little meanings"

(Serres [1982] 1995, 67), they remain as almost imperceptible fluctuations or surges from which new forms may eventually emerge. In mythic narratives, insoluble existential oppositions or contradictions are defined in terms of metaphorical equivalents from a cosmological realm with central figures or terms acting as mediators, such as "Asdiwal, the earth-born master of the hunt," who is introduced in Lévi-Strauss's "The Story of Asdiwal" (2010, 16). In stand-up comedy, particular problems are worked out with the performer as the mediating agent, who allows for some Hermetic passage between the principal contradictory terms. In the present discussion, the oppositions that are established in the comedic bit (e.g., the need for social relationality versus exaggerated sense of self-righteousness) are subordinate to or structured by the principal guiding contradiction between form (need for laughter) and content (need for honesty), which is what eventually ends up disrupting the internal consistency of the bit. Following from this, it seems reasonable to suggest that the functionality of the sequence of oppositions in stand-up comedy is analogue to the processual form of cosmological tropes in mythic narratives. In both cases, the transposed terms are metaphorical equivalents to the societal or existential problem that the narratives (unsuccessfully) seek to work out.

Inspired by Serres's reflections on the Hermetic "intermittent" emergence of new forms of fluctuating meaning, however, I think that there is something else going on in stand-up comedy. Rather than simply reaching an inconclusive finale where the insoluble mythic paradox is acknowledged, contradictions are intensified and deepened almost as if the comic is performatively digging his or her own comedic grave. Whereas mythic narratives advance through a sequence of tropic transformations where one semiotic relationship "dies" and is "differentiated" into something new (Wagner 1979, 24), the principal contradiction that is activated at critical inflection points seems to gradually acquire more solidity through the series of incongruous oppositions that the joke makes use of in order to move from setup to punch line. This comedic life beyond the moments of principal contradiction, as it were, is given support by the performer's insistence on the pertinence of an aesthetic form that is increasingly (albeit comedically) problematic: *Is the alleged manic depression for real? Did he actually approach a stranger?* As we shall see in the last section, this perseverance of the principal comedic contradiction is both the strength and potential problem of stand-up comedy. Moving toward a conclusion, I will therefore suggest that the fluctuations of the principal contradictions may also point toward an ethnographic Serresian analytics beyond (or parallel to) Serres's own thinking.

Misdirection: Liberating the Humor from the Joke

If stand-up comedy can be considered as the presentation of a series of contradictory relationships that reaches its endpoint with the punch line, it is strikingly similar to the sequence of transformations in myths where a degree of closure is reached when the narrative returns to its point of origin in a transformed state.[8] In *The Fire of the Jaguar* (Turner 2017), Terence Turner makes a comprehensive analysis of the processual structure of one myth on the origin of fire among the Brazilian Kayapo people. By considering the structure of a mythic narrative not as an invariable relation that connects different contents but as "a demiurgic power of self-creation" (1991, 155), Turner convincingly shows how the Kayapo myth ultimately becomes a dynamic scheme for dealing with society's internal contradictions in relation to the necessary socialization of its members. In Fidance's stand-up comedy bit, we end up with a series of reflections on the "qualitative mathematics" (Lévi-Strauss 1954) of the one and the many for someone struggling with the challenges of social relationality. When the myth returns to its point of departure and, equally, when the final punch line has been delivered by the comic, we are left with unique insights into particular existential or societal dilemmas that have been mapped out and progressively explored. In both instances, the narrative seems to be held in check by the need for closure. Looking closer at the stand-up comedy bit, however, it is less in check than what might be assumed. With Serres ([1982] 1995), we could say that the punch line is less an endpoint than it is a moment of turbulence from which new patterns and relations may potentially arise.

In a stand-up comedy bit, the principal contradiction (here between form [laughter] and content [honesty]) is deepened rather than obviated. Consequently, if we consider the comedic bit as successive movements held in place by crucial inflection points, the punch line does not function as the definitive response to some initial target assumptions. It is rather the ultimate signpost that the aesthetics of the comedic bit has exhausted itself through the increasing fluctuations that disrupt the relationship between form and content. The comedic bit therefore does not resolve the principal contradiction that it has introduced, and it also fails to articulate—through the now exhausted form— that it is incapable of delivering a finality to its own narrative dilemma. In effect, the transformational work of stand-up comedy does not necessarily end with the concluding punch line. If the joke could play a joke on itself, it would be that the punch line liberates the humor rather than fences it in.

What might the effect be of a joke that never manages to reach its point of destination? With Serres, we could hypothesize that if the initial setup is not resolved by the punch line, it is Hermes, the "god of the crossroads" (1982b, 14), who is no longer capable of weaving connections between disparate knowledges. I will suggest, then, that a comedic dilemma that is not resolved by the punch line, and therefore effectively escapes the confines of the joke, endures as a vehicle of continuous metaphoric transformation but without the means to articulate what domains it might bring together.

Peter Young is a New York–based comic whose material centers on contemporary political issues. During an interview shortly after the 2016 presidential election, we discussed whether political jokes might have an impact on the audience's sense of civic engagement. "I often doubt that," Peter told me. "Trevor Noah said in an interview I heard recently that his audience often feel that they have been politically active simply by laughing and then carrying the sense of 'hey, something is wrong' around with them. But they rarely do anything at all."[9] Maybe this is the major paradox of comedic transubstantiation. While it may implant in the listener an acute sense of transformative agency, it cannot articulate what the object of this transformation is.

Punch Line

In his inspired essay on Molière's Don Juan, Serres acknowledges the crucial influence of Marcel Mauss: "Now open The Gift. . . . There you will find match and counter-match, alms and banquet, the supreme law which directs the circulation of goods in the same way as that of women and of promises; of feasts, rituals, dances, and ceremonies; of representations, insults, and jests. There you will find law and religion, esthetics and economics, magic and death, the fairground and the marketplace—in sum, *comedy*. . . . But could we ever have read Molière without Mauss?" (Serres 1982a, 13).

According to Serres, comedy is a theatricalization of the law of exchange—the transaction of gifts during a feast, for instance, which ends up confirming a conservative circulatory system. By shortcutting the sequence of equal transactions, then, Don Juan parasitically rejects the rule of reciprocity (alike for alike), which nevertheless results in a further potentialization of the system. As such, comedy allows for a universal overview, "a god-like presence that assumes the integrity of the various exchanges in terms of their determining structural casualties" (Hertz-Ohmes 1987, 245).

In this essay, my ambition has been to ethnographically probe Serres's analytical insights about comedy. But I have done so by slightly extending the parameters of his analysis. It seems to me that Serres approaches comedy straight on. He investigates the social universe that is so beautifully (and co-medically!) brought to life by Molière in order to make a series of general in-ferences about the relational infrastructure that it is built upon ("What does one do at a feast if not to exchange?" [1982a, 6]). In so doing, Serres engages his conceptual arsenal (parasitic disturbances, Hermetic circulations, the ten-sion between system [global] and practice [local], etc.), which effectively makes Molière's comedy of Don Juan into a drama about Serres's philosophical proj-ect. Moving beyond Serres's own inferences about Don Juan, I have therefore suggested to consider Serresian comedy, as it were, as a topological space for staging the tenuous relationship between the global and the local, that is, between the drive toward a universalizing theory and a simultaneous and careful attention to the singular forces that resist such encompassment. In order to ethnographically test this idea, I have moved laterally from Serres's own work to my empirical material on stand-up comedy by asking: What would a version of Serres's mythodological structuralism look like if it were to be applied to ethnographic data? This involved an explicit attempt at stay-ing loyal to and operationalizing Serres's two interrelated analytical tenets that myths go beyond the limits of scientific knowledge in providing insights about the world and that structuralism is the methodology for eliciting these (Webb 2012, 54).

We may rightfully ask ourselves to what extent the comedic transubstan-tiation of form onto content constitutes a confessional and inherently real sacrificial exclusion of intimacy for the sake of laughter. Are inwardness and personal privacy sacrificed by the comic in order for the audience to be mo-mentarily bound together in the enjoyment of "all against one"? To be sure, the eversion of all imaginable personal information during a comedic performance does indicate a Foucauldian "confession of the flesh" (Foucault 1980), as it were, where the comic surrenders his or her stage persona to the disciplinary paradigm of organized unity that is enacted through a popular cultural event. But in this regard, it seems that the stand-up comedy performance has a final joke to play on itself, for it is never quite clear whether the sacrifice of intimacy is real.

As Ian Fidance and many other comics performing on the New York com-edy scene seek to truthfully expose their intimate personal weaknesses, it is the stretching of the porous relationship between form and content that gives their performances a unique comedic vibrancy. In a nutshell, as audi-ence members, we constantly wonder whether their personal anecdotes will actually be funny. Did the comic *really* mean what was said from the stage?

Is stand-up comedy *really* about being as honest as possible? It is the tension between comic and comedy, performer and performance, in other words, that exaggerates the internal complexity of the bit and ends up challenging the coherence and reliability of the narrative. As I have tried to show, what happens is a decoupling of the narrative's form from its content so that the former appears to operate on the latter as a vehicle of transformation. It is through this process of comedic transubstantiation, I will argue, that the terms of the narrative, its meaningfulness, are brought into question but without the narrative losing its authority. In fact, the very questioning of its substance is the structurally motivating factor for the continuation of the narrative. When stand-up comedy allows for comedic transubstantiation to define its performative aesthetics, it potentially becomes a highly efficacious communicative vehicle for creative (moral) disruption, the only downside being that it does not also produce a target. In effect, the audience is stranded with a tension between form and content that cannot be resolved by the punch line. The global form has (at least ideally) allowed for a number of laughs caused by the deviations away from the setup by the punch line, but it has also disrupted the audience's appreciation of the comic's quest for honesty.

In her wonderfully thoughtful discussion of Serres's reading of Levy's *The History of Rome*, Maria L. Assad emphasizes that Serres does not simply dwell in the "endless violence inherent in the triumph of order" (1991, 288). By insisting on the importance of the "excluded third," Serres uncovers how the seeming order of Rome contains within it an "ocean of multiples" that may allow us to "refill our 'fountain' of possibilities" (Assad 1991, 288). Taking our cue from Assad, we may speculate that although comedic transubstantiation seems to suggest the crystallization of a paradigm of organized unity, what we end up with is an "ocean of multiples" that is constantly threatening to overflow the systematicity of the narrative aesthetics of the stand-up performance. What is being sacrificed is not intimacy onto laughter but, rather, the consistency and mutual contingency of both.

Returning to Serres's analysis of comedy, we end up with a different and perhaps more volatile understanding of Hermes, the messenger God, than what was initially presented to us in the essay on Molière's Don Juan. By tracing his parasitic movements across a comedic structural terrain that is constantly being destabilized by the fraught relationship between the global and the local, Hermes now appears to be always at risk of losing the capacity to operationalize the "most profitable positions, at the intersection of relations" (Serres [1980] 1982, 43). While he may be the "reconnector of an explanatory system—myth" (Harari and Bell 1982, xxxii), this potent capacity does require that fragments and knowledges can be synthesized into clusters of meaning. A scalar differentiation between system and element, in other words. However, when a global form exhausts itself and comes to work on the local content, it is revealed to us that the relational power of Hermes is never over and above the turbulent chaos across which he tries to find new passages.

What may Serres's "philosophy of circumstances" (Herzogenrath 2012, 9) mean to ethnographic explorations of the porosity of social life? How may we productively engage with Serres's insistency on the importance of a synthetic weaving together of different and often discordant knowledges and imageries? In this essay, I have suggested to laterally operationalize Serres's ideas and to remain loyal to his main analytical principles. Consonant with the Hermetic credo of always listening to the "cacophony of turbulence below" (Assad 1991, 281), however, to remain loyal to Serres will invariably mean to allow for other turbulent surges to arise. By tracing the exhaustion of a narrative form widespread among stand-up comics in New York, I have therefore also—tentatively—alluded to a productive exhaustion of a Serresian analytics. What lies beyond? Hermes will show us.

NOTES

1   In the English version of Molière's Don Juan that is cited later, *tobacco* is translated as "snuff" (Molière 2021).

2   The doctrine of transubstantiation was based on the assumption of the essential reality of conceptual categories. The "type-essence" was considered as *substantia* whereas the perceptible and sensuous aspects that distinguish one thing from another were called *accidentia*. Hence, "the idea of transubstantiation is that in the consecration of the elements the *substantia* change but the *accidentia* remain the same. The *substantia* of the bread and wine become the *substantia* of the body and blood of Christ. The *accidentia* remain the same, and the *accidentia* are all that remain of the original bread and wine" (Barclay 1967, 72). Inspired by Braverman, I consider comic transubstantiation here as the comedic dramatization of form onto its content.

3   Humor research abounds with often contradictory definitions on what constitutes a joke. Here I follow Sherzer, who simply considers a joke as a short narrative that ends in a surprising punch line (2002, 38).

4   Joke made by New York–based comic Rachel Feinstein at the Comedy Cellar, March 2, 2017.

5   Los Angeles is another key center for stand-up comedy in the United States. In contrast to the efficiency of words that characterizes New York–based comics, Los Angeles is known for its "act-outs," that is, physical performances where comics act out the characters and voices in the jokes.

6 The "open mic" comedy scene is an umbrella term covering all the stand-up comedy shows throughout the city where anyone can perform on stage either for free or by paying a small fee.

7 For a recent apt discussion of Serresian surges in relation to the "anti-predictability of breaks in space and time," see Knight (2022).

8 Roy Wagner alludes to this relationship himself: "The point of closure in a myth—the punch line of a joke" (Wagner 2010, 106, in Dulley 2019, 67)

9 Trevor Noah is a South African–born comic who hosted *The Daily Show* on Comedy Central. I have since—unsuccessfully—tried to find the interview that Peter mentioned. Peter also could not remember where he heard it.

12

MICHEL SERRES, WISDOM, ANTHROPOLOGY

Introduction

Anglophone anthropologists (and other social scientists) have made something of a habit of looking to the European continent, and particularly perhaps to France, for theorists. Levinas, Foucault, Deleuze, Latour . . . in each case, the name as perceived from within anthropology stands for a body of theory and/ or a bundle of concepts. These theorists are understood to fertilize anthropological thinking from the outside, bringing terms, notions, and concepts that can be borrowed to make sense of ethnographic material.

Adding Michel Serres to that list challenges the formal pattern, forces us to rethink what the move of looking to "theorists" is about, for Serres's work, as the introduction to this volume makes clear, is peculiarly difficult to treat in the way we normally treat "theory." His writing, as has often been pointed out, is enthusing, fascinating, and difficult in equal measure. Serres has occasionally been deployed as a provider of concepts, in the manner of other continental philosophers—his coinage of "the parasite" (Serres [1980] 2007) has been particularly significant in this regard as a number of contributions to this volume attest. And yet he himself insistently refuses to describe his own work as a matter of concept creation, preferring to express his thought through "characters" (Thumbelina, the hominescent, Hermes, angels; cf. Serres, Legros, and Ortoli 2016) that resist the kind of transportability and applicability that we usually look for in the work of those we designate as theorists. One can find in Serres's works extremely sophisticated and technical arguments that prefigured some of the main advances in the social studies of science and can provide radically new ways of envisaging a range of topics, and yet much of his writing seems to flow with the unsystematic verve of literary evocation, deliberately shunning the standard apparatus of academic discussion. There are other paradoxes too: Serres's deep-rooted suspicion of authoritative discourses alternates with his readiness, as this volume's introduction puts it, "to quote Christian scripture at the drop of a hat"; his deep commitment to working from and through examples and cases is belied by relatively frequent bursts of what looks very much like grand-historical and cultural generalization.

Faced with this dilemma, one solution is to treat Serres as a theorist anyway—either by cherry-picking borrowable concepts and leaving the rest or by engaging the corpus as a whole and deploying the gargantuan second-order work required to produce something like a systematic account of Serresian thought (see, for instance, Watkin 2020).[1] It is not clear whether this is something Serres himself would have welcomed. Not unlike his contemporary Michel Foucault, when Serres provided something like a summation of his work, this was in the fluid and ever-renegotiable form of dialogues and interviews (Serres and Latour 1992; Serres, Legros, and Ortoli 2016). But thought is there to be used, and there is of course nothing wrong with seeking to "theorize" Michel Serres, either wholesale or piecemeal. There are, however, alternatives.

One alternative, suggested by the editors to the present volume, is to think of Serres as an "informant." The present chapter suggests another alternative. What if, instead of thinking of Serres's writing as "theory," one were to approach it under the rubric of "wisdom"? The suggestion might raise hackles—isn't wisdom a woolly, evaluative term that has no place in serious academic discussion? Serres himself seems to think otherwise, as we shall see, and enjoins us to take wisdom seriously. As it happens, this is something that anthropologists have in fact started to do, as I will outline in the second part of this chapter. Indeed, suggesting that we understand Serres's writing as a form of "wisdom literature" (cf. Kaufmann 1996) might help us not only in approaching Serres but also in clarifying the comparative stakes and the potential traps and pitfalls of the category of wisdom, currently a thorny subject for anthropology. The second part of the chapter gives a whirlwind tour of some key themes in an emergent comparative anthropology of wisdom—what I call below a *comparative sophiology* (cf. Faubion 2020). These themes include the fraught relation between wisdom and knowledge, the interplay of exemplarity and exception in characterizations of wisdom, and the paradoxical tension between wisdom, difference, and "culture."

This foray into the comparative anthropology of wisdom—a very different project to Serres's, of course—will be used to distinguish two angles from which Serres's own writing might be understood as wisdom literature. The first is the substance of his proposals for approaching the pressing ecological, geopolitical, and technological concerns of the late twentieth and early twenty-first centuries. The second is to look for formal characteristics of wisdom in the nature of Serres's method and mode of thinking. The former—Serres's self-conscious invocation of wisdom as a corrective to Western modernist hubris—is familiar in ways that anthropologists might find uncomfortable. The latter—what one might think of as the methodological structure of Serres's wisdom—echoes some broader intuitions that emerge from the comparative anthropology of wisdom. These commonalities help us make sense of some fundamental ways in which anthropology itself is often conceptualized by its practitioners. Reading Serres's work as a form of "wisdom literature" helps us tease out and think critically but also constructively about anthropology's own investment in that genre.

Wisdom according to Michel Serres

Bruno Latour (1987, 97) notes that Serres's notably "difficult" writing stems in part from the fact that it rests on the assumption that readers are already familiar with the material Serres is commenting on. Serres doesn't introduce or summarize and is pointedly allergic to citation (Watkin 2020, 163). As a result, the reader who doesn't already master Lucretius, La Fontaine, or mid-twentieth-century mathematical paradigm shifts (or indeed all of these) is likely to find some passages of Serres deeply obscure. Other passages will seem, for the very same reason, misleadingly transparent—it is easy to miss a depth of reference in Serres's offhand mentions of texts, theorists, or arguments.

Serres's engagement with Stoicism is one such case in point. In the final interview with Latour in the book *Éclaircissements* (1992, 243–94), titled "Wisdom," Serres articulates his vision of philosophy as (returning to its etymological roots) the search for wisdom. Serres makes a single passing reference to the Stoic philosopher-emperor Marcus Aurelius (Serres and Latour 1992, 253), which readers unfamiliar with the writings of the Roman Stoics might easily sweep past as just another instance of Serres's fabled eclecticism. Yet that single mention is a clue to the fact that the whole interview in fact begins with a deep engagement with and rethinking of the fundamentals of Roman Stoic philosophy. Readers familiar with the corpus will have pricked up their ears at Serres's earlier claim that the key to wise conduct has always been the fundamental "distinction between the things which depend on us and the things which do not depend on us" (Serres and Latour 1992, 246).

This contrast that Serres begins from and that he then re-elaborates, as we shall see later, is a technical distinction fundamental to the particular period of Stoicism associated with Marcus Aurelius and Epictetus: distinguishing between that which is within our power (*ta eph'hemin*) and that which is not within our power (*ta ouk eph'hemin*) is fundamental to the Stoic practice of wisdom.[2] What depends on us, for the Roman Stoics, is our inner attitude to external things and events. Although sometimes located in "the mind," this inner space is more precisely understood by the Roman Stoics as the "ruling principle" (*hegemonikon*) as constituted by three fundamental capacities: judgment (*hypolepsis*), desire (*orexis*), and impulsion (*horme*). Everything else—in Epictetus's memorable summary "our body, our property, reputation, office, and in a word, everything that is not our own doing" (1989, 483)—does not depend on us.

In their starkest and most intentionally paradoxical pronouncements, Roman Stoics sometimes cast this distinction as one between absolute inner freedom and total external necessity. Stoic wisdom would thus be reduced to an injunction to modify your internal experience of a world you cannot in any way control. As Pierre Hadot (1992) shows, however, the Roman Stoics deployed their core distinction in practice in a way that was far more subtle and psychologically realistic than some of these extreme pronouncements might suggest. On Hadot's reading of Roman Stoicism as enjoining a realistic way of life, the core point was that we can of course effect some changes in the world, care for our body, our property, our office or reputation, yet the outcome of our efforts is always ultimately in the hands of fate. Conversely, while most people cannot, at the drop of a hat, fully and straightforwardly control their reactions to external events, quell their fears, and redirect their desires, these are things that can be worked on and developed. Wisdom, therefore, consists of the proper application of our efforts: honing our judgment on the distinction between what is within our power and what is not so that we can resist the appearance of external goods and ills, orienting our impulses toward actions that are just and good for the community while accepting patiently and even welcoming those things that are out of our control. By contrast, for the Stoics, most people direct their efforts incorrectly—their desire is aimed at the outcome of their actions and at things that are entirely out of their control, their impulses are left to roam and range without particular purpose, and their judgment accepts every semblance of good without proper discernment.

In a characteristic mash-up of the ancient and the (post-)modern, Michel Serres takes the basic Stoic distinction between that which does and does not depend on us and reads it through one of the recurrent themes of his writing: the hubristic expansion of (Western, modern) science and technology in the twentieth century. The problem, as Serres articulates it, is that this expansion has pushed back the bounds of necessity and seemingly extended the sphere of liberty indefinitely. Whereas Marcus Aurelius—given the limited effectiveness of the medicine of his time—could conceive even of his own body as outside his control (Serres and Latour 1992, 253), Western moderns have to reckon with the fact that their technological advancements have brought not only bodies but the continued survival of life on Earth itself within the sphere of things that, for better or ill, depend on us. "The old saying changes and becomes: *everything depends on us or will depend on us one day*" (Serres and Latour 1992, 248).

The result, Serres argues, is "a curious and important reversal of the image of the wise" (Serres and Latour 1992, 245), for what may have struck the first generations of moderns as an immense extension of their freedom and the final defeat of necessity is now revealing itself as in fact quite the opposite. Necessity reappears precisely where (for the Stoics) freedom used to lie: in our own powers of action, judgment, and decision. If everything now depends on us, nothing can be left to fate. Our fate is, in effect, to be responsible for everything: "As we dominate the planet, we become accountable for it; as we manipulate death, life, reproduction, the normal and the pathological, we become responsible for all this. *We will have to decide about everything*" (Serres and Latour 1992, 252, original emphasis). The Stoic distinction between that which depends on us and that which does not depend on us is in effect abolished or, rather, turned inside out: "*It no longer depends on us that everything depends on us*" (Serres and Latour 1992, 250, original emphasis). Our freedom is also our necessity. The wise, in this "reversed" image, are not free to retreat into their virtue and leave fate to its own devices. Rather, the wise must now rely on the powers of science and technology to guide and shape the world—their freedom to do so also necessitates that they do so. Is and ought, duty and fact, have been fused: "Why must I behave like this and not otherwise? So that the Earth may continue to exist, so that the air may remain breathable, so that the sea remains the sea. . . . *Duty equals fact*" (Serres and Latour 1992, 255).

These pages contain one of the most explicit accounts of the underlying logic of a tension that readers can detect throughout so much of Serres's oeuvre, namely his simultaneous admiration for and distrust of the power of science and technology. When it is stated so clearly, however, Serres's position raises some questions and concerns, some of which Latour articulates in the exchange. Surely, Serres is exaggerating the purported "mastery" of modern science or technology? And who, after all, is this "we" to whom Serres is constantly referring? What about the many people(s) of the world who have little if any control over the conditions of their life? To these questions, Serres responds by pivoting from generalizations about modern humanity to a binary vision setting on the one hand powerful Western moderns and on the other the "almost universally miserable world":

> The exponential growth of misery, famine and illness, new and residual, which ravage the third and fourth worlds, mark the return or persistence of the old necessity. Those, to whom I belong, who inhabit the shining core which trails and multiplies these many miseries—like the tail of a comet—are accountable for them and seek this wisdom. This is the second responsibility, the new obligation, the new conditions resulting from our own actions, the last blow against the collective narcissism of the rich nations. . . . Masters of the Earth, we are building an almost universally miserable world which becomes the basic, objective data of our future. (Serres and Latour 1992, 256–57)

There is much that is problematic here, as I outline later. But let us continue to follow Serres for a moment. This second kind of responsibility, he argues, requires another kind of knowledge alongside the scientific and technical knowledge discussed previously. The humanities are important, he argues, because they represent "the continuous cry of suffering, the multiple, universal expression, in every language, of human suffering" (Serres and Latour 1992, 261). Without an attentiveness to this testament of human frailty and tragedy through the ages, experts and deciders shaped by science alone and its hubristic short-term powers would become truly "inhuman" (Serres and Latour 1992, 264).

In sum, wisdom according to Serres rests on two pillars: on the one hand, the post-Stoic—or rather anti-Stoic—reversal of the relationship between freedom and necessity forces the philosopher (the one who would be wise) to attend to and to take seriously the facts and powers of science and technology.[3] On the other hand, the philosopher must look to the humanities for a corrective to the hubris of modernist knowledge and power. In a set of striking oppositions that take shape in the rest of the interview, Serres elaborates a vision of wisdom built from the crosswise light shed by science and the humanities, which stand respectively for power and frailty, knowledge and suffering, the young knowledge and the old (a theme Serres also develops in Serres [1982] 1997, 33ff), confidence and tragedy, the short and the long term.[4] The philosopher, the wise decider, poised between the knowledge of science and the attention to suffering born of an encounter with the humanities, must make the weak their "elective other." Philosophy, for Serres—which is to say the search for wisdom—is tasked with bringing the humanistic witnessing of human suffering in the court of absolute scientific knowledge.

As often in Serres's writing, lyrical prose hides or perhaps transfigures an underlying technical precision (Latour 1987). Here, grand statements about wisdom articulate the logic of a very distinctive set of methodological devices that Serres uses throughout his work to disrupt and reengineer the relationship between scientific and humanistic knowledge (many of these have been detailed by Watkin 2020). But the spirit, style, and affective charge also matter. Reading these pages, it is easy to see why Serres became such an important figure in science studies. Beyond the obvious ways in which Serres's work inspired Latour's articulation of Actor Network Theory (2005), the interdisciplinary sensibility Serres articulates here also dovetails neatly with the type of vision entailed by Donna Haraway's search for a situated feminist objectivity that doesn't refuse science but rather brings the care and concerns of the humanities into the very heart of scientific controversies (see, for instance, Haraway 1988, 1989, 1997). This vision and its various avatars and mutations in the tradition of feminist technoscience (Stengers 1993; Despret 1996) have also been profoundly influential on the anthropology of science (see Candea 2023). Serres's particular vision of wisdom at the confluence of science and the humanities has many broader echoes for anthropologists as well. Not all of these echoes, however, are comfortable.

The Problem with Wisdom

Serres has more to say about wisdom, and I return to it later. But I will pause at this point to register a discomfort that I suspect some readers will already have felt. Serres's vision of privileged Westerners as "Masters of the Earth," ultimately responsible both for the ills of the world and for seeking the wisdom to remedy them, is likely to raise some eyebrows at least. A critic might say that far from puncturing the collective narcissism of the West, this vision rather comforts it. Supremely powerful, supremely accountable, and first in line in the search for wisdom, the West once again steals the show even as it makes of the Rest its object of care and concern.

This is one of those moments when, as the editors of this volume suggest, one would be well inspired to think of Serres as an informant. However much one might disagree with his description here, it surely echoes a pervasive Euro-American mythological sensibility. More pointedly, one might suggest that Serres is articulating here the affective core of an orientation that anthropologists have not themselves sufficiently guarded against in their own appeals to comparisons between "us" and "them" (cf. Candea 2016, 2019). This feeling of seeing anthropology's own mythologies as in a distorted mirror is perhaps strongest when Serres—as an illustration of what is required of a would-be sage—embarks upon a lengthy evocation of his own personal travels through every country and station: "I have known and loved Korean men and Japanese women . . . I have loved because I knew them in their own places, Africans from the North, Center and East . . . peasant, I have worked in the fields, labourer, I have worked in the factories . . . I have rubbed shoulders with ambassadors and nuns . . . I have even visited the Indians of South America . . ." (Serres and Latour 1992, 270–71).

This breathless personal tour of humanity issues forth in a general characterization of the fundamental unity in diversity of humankind: "in sum, unhappy, and statistically, generally, globally, essentially, ontologically, objectively *pitiable*" (Serres and Latour 1992, 271). Wisdom, argues Serres, is a human universal (hence "homo sapiens"; Serres and Latour 1992, 243, 269–70) because it is rooted in this suffering, weakness, and frailty, universal characteristics of the human condition (Serres and Latour 1992, 270). The passage is also intended to characterize a sort of moral education of the wise, "the sage who knows but who pities, [who] belongs not only to our time, the period when the winners, producers of reality and of men, play the game of 'winner loses all,' but rather belongs to all of human time and history, since it is weakness which makes time" (Serres and Latour 1992, 272).

This passage makes for particularly uncomfortable reading because it condenses—one might almost say caricatures—some temptations that anthropology has faced throughout its history. This might seem paradoxical. After all, Serres's universalizing vision of suffering humanity and his religiously inflected appeal to "pity" might seem to find few direct resonances in anthropology—except perhaps in the "anthropology of suffering" as memorably characterized by Robbins (2013). But what is much more familiar is the ambiguous move through which the Western intellectual goes elsewhere to find a corrective answer to Western modernity gone wrong. The move is intended as humble and self-critical: wisdom is elsewhere. But Serres makes (unintentionally) evident the unpalatable aftertaste of that move: a heroic, self-aggrandizing view of one made wise by an experience of the difference of others. Few individual anthropologists would claim the kind of universal experience Serres evokes in this passage. Each instance of ethnographic enlightenment might seem to be about particularity and difference, not universality. But this is partly an effect of scale. The discipline as a whole, as a summation of these individual efforts, can seem to come uncomfortably close to "saying" something much like what Serres is claiming of himself.

These temptations are particularly evident when anthropologists themselves approach the question of wisdom. Johannes Fabian once argued that "primitive" was a category, not an object of Western thought (Fabian 1983, 18). Much the same might be said for "wisdom" and anthropology. With a few notable exceptions I discuss later, anthropologists have not extensively explored "wisdom" as an object of analysis, comparison, or elaboration. As a mostly implicit category of anthropological thought, however, wisdom has loomed large throughout the history of the discipline.

The figure of wisdom is often lurking behind anthropological moves to destabilize "Western" or "modern" epistemological, cultural, and political assumptions by confronting them with non-Western alternatives—the form I have described elsewhere as "frontal comparison" (Candea 2016, 2019). There is something rather uncomfortable about making this move explicit, as in the title of Maybury-Lewis's volume *Millennium: Tribal Wisdom and the Modern World* (1992)—and that discomfort is in itself telling, as I will argue later. Yet the implication is unmistakably there in classic works such as Sahlins's *Stone Age Economics* (1972), in which the wisdom of an "original affluent society" is counterposed to the folly of capitalist modernity, or in Pierre Clastres's influential portrayal of Amazonian "stateless" societies as pragmatic anarchists who have wisely eschewed the trappings of the centralized state (1990). Some anthropological work that counterposes non-Western understandings of and care for the environment to the unreflexive rationalism of "metropolitan" science partake of this sensibility (see, for instance, Cruikshank 2005).

In these ways, throughout much of the history of the discipline, anthropologists have been beholden to and have also fed the vision of "wisdom" as something out there, a corrective to modernist hubris that can be found in distant times and places. This is itself, of course, a distinctly "Western modern" vision of wisdom with at least part of its roots in nineteenth-century romanticism and that has also been given a striking new coat of paint by discussions of degrowth and environmental catastrophe. It is, at heart, the dynamic entailed as well in Serres's discussion of wisdom earlier—even though Serres himself is not foregrounding comparison as his core device. The discomfort one might feel at titles such as Maybury-Lewis's speaks to this sense that the search for wisdom, however well meaning, might be structurally linked with anthropology's heritage of othering and "denial of coevalness" (Fabian 1983). It is not therefore surprising to find anthropologists who explicitly reflect on wisdom, such as Kao and Alter, seek to guard against "the discursive colonization associated with patronizing primitive society and their folk wisdom" (Kao and Alter 2020, 13). Yet one might be forgiven for worrying about a similarly patronizing implication when one of the contributors to Kao and Alter's volume claims that "a unique and globally pivotal task seems to have fallen upon today's ethnographer: that of becoming a spokesperson for apparently peripheral wisdoms" (Van Binsbergen 2020, 184).

One typical anthropological solution, of course, is to double down on this othering dynamic by using ethnography to demonstrate that wisdom itself can be something other than what "Westerners" believe it to be. For instance, Clark Chilson (2020) in Kao's volume argues that whereas Western understandings of wisdom today tend to be fundamentally oriented toward the self and the question of knowledge, Japanese *chie* is oriented toward social relations and the question of discretion and emotional sensitivity. But however far one leans into it, the dynamic of frontal comparison remains somewhat self-defeating: after all, the vision that wisdom is not what the West believes it to be is itself a classic Western vision. Indeed, readers of Serres will see how easily the putative contrast drawn between "West" and "East" can be encompassed by a narrative of the West's own internal duality—between the knowledgeable sciences and the socially and emotionally intelligent humanities, for instance.

This leaves anthropologists who would try to think comparatively about wisdom with a sharp set of difficulties that Kao and Alter articulate but do not fully resolve—in part because, as I have begun to suggest, the problem is structural:

> To locate the emplacement of wisdom, we ask ourselves: Where exactly is wisdom? Is it in culture, in our heads, in our bodies, in public representations? The danger in showing these cultural accounts is the implication that wisdom is something fixed and universal, and that all cultural anthropologists have to do is to comb the earth for wisdom's different ornamentations and expressions. These cautionary words are not meant to pander to cultural relativism, nor to take anything away from the search

for "wisdom" across human societies and history (see also Curnow 2010; Katz 1999; Maybury-Lewis 1992; McConchie 2003; Radin 1957). Rather, they show the importance of recognizing that wisdom is constituted by differing ideas of the world and of the person, which can be very different from conventional understandings arising from Western cognitive science. (Kao and Alter 2020, 12)

Kao and Alter in this passage and elsewhere seem to tack back and forth between the impulses of universalizing and relativizing wisdom—much like Serres himself, in fact—without finding a comfortable place to rest. This familiar problem cuts particularly deep in the case of wisdom because of a fact that the authors note but whose significance they do not perhaps sufficiently highlight: wisdom's peculiar relationship to "culture" in the Euro-American imaginary. The authors cite approvingly the observation of evolutionary biologist Jeffrey Schloss that "[wisdom] is a quaternary heuristic that is both contained within and partly free to transcend—even oppose—the cultural heuristic. This capability is perhaps reflected in the Pauline exhortation to 'be not conformed to this world, but be transformed from within by the renewing of your minds'" (Schloss 2000, 166; quoted in Kao and Alter 2020, 19). This intuition rejoins the fact that in many if not most of the contexts in which anthropologists have sought to describe "wisdom," as we shall see later, this is characterized locally as a matter of exception and personal excellence, one that marks certain people out from others in their society.

There is thus an implicit tension between wisdom and culture (understood as a complex of common practices, conventional attitudes, and shared experiences) in the conceptual tradition in which anthropology and its comparative visions emerged. In seeking to think comparatively about wisdom, in seeking to situate different wisdoms in different cultural contexts, anthropologists are faced with the problem that the vision of wisdom they are often implicitly working with is defined precisely through its ability to break with common notions and to transcend mere convention. Like "nature," albeit in a different way, wisdom (in a certain Euro-American "conventional understanding" from which anthropology itself stems) is an other to culture. Like nature, therefore (Strathern 1980; Viveiros de Castro 1998), wisdom puts distinctive pressures on a comparative imaginary that still uses as its key operator something like "culture" (e.g., globally varied and locally shared "ideas of the world and the person"). Wisdom needs to be more than conventional. This is why appeals to "folk wisdom" or "popular wisdom"—however well meaning—so often sound patronizing and demeaning. On this implicit view, wisdom which permeates a whole culture is as meaningless as the fabled town where "everyone is above average."

Added to these knotty comparative problems is the difficulty—so clearly illustrated in the preceding passage from Serres—of where the commentator is placed in such articulations of wisdom. The intuition that wisdom is "elsewhere" has a particularly close link to the value of self-critique, be it individual or collective: if believing oneself to be wise—and therefore exceptional—is a sure sign of hubris and folly, realizing that wisdom is elsewhere and operating one's (individual or collective) self-critique in relation to it is, conversely, the first sign of wisdom. The dynamic is, however, paradoxical, for as soon as one takes up this pivotal position—whether as "spokesperson" for the "peripheral" (Van Binsbergen 2020) or as the one who connects and straddles every human register, as in Serres's preceding passage—one seems to be describing oneself as wise and thereby revealing one's hubris. In either case, the "self" in self-critique seems to have shifted slightly away from the self of the commentator.

Conversely, and to be fair to Michel Serres, this detour through some of the inherent difficulties of an anthropology of wisdom suggests that where Serres's writing seems jarring, this is at least in part a structural feature of the very attempt to evoke wisdom and place oneself in relation to it, in a certain Euro-American conceptual tradition to which Serres and anthropology are both (despite their self-critical bent) still beholden. The problem comes with the territory as it were, and it would not be fair to single Serres out for it.

Comparative Sophiology

One route out of some of these paradoxes is, building on Kao's observation that "wisdom is constituted by differing ideas of the world and of the person," to draw a distinction between the variability of these ideas and the variability of the wisdom(s) they inform or constitute. Anthropology has a well-established language for thinking about the former—epistemological and ontological variability. To this it would be useful to add another term, coined by James Faubion (2020): *sophiology*. Just as comparative epistemology seeks to describe not merely what different people(s) know but what they understand knowledge to be, comparative sophiology describes not people's different access to wisdom but rather their different understandings of what wisdom is. These sophiologies certainly entail and are constituted by epistemologies and ontologies, which may themselves be familiar or unfamiliar. But sophiologies are not reducible to these epistemological and ontological differences. Ethnographic accounts of different understandings of the person, the body, or the nature of knowledge in any given setting do not in and of themselves necessarily tell us whether something like "wisdom" is a locally relevant category and what its content may be.

To clarify my argument here it might be useful to compare and contrast it to Robbins's famous argument about the need for an anthropology of the good (2013). In order to understand what "the good" might be in any given setting, Robbins argues, one must attend to often deep and uncanny ethnographic variability—but this variability doesn't negate the broader fact that people everywhere have some sense of "the good," some positive horizon of striving. I am making an analogous move in detaching the question of ontological and epistemological variability from the question of how people envisage wisdom (what one might call sophiological variability). I do so, however, with the caveat that people everywhere do not necessarily have "wisdom" as a relevant category or horizon of striving. "The good" in its generality and abstraction may be a convincing candidate universal (I leave that to others)—"wisdom," it seems to me, is rather too specific to play that role. The failing of some anthropologies of wisdom has been precisely that of trying to find it everywhere. In that sense, the comparative sophiology I am interested in here is less like a comparative anthropology of the Good as such and more like a comparative anthropology of certain more modest "goods," such as fasting (Fadil 2009), feasting (Humphrey 2015), or hospitality (Candea and da Col 2012): relatively concrete positive horizons that may be central and highly elaborated in some settings but not really a matter for much concern or interest in others. The question of whether hospitality is (in some form) a human universal may be interesting in its own right. What interests me as a social anthropologist, however, is what hospitality is and does in the contexts in which it is a matter of highly reflexive concern. The same goes for wisdom.

Clark's preceding discussion of Japanese *chie* is one instance of comparative sophiology. In the same vein (but perhaps less self-consciously concerned with a frontal contrast to "Western wisdom"), Keith Basso's wonderful book *Wisdom Sits in Places* (1996) provides an elaborate account of the form, structure, and implication of Western Apache wisdom, based on his own ethnographic observations but also crucially on the glosses and explanations of Dudley, a key informant, concerning "*'igoyá 'í*"—which Dudley himself translates in English as "wisdom." Similarly, Peristiany (1992) investigated the meaning of wisdom in mid-twentieth-century Cyprus by building on his ethnography and on accounts given by key informants of the nature of *sophrosyne*, which speakers of classical or modern Greek have no trouble translating as "wisdom." Cook's elucidation of distinctions made by Thai Buddhist nuns between wisdom and knowledge (Cook 2010, 96–116) might also be said to belong to this genre.

In all of these cases, "wisdom" is scaffolded by very diverse ontological and epistemological assumptions. Apache *'igoyá 'i* entails a detailed theory of mind focused on mental "smoothness" and an understanding of the formative power of specific "places as durable receptacles and the knowledge required for wisdom as a lasting supply of water resting securely within them" (Basso 1996, 134). Progress in wisdom is effected in great part through the repeated and mindful inhabitation of a landscape full of particular places, landmarks associated to particular stories that exemplify wise habits of mind. Wisdom, like water, "sits in places" and soothes the mind. Like water, you cannot live without it, for wisdom is fundamentally an instrument of "survival" (Basso 1996, 134): "Stated in general terms, the Apache theory holds that 'wisdom'—*'igoyá 'i*—consists in a heightened mental capacity that facilitates the avoidance of harmful events by detecting threatening circumstances when none are apparent" (Basso 1996, 130).

Much of this discussion would have seemed strange to Peristiany's informants from mid-twentieth-century Alona in Cyprus. In Peristiany's account of Alona, the *sophron* (wise man) is primarily situated and bound by community (and gender), not mnemotechnic landscapes. While "prudence and commonsense are not the preserve of the wise" (Peristiany 1992, 114), a *sophron* is first and foremost a social role, one of mediator and fixer who acts to resolve individual strife "for the community as a whole." As a person can have only one best friend (*bisticcos*), so a village can have only one *sophron*, who is the *bisticcos* of the village (Peristiany 1992, 114). The wise, in this formal sense, has to be structurally independent from existing hierarchies and ties—thus, a priest, ensconced in a church hierarchy, fits the bill rather poorly. So does a man younger than sixty, the age at which one begins to hand over matters of property and family management to younger kin. Far from being focused on the achievement of personal survival, *sophrosyne* is associated with renunciation. Indeed, a key characteristic of the status of *sophron* is that it should ideally follow on from a life lived in full enjoyment of sexualized masculinity (the *sophron* would ideally have been "*archatos*, heavy-testicled, as a young man" [Peristiany 1992, 115]) and a certain measure of local notoriety and power. *Sophrosyne* comes with a renunciation of these vital pursuits and advantages. Thus, Peristiany reports the saying: "Our old men have the wisdom of the cicadas. Having had their fill of everything they now sing 'hosannas on an empty stomach' [a reference to the meager diet of dependent elders]" (Peristiany 1992, 122).

While places are a vehicle for wisdom in Basso's case and sayings are such a vehicle in Peristiany's case, in Cook's ethnography of Thai Buddhist nuns, wisdom is closely linked to a contrast drawn between two languages: Thai and Pali (Cook 2010, 102ff). Pali, unlike Thai, is understood to speak not only to the brain (*samong*) but also directly to the mind/heart (*jai*)—the sixth internal sense base in Buddhist philosophy, located not in the head but in the chest. Pali chanting can thus be a vehicle for wisdom even for those who do not understand the meaning of the words. Rather, wisdom is conveyed through enjoining practitioners to work on their *jai* in order to register the effects of Pali sermons—in a way that is consistent with the broader telos of mindful meditation: "It is through the cultivation of mindful attention that the practitioner comes to 'hear' the wisdom of the Pali words—a form of understanding that is not exclusively executed through conceptual clarity" (Cook 2010, 104).

In just three examples, we can see a variety of ways in which wisdom can be understood to interface with very different ontologies and epistemologies of human bodies, minds and desires, places, language, and gender and social forms. And yet a number of key similarities traverse these and other ethnographic attempts to characterize sophiologies—people's diverse understandings of wisdom. I will come in a moment to what conclusions we might draw from these similarities—suffice it to say for now that I am not claiming these are universal features of wisdom, but something rather more complex. Let us characterize these, for now, as a set of "intuitions" that emerge from these ethnographies.

One set of intuitions circles around distinctions drawn in all of these studies between wisdom and (merely theoretical) knowledge. The phenomena identified as wisdom in the preceding ethnographies all share a number of features in this respect. Firstly, in all of these studies, *wisdom is practical as well as conceptual*. "Mere knowledge pertains to 'professors,' a most derogatory term. . . . Knowledge joined to the capacity to apply it is a constituent of wisdom. Pure knowledge is an ingredient of foolishness. . . . [Wisdom is] manifest in our daily dealings as a combination of transcendent enlightenment and practical virtuosity" (Faubion 2020, 157).

Closely related is the intuition that *wisdom works in cases and exempla, not rules*, or at least in the skillful interplay between cases and rules, the particular and the general: "One of the main criteria of the sophron's wisdom consists in applying general principles to particular cases in such a manner that the general is not seen to bend so as to serve practical ends. . . . The good doctor, say the Aloneftes (and the sophron is a doctor for social ills), uses as his starting point a particular patient's illness and then determines the cure" (Peristiany 1992, 105, 110; see also Højer and Bandak 2015; Faubion 2020, 164; Kaufmann 1996). The "places" in which Apache wisdom "sits" are also, in this sense, cases, or exemplars: mnemotechnics for striking stories: "The knowledge on which wisdom depends is gained from observing different places (thus to recall them quickly and clearly)" (Basso 1996, 20).

This in turn relates to a set of intuitions about the expression of wisdom. In its verbal articulation, wisdom is often presented through *a paradoxical interweaving of clarity and difficulty*: as we have already seen from in the preceding examples, wise speech is often either deceptively simple, hiding deep implications in seemingly straightforward metaphors, or articulated through Delphic paradoxes that require interpretation and unpacking, places that require repeated memorial inhabitation, a language that one must work to hear properly.

Associated with this bundle of insights is another set that matches wisdom to persons. If wisdom is often so hard to express abstractly, it is because it is, first and foremost, *exhibited in exemplary conduct* (cf. Bandak 2015; Robbins 2015; Zagzebski 2017), in "a certain kind of poise" (Kao and Alter 2020, 11)—wise speech in this sense is merely an instance of wise conduct. "As Apache men and women advance farther along the trail of wisdom, their composure continues to deepen. Increasingly quiet and self-possessed, they rarely show signs of fear or alarm. More and more magnanimous, they seldom get angry or upset. And more than ever they are watchful and observant. Their minds, resilient and steady at last, are very nearly smooth, and it shows in obvious ways" (Basso 1996, 139).

Wisdom is difficult to master and requires effort, apprenticeship, and personal transformation. "Disciplined mental effort, diligently sustained, will eventually give rise to a permanent state of mind" (Basso 1996, 138). Wisdom "is gradually revealed in speech and action consistent with the image of a selfless and judicious man who applies these qualities to the welfare of his community" (Peristiany 1992, 114). "It is only through the cultivation of the ethical self that such knowledge becomes possible" (Cook 2010, 104).

A final set of intuitions relates to the exceptionality of wisdom and the wise. Wisdom is, in almost all of these ethnographies, a matter of exception and personal excellence, one that marks certain people out from others in their society. This is directly related to the fact that wisdom, as I noted previously, often involves breaking with or at least taking a partly external perspective on locally accepted notions, approaches, and ways of being, common to other members of the society in question. This may set one up as a local arbiter and mediator or, on the contrary, cut one away from the everyday life of the surrounding society—or in some ways both. In any case, wisdom is often locally unconventional just as the wise are locally exceptional. This is true even in Peristiany's account despite the central concern with social structure and cultural totality derived surely at least in part from Peristiany's own theoretical interests. The *sophron* may be "the unquestioned mouthpiece of public opinion" who "consolidate[es] communal belief in shared ideals" (Peristiany 1992, 114). And yet this is precisely because or insofar as he is able to take a partly external position on these ideals and to step outside the fray in all of the ways Peristiany describes.

What are we to make of these shared "intuitions" about wisdom across otherwise varied settings? An older, less self-conscious anthropological imaginary might have taken them as evidence of broader, more general, or even universal structural features of Wisdom per se. Conversely, the more self-regarding anthropological heirs of postcolonial, feminist, and writing culture critiques would likely see these shared intuitions as evidence of a different sort—evidence of the (definitionally unwarranted) projection onto a range of cultural others, of the assumptions of anthropologists, themselves beholden to particular Euro-American visions of wisdom.

Conclusion: Wisdom as Method

For the sake of my argument here, however, it is sufficient to note that, judging by the phenomena they pick out in their comparative sophiology, *anthropologists* at least seem to share the following intuitions concerning wisdom. Whether or not these intuitions reflect something general about wisdom in the world, they certainly characterize the vision of wisdom that emerges from anthropology's efforts at comparison. To summarize the previous section, one might deduce from the preceding comparative literature that wisdom (as seen by anthropologists)

1   is practical as well as theoretical
2   works through cases
3   is "difficult" and/or paradoxically "simple" in its expression
4   requires personal transformation
5   relies on exemplarity
6   is conventional yet also more than conventional

One striking fact about this list is how easily it also could be taken to characterize anthropologists' vision of their own discipline. It is to features like these that anthropologists tend to turn when asked to define the distinctiveness of their discipline within the chorus of the social sciences and humanities. The distinctiveness of anthropology is frequently articulated at the intersection between the practical engagement of fieldwork and the abstractions of theory.

Its reliance on ethnographic detail and casework, on the slow and painstaking monographic form, is often hailed as what makes it distinctive but also difficult, troublesome knowledge, knowledge that resists cutting to the chase of a simple, portable conclusion and dwells in the complexity of the particular. Anthropological writing can sometimes seem "merely" descriptive, like a story engagingly told—yet these seemingly accessible or transparent descriptions can carry deep theoretical meaning. Despite repeated critiques of the "Malinowskian" model of ethnographic fieldwork, anthropology also continues to be popularly described and understood as a discipline requiring deep personal involvement and transformation, a shaping of experience and embodied knowledge and intersubjective apprenticeship that is not a mere matter of "methods training." Ethnographic evocation, for better or for worse, continues to tie the exemplarity of the ethnographer in with the exemplification of the case—a point brilliantly elaborated in the introduction to the edited collection *The Power of Example* by Lars Højer and Andreas Bandak (2015). As for the final point—wisdom's paradoxical relation to convention—it echoes anthropology's deep investment in comparison, which places every account both inside and outside a setting, both situated and more than situated (Candea 2019). This provides a different spin on the various contortions through which anthropologists of wisdom have sought to challenge both "Western assumptions" and the inherently Western orientalist assumption that "wisdom is elsewhere."

This parallel between anthropology's intuitions about wisdom and anthropologists' intuitions about anthropology could be read in various ways. A cynic might say that anthropologists are merely recognizing as "wise" those practices they have learned to value in their own disciplinary training. A more generous interpretation might be that anthropology has borrowed, in its own disciplinary constitution, some widespread Euro-American intuitions about wisdom as method, some of which might even reach beyond merely Euro-American settings.

That more generous interpretation is supported by another interesting parallel, for after all, the preceding list also recalls some of the key features of Michel Serres's writing. Serres—in his own discussion of wisdom and elsewhere—self-consciously interweaves abstraction and a material concern with the world. He doggedly refuses concepts and laws, replacing them with cases and exemplary characters—"that is the whole of my philosophy: there are no concepts, there are examples and events, that's all" (Serres, Legros, and Ortoli 2016, 81). His writing alternates passages of excruciating conceptual difficulty, with flows of deep insight hidden in what a careless reader might mistake for mere descriptive prose. Serres also *exemplifies* his thinking through his own writing

270 / MATEI CANDEA

and method more than he explains it, as Watkin (2020) and others have noted. And when he does explain it—as in the two book-length interviews that have weighed so much on the interpretation of Serresian thought (Serres and Latour 1992; Serres, Legros, and Ortoli 2016)—Serres explains his thought by and through an account of his own person, his conceptual transformations, his life story. Serres theorizes this explicitly in his discussion of wisdom with Latour: "One never invents an abstract wisdom without first trying to shape a real, living sage. What does it matter if I am [wise] if my successors do not become it?" (Serres and Latour 1992, 266). These are no throwaway observations. Twenty-two years later, interviewed by Legros and Ortoli, Serres is still similarly concerned with "philosophy as a way of life" (Serres, Legros, and Ortoli 2016, 358) and with his own ability to effect a kind of exemplary transformation in others (cf. Serres [1991] 1997). Serres reflects with pride on students recalling that they used to take notes during the first half of his lectures and then just write their own texts during the second half while he was still talking. This exemplifies the way something like a breath of inspiration "passes" from teacher to student, from one thinker to another (Serres, Legros, and Ortoli 2016, 357): "For in the end, what do we need in life? We don't need ideas or quotes—I couldn't care less that Mr Such-and-Such thought this or that. On the other hand, if he can pass me something I really need . . ." (Serres, Legros, and Ortoli 2016, 356).

We might therefore contrast Serres's explicit and in some ways problematic discussion of the nature of wisdom in late modernity, outlined in the first part of this chapter, with a more pervasive sense of wisdom as method, which runs through his work. The two are not unconnected, of course, and Serres's explicit account of wisdom provides as it were the ontological and epistemological underpinnings of his sophiology. But whereas anthropologists may well find themselves at odds with aspects of Serres's ontology and epistemology, there is a clear elective affinity between his sophiology and that of our discipline.

Reading Serres's work as "wisdom literature," then, means staying attuned to this shared set of assumptions and intuitions. Serres's example might, on the one hand, encourage anthropologists to be occasionally more careful in moments when they might be tempted to think themselves too wise and remind us to ask each time, on whose definition and at whose expense? But there is a more positive lesson there too. Serres's unabashed adoption of wisdom as a descriptor for the core purpose of his oeuvre might be taken as a model for anthropologists, many of whom, I suspect, do think in their heart of hearts, like Serres, that wisdom is a rather apposite horizon for which to be setting sail (for some explicit statements to this effect, see Eriksen 2018; Stoller 2020). At the very least, making that ambition more frequently explicit, and thinking through its implications, would spare us the return of fruitless arguments over whether anthropology should aim to be more like "science" or more like "activism." That is an alternative that Serres at least wisely eschewed.

NOTES

1   Watkin, agreeing with a line of previous commentators (2020, 203), notes that doing this requires reading *all* of Serres's oeuvre—a feat given his prolixity.

2   Not unlike Serres's own writing, that of Epictetus and Marcus Aurelius often belies its own technicity. In the following discussion, I am drawing not only on the original texts but also on the elaborate reconstructions of the structure of Stoic wisdom by Pierre Hadot (1992). Hadot's influential interpretations of classical philosophy as a form of lived praxis, *"une manière de vivre,"* rather than a mere construction of theoretical systems (Hadot 1995, 2003) inspired Michel Foucault's late concern with *technologies of the self* (Foucault 1988). Serres, as we just noted, doesn't cite. However, the following passage suggests that he may have been familiar with Hadot's work, at least by 2016: "Philosophy isn't just books, a series of theories, it's also a way of life [*une manière de vivre*]. That is the great, the immense lesson of Classical philosophy" (Serres, Legros, and Ortoli 2016, 358).

3   Indeed, a Stoic might argue that Serres has rather misread the import of the Epictetan distinction: just because some material things are *partly* under our control—and that is surely all one can say, hyperbole aside, of our bodies or the fate of the Earth despite the averred powers of science, medicine, and technology—doesn't make them into moral ends. After all, "body, property, reputation, office, etc." were always things that could be shaped or influenced by human actions. Add to this the profoundly Christian-inflected grounding of wisdom in human suffering, and you find a reconfiguration of wisdom that, while it starts from a Stoic distinction, is entirely at odds with both the spirit and the letter of Roman Stoicism.

4   I am grateful to Andreas Bandak for the parenthetical observation.

JANE BENNETT, ANDREAS BANDAK, AND DANIEL M. KNIGHT

AFTERWORD

Conversations with Jane Bennett

Michel Serres favored conversation as a malleable means for forming and disseminating porous knowledge. Ideas in flux, pathways partially taken and others left unexplored, piecing together new connections across minds, senses, and imaginations. In his oeuvre, Serres is in conversation with an array of traveling companions—from Lucretius to Leibniz, Venus to Verne, Plato to Picasso, the Bible to NASA. Perhaps his most accessible book also took the form of conversations, with his friend Bruno Latour.

And so we choose to close this book—itself a collection of entangled exchanges between scholars, theories, research participants, and a late philosopher—with a conversation. As a theorist across the humanities and social sciences who has engaged extensively with Serres's ideas, it could be said that Jane Bennett has traveled with Serres on a grand tour of the cosmos. Across strata of existence that far exceed the human, on their adventures they encounter matter, atmospheres, senses, and aesthetics. In the register of friends chatting over coffee, our conversation took place through email over the course of six months (and continued later during an encounter in Copenhagen), with a concise brief of conversing about where conversations with Serres might lead.

Close Encounters

AB/DK: Many contributors to this collection and people we have spoken to about the project have recited their first "encounter" with Serres's work. Whether in the dusty depths of a small secondhand bookstore, on a graduate program reading list, or by way of an obscure reference to one of his main conceptual hooks, such as the parasite or noise, that initial encounter left an indelible impression. Can you recall your first encounter(s) with Serres and your immediate feeling?

JB: I came across Serres sometime in the late 1990s when I was trying to learn about Bruno Latour's actor-network theory; it was in the book called *Conversations on Science, Culture, and Time* by Serres and Latour. I didn't understand what they were talking about!

AB/DK: That is our experience of our first encounters too—Daniel also with *Conversations*, Andreas with *Genesis*, and there seems to be this sense from most of our contributors. There was a surge of excitement, but we really didn't understand what we were reading or why we felt so invigorated . . . Serres was *different*. After this initial confusion, in your reading of Serres, have there been any specific works, concepts, or key terms that have especially appealed to you and helped you make sense of your research material? Noise, assemblage, and vortices, central themes in *The Birth of Physics* and *Genesis*, seem to have been particularly powerful influences. Is this the result of an instantaneous "connection" with Serres?

JB: I have come to love Serres's bold efforts to name the very heart or essence of processes—for example, turbulence, *noise*, Venus, Hermes—terms that are evocative, generic, mythic. I am amazed by his ability to direct attention to an "archaic" (or is it "cosmic"?) level of our existence. When I read him, I am reminded that, in my everyday life, I am participating not only in strata of existence that are biographical, biological, and cultural but also those that are more-than-human (of physics, the realm of matter-and-void, energies and atmospheres or spirits—it's hard to know what to call it). For example, changes in barometric pressure are insinuated into our moods and desires. Serres puts into words that language-resistant but weirdly ubiquitous dimension. His books strike chords between one's "own" experience and the ongoing universe, one could say!

AB/DK: That is very true. The way Serres captures atmospheres or aesthetics beyond everyday dimensions of knowing the world does seem to speak to much that evades capture in general descriptive concepts of written and spoken word. Anthropology tends to still be constrained to ethnographic narrative, and we have been thinking about how to start addressing other forms of knowledge—feelings, textures, atmospheres of life. Could you perhaps elaborate a little on how Serres's poetics and ability to audaciously pin down cosmic existence resonate with your own experience of the world—the uncanny, the unforeseen, the "something that makes the world make sense"? Or, perhaps even, "what makes matters vibrant," if you like?

JB: In rereading *Conversations* recently, this line hit me: "No one can think without somewhere depending on an invariable that underlies variations" (Serres and Latour [1992] 1995, 195). I guess my heuristic invariable is "vibrant matter": it has helped me to think about how my possessions possess me, how my achievements are profoundly circumstantial, and why the not-quite-human world is able to fascinate not just by way of beauty but also via a repellent violence and cruelty (Bataille's essay "The Big Toe" comes to mind). Serres's "invariable" may be turbulence. He spins out an onto-tale of a churning, creative-destructive process, within which human beings are presented as but one of the astonishing congealments, alongside things like the "thalweg," wind, mud, parasites. His process-centric tales (often invoking classical Greek gods and goddesses who are themselves simultaneously natural forces and personalities), have the effect of highlighting the *impersonal* subsisting within mortal persons—how our constitution and our efforts include, for example, those of electrical pulses, gravitational, magnetic and other attractions, and so on. Serres marks the impersonal within, as in his extraordinary discussion at the end of *Conversations* about evil as an objective force. The impersonal is anything but inert for him; it is both more active and more replete with pluripotency than the figure of "inorganic matter" that still circulates in the social sciences. The (unstable) strata of our objecthood or it-ness is not defined by Serres in contradistinction to an active and creative subjectivity. Serres's impersonal does things (but not necessarily consciously so), and so do we. Even as we also are, like every other thing, susceptible to being acted upon (maybe I am Spinozaizing Serres here?).

Serres's turbulent, asubjective vitality is different from Freud's uncanniness, primarily because the former is less a psychic phenomenon than an ahuman force-presence capable of a life outside the human body. I talk about this in *Influx and Efflux* in the contrast between, on the one hand, experiencing doodling as the channeling of the graphic line's own trajectory (here I am thinking of Paul Klee's notebook writings and Tim Ingold's *The Life of Lines*) and, on the other hand, experiencing doodling as the Surrealists did, in other words, as an expression of the (human) "unconscious." Doodling responds to variations in cosmic forces as it saunters along.

AB/DK: Serres notoriously draws together scholars of disparate eras, disciplines, and cosmologies, often through comparison, analogy, and messages. How does Serres fit in with the ensemble of scholars that have influenced your thinking?

JB: Serres is a subtle eco-philosopher or eco-poet. I first recognized this in his Lucretius book when he affirms an Epicurean ethos of making the most of the least. Lucretius: "the greatest wealth is living modestly, serene, content with little" (Lucretius, *De Rerum Natura V*, line 1117, in Lucretius 1968, 191).

This logic of little-is-enough is a revolutionary alternative to the insane capitalist fantasy of perpetual economic growth. Serres expressly links the Lucretian ethos to its physics, to, that is, the extreme slightness of the clinamen:

> ... [a tiny flower's worth] of a little wine, or a little jar of cheese to make a great feast.... In the absence of wine, water is enough, whatever's there..... Just a little and no more: ne plu quam minimum, this is the definition of the clinamen. Tantum paulum: as little as it would be possible to say, still, that movement has occurred. As little as it would be possible to say, still, that my desire may find itself satisfied.... The movement of the soul is differential, it is ... the same deviation ... as that which ... changes the cataract of the atoms.... The wise man inhabits this minimal deviation, this space between little and nothing.... Beyond there is only vain and superfluous growth.... (Serres [1980] 2001, 183)

It is in this liminal space or interval that new turns begin and creative thinking can show up.

Just Relax . . .

AB/DK: Earlier you said that when first reading *Conversations* you didn't understand what Serres was talking about. In that book with Latour, they engage in a vast array of disciplines, methods, and timespaces. And *Conversations* is widely considered one of the easier introductions to Serres's thought! Is there a point, do you think, where this disciplinary porosity becomes overcomplicated, too dense, and shuts down "conversation"? You say that little is enough, less is more . . . but that seems almost the antithesis to what Serres provides his readers!

JB: Yes, indeed! I'll take your comment as an occasion to reflect more generally on Serres's mode of inquiry ("method"). The first thing I noticed was its mood: the writing seems to flow from a comportment of *relaxation*, of slow and steady breath. Serres is in no rush. I envy that ease, calm, and slowness, since what is more typical for me and many of my friends as we sit down to write is a humming anxiety—"Do I really have anything to say? Hasn't everything already been said? And if I do have something to add, can I call it forward and type it fast enough before it slips away?" I like Serres's solution to this anxiety—a "retreat house" (I wish!). He says that "new ideas come from the desert, from hermits, from solitary beings, from those who live in retreat and are not plunged into the sound and fury of repetitive discussion. The latter always makes too much noise to enable one to think easily. All the money that is scandalously wasted nowadays on colloquia should be spent on building retreat houses, with vows of reserve and silence" (Serres and Latour [1992] 1995, 81–82).

He goes on: "We have more than enough debates; what we need are some taciturn people" (Serres and Latour [1992] 1995, 82). But as you note, Serres is hardly taciturn! His style is "on and on." It often works to startling and profound effect; it also can drift into a (highly educated, philologically sophisticated) stream of consciousness. My partner Bill Connolly and I both confess to a sense of "Enough already!" if we have read too many pages at one sitting.

AB/DK: In our experience of reading Serres, it is the inspirational "ah-ha!" moments that captivate us and keep us returning to his work. But we also acknowledge that the reader must do a lot of hard graft, through repetition, messy logics, and endless crisscrossing stories, to find the passages that appeal to them. Is this, perhaps, part of the reason why anthropologists and social theorists have found it more challenging to engage with Serres than his ever-popular contemporaries? Perhaps this writing style could be seen as indulgent in an era of rapid citations, commercial publishing, and audit culture?

JB: Yes, you put it perfectly. I agree. Slowness is not the hallmark of our time—too much anxiety and too many volatile conditions make it extra hard to practice (Epicurean) ataraxy.

AB/DK: In this volume, we start from the premise that, in anthropology at least, engagement with Serres regularly constitutes a footnote or a passing citation, often to two or three of his most widely known ideas. But then the analysis goes no further; Serres remains a "sound bite." He does not seem to garner the same detailed, some might say exhaustive, attention as his contemporaries or students, such as Gilles Deleuze, Bruno Latour, or Jacques Derrida. Would you agree, and if so, why do you think this is?

JB: That is a really good question. I agree that while many people read and love Serres's books, there hasn't emerged the phenomenon of Serresians in the way there are Deleuzeans, Derridians, Latourians. I'm tempted to say that Serres's mode of writing/thinking resists systematization (even more than the others on that list), even as he explores, against the grain of much contemporary theory, a philosophy of a shared ("universal"?) archaic pulse.

Or maybe it's less that Serres actively "resists" a system than he just goes elsewhere and elides the particular kind of systematicity that (a French, twentieth-century tradition of) philosophy pursues. Serres really takes seriously the fact that the (very big, more-than-human picture of the) "order of things" just isn't the same kind of "order" as that which (a French, twentieth-century tradition of) philosophy (of "solids," not "liquids") has tended to seek. I'm not sure about what I just said, though . . .

AB/DK: We think that your previous response shows the searching spirit of Serres—he takes us on a tour of eras and spaces—not knowing the questions to pose from the start, not searching for an ordered philosophy of things, but exposing us to options. We detect in your responses thus far some hesitancy to claiming to *know* Serres, some vulnerability in the way you present knowledge, which is itself very Serresian. He offers us pathways, allows the reader to make connections, search with him for cosmic relationships. Is this perhaps part of how Jane Bennett sees the world—resisting systemization, taking us on a grand tour of the more-than-human?

JB: I hesitate also to be included in a comparison with Serres—but thank you, yes, I do like the grand tour. And I do like to dig around in many plots at once. Maybe because it seems that every identifiable process or system or assemblage is marked by pulses and energies that exceed it.

An Intuitive License

AB/DK: In your own work, you do an impressive job of connecting disciplines, navigating between and along concepts that link but also potentially separate epistemological canons. We might suggest that this is a form of "hyphenation," in Serres's terms. How might engagement with Serres encourage social researchers to embrace connections across disciplines?

JB: I love that idea—yes. But I don't know where Serres talks about hyphenation. Please advise . . .

AB/DK: Sure! In *L'Art des Ponts: Homo Pontifex* (2006, 77), Serres likens the hyphen to a bridge between the "soft empire of signs" and "the hard realms of physics and biology," allowing communication to transcend two planes. The result is the production of newness (like the harlequin figure that is born of the mixing of the hard sciences and humanities in *Troubadour*). The hyphen allows alignment and conjunction of disparate parts, indicating a branching out of expertise into new collaborative domains (centrifugal) but also the drawing together of units of knowledge (centripetal) that do not obviously fit together. Serres also uses hyphens to explain complex concepts, regularly by way of etymology—like com-putare in *Thumbelina*. The hyphen facilitates conjunctions and a general idea of connectivity. In Maria Assad's reading of *Troubadour*, she concludes that "the hyphen imprints diacritically the meaning of the middle-ground . . . neither posing (thesis) or opposing (antithesis), the exposing middle (or third) is the same as the non-present present" (1999, 133). This allows for connection across hard and soft concepts, porosity between disciplines and bodies of knowledge. We might suggest that you "hyphenate" in the way you allow so many actors (human and nonhuman) into your academic conversations. Just a thought!

JB: Yes, now I see what you mean. Yeah, once you start looking, you can't help but find yourself in the presence of a cascade of similarities and relations all hyphenating with each other. The best grad students are often the ones who intuit these affinities without even trying (and then the painful, falsifying task is to create a border wall for the dissertation!). These resonances are intuited rather than seen or known. Which brings me to another aspect of Serres's mode of inquiry that I admire, his intuitive license. Serres: "Aren't the great inventions, even the conceptual ones, based on an intuition? It always makes the first move" (Serres and Latour [1992] 1995, 39).

Serres figures intuition as a "dazzling, obscure, and hard-to-define *emotion*" (Serres and Latour [1992] 1995, 99, my emphasis). The feeling of intuition is a response to goings-on outside of the self. It is prompted by signals emitted from elsewhere, signals that are fleeting and vague. They form "a confluence not a system, a mobile confluence of fluxes . . . overlapping cyclones and anticyclones . . . wisps of hay tied in knots . . . clouds of angels passing" (Serres and Latour [1992] 1995, 122).

Intuition marks, for consciousness, the presence of unnoticed activities and relationships or proto-relations. Serres follows up on these hints, wades around in them, and then writes them up (or stitches them together)—in ways provoked by the signals but not quite identical to them. Serres explicitly links apparently disparate events—Lucretius-Balzac-complexity theory or war-Hermes-*Challenger* explosion. The crumpled nature of time allows this.

Scales and Tales

AB/DK: If we may turn for a moment to anthropology specifically, from where we know you have taken a lot of recent inspiration. The discipline has lately been dealing with the issue of scale, from the level of the individual, through community, nation, to planetary events. This has seen anthropologists move away from the classic focus on small-scale societies and singular "domains" of analysis (kinship, economy, ritual, etc.). Serres seems an ideal way in for anthropologists to start moving across and between scales given that he talks of how localized knowledges, truths in their own right, clump together as building blocks toward greater paradigmatic realities. What are your thoughts on this, particularly given your own work on keeping local viewpoints on global realities?

JB: My feet are in Baltimore, but I tilt, as I think Serres does, toward the cosmic. I am drawn to that strata of experience, and I like to decorate it with words. This spinning of onto-tales is a pleasurable aesthetic practice that helps also to orient more local- or problem-centric forms of thinking. Serres too leans grand, and sometimes even goes further, from onto-*tales* to "the universal": "Never forget the place from which you depart, but leave it behind and join the universal. Love the bond that unites your plot of earth with the Earth, the bond that makes kin and stranger resemble each other" ([1990] 1995, 50).

I think it's safe to say that Serres is more fabulator-of-the-generic than ethnographer-of-the-unique-locale, even though he insists on the value of both endeavors. It is very helpful to put the point, as you do, in terms of *scales* of analysis. But maybe there is no best (methodological) way to cross or connect scales. Maybe it's okay that such a practice be idiosyncratic? I haven't given this much thought and would like to hear more about what you think . . .

AB/DK: For us as anthropologists, it is helpful to work with Serres on maintaining local knowledge while building toward universal or planetary truths—what he calls algorithmic thinking. This seems to be a challenge for the anthropological method, particularly given the increasingly vocal calls for anthropology to develop its applied dimension. We have to deal with the planetary, perhaps the universal, and we on the whole want to. But we also want the local, the relative, embellishment to still glow through. So we might ask, is there really relevance in Serres for current affairs, particularly on how social researchers confront contemporary crises, conflicts, and trends in thinking beyond the human?

JB: One of the difficult lessons I've learned from Serres is the value of *postponing* the desire to identify "contemporary relevance," especially when there is reason to suspect that the terms of analysis/categories of experience/regimes of perception of contemporary life are themselves producing bad effects (bad in the sense of cruel, ecologically destructive, unjust, and ugly). This elision of "relevance," however, will make you subject to (academic and political) criticism—charges of moral indifference or worse. It is sometimes still worth the risk . . . though it's really hard to make the political-ethical call.

AB/DK: We take your point on the breakdown of both "contemporary" and "relevance." Well, perhaps some of the "contemporary relevance" we refer to is, for instance, in the apparent need for a new natural contract. This is something our contributors have picked up on—have we outgrown Serres's version of the natural contract, and does it require addendums, whether in the form of machine technologies or in light of recent perspectives on the Anthropocene and climate change? This is a political-ethical call (although we could debate both its contemporaneity and its relevance). What is your take on the state of the natural contract in the 2020s?

JB: I just reread *The Natural Contract*. As many others have noted, framing the problem in terms of "contract," even in the broad way Serres uses the word, may still be too limited for the complexity of the events known as climate change, which now include a patho-psychology of denial. It may also be somewhat at odds with Serres as a process philosopher, in that a contract can readily imply a prior *separateness* of parties who *then* engage in relations. In other words, "contract" seems to posit, to use Karen Barad's terms, inter-action rather than intra-action.

But let me also nominate two aspects of *The Natural Contract* that seem to me to be as valuable now as they were when the book appeared.

The first is its effort to present a specifically ecological *ideal-of-self* with the power to attract, inspire, and convert readers. We need that if the political will to disassemble the relentless political economy of waste and exploitation is ever to catalyze. What sources does Serres draw upon to compose this (counter-)ideal? While I think he would agree that non-indigenous peoples have much to learn from indigenous peoples' ways of thinking/living/acting in and as Earth, his primary tack is to take sustenance from European subcultures whose participants do not for the most part identify as indigenous. These are the Earth-and-sea-loving "peasants and sailors" whose ways of life are now ignored, denigrated, and almost eliminated. Serres invokes these exemplars and writes up a modern, modified version of these exemplars, which he calls the Sage or the "Instructed Third."

This "free spirit and damned good fellow" is radically trans-disciplinary and trans-temporal, and is a "traveler in nature and society . . . fascinated by different gestures as by diverse landscapes . . . navigator of . . . waters where scientific knowledge communicates, in rare and delicate ways, with the humanities; conversely versed in ancient languages, mythical traditions, and religions . . . sinking his roots into the deepest cultural compost, down to the tectonic plates buried furthest in the dark memory of flesh and verb; and thus archaic and contemporary, traditional and futuristic, humanist and scientist. . . ." (Serres [1990] 1995, 94–95).

The second item on my list concerns the brave effort of *The Natural Contract* to mark the *aleatory* dimension of the turbulent physical-cultural processes of existence. Serres talked about this in *Genesis* as a pervasive logic of "it depends"—sometimes the turbulence (of the sea but also of a crowd) will repeat itself in a way capable of producing a slightly more regular "fluctuation" that could, maybe, become further consolidated as a distinguishable event. But maybe not. In *The Natural Contract*, this aleatory logic appears in a discussion of whether the crowd or swarm of 1990s environmentalists will congeal as a collective *agent*. Will they compose in such a way as to be capable of producing a significant effect? In his ontology of turbulence, groups are continually amassing, many times nothing comes of it, but sometimes a spark ignites and binds them together such that a group becomes capable of action. Will the green people convening mobilize to act? You just can't know in advance, says Serres: "The crowd is massing like clouds before the storm, which may or may not break" ([1990] 1995, 5).

I'm not sure what to do with this logic of "it depends." It rings true to me. But what are its implications for engaging the eco-crisis, for changing our course? Is the task to try to discern what exactly the "it" of "it depends" *depends on*? Does Serres call for experiments, in the lab and in the polity, to find out? Does Serres's accent of the ever-presence of an aleatory element encourage a shrug? The storm of an ecological revolution (in economy and society) didn't happen in the 1990s, and it still hasn't. How does cognizance of an irreducible element of chance get factored into, say, Connolly's climate crisis "politics of swarming" (in *Facing the Planetary*)?

AB/DK: That turbulence, a swelling that may or may not break, a surge in tension that may rupture into an event, is so often there in Serres, from *Genesis* through to *Branches*. And this also seems pertinent to anthropological interests in social/eco-movements and their political/politicizing potentialities. We guess the *movement* aspect of social/eco-*movements* would be important to Serres, the momentum of a crowd of clouds and its potential to bring forth strikes of lightning rather than getting benignly blown out to sea.

The natural contract is of course closely linked to Serres's ideas of parasitism. Have we gone past the threshold where the human can be anything other than a parasite on the natural environment? How do you perceive this relationship of hospitality?

JB: I don't think that threshold has passed. I'm not even sure how or if (short of human extinction) one could ever tell where such a final threshold would be located.

AB/DK: We would like to turn from concepts to method for a moment. The natural contract and parasitism seem to be about navigation. Serres points to the need to engage in conversation with friends and enemies to navigate a third way. We think that this has metaphorical applications as well—conversations between human/nature, hard/soft academic disciplines. We mentioned before that this might be the work of the hyphen—and you alluded to this previously in talking about contracts bringing separate parties to a conversation. For Serres, we feel that bringing bodies of knowledge together makes a stronger—and novel—third way through symbiosis. In your answers thus far, you seem very open to what you don't already know—do you see methodological similarities between your own work and Serres?

JB: No one really knows where they are going or what they need to know or feel or be in order to arrive. So it seems prudent to engage in the pleasures of learning about many different things, even if you never become expert in any. I admire Serres's mode of inquiry, especially the part about being willing to follow a murky intuition, at the risk of looking and sounding silly, ridiculous, weak, and "feminine." One of the best compliments I've been given is to be called, in the preface to this volume, a polymath. Thank you!

AB/DK: Well, we are delighted to have you in conversation as someone who really does travel the cosmos and takes us all with you. On conversations, Serres is quite clear when he is speaking to Latour that conversations do have their limits. At first, he seems unconvinced, saying that one has to carefully select which conversations to engage in. Perhaps you could pass comment on the limits of conversation.

JB: I see what Serres is saying: conversations are good, but only, as my friend David Howarth is fond of saying, "up to a point." Especially written ones, where the affective and bodily components of communication are minimized. (Plato's *Phaedrus* makes this point so well.) Talking and writing are what we do as teachers and scholars, but Nietzsche was onto something when he said: "Man, like every living being, thinks continually without knowing it; the thinking that rises to *consciousness* is only the smallest part of all this—the most superficial and worst part—for only this conscious thinking *takes the form of words, which is to say signs of communication . . .*" (Nietzsche 1974, 354). Whitehead too says that so much of what is communicated across bodies happens at a visceral level below the radar of both cognitive and sensuous detection.

AB/DK: Yes, and our, written, conversation may look, feel, and be recalled differently if we really were out for coffee in Baltimore, Copenhagen, or St. Andrews! Conscious of these limitations, are there times when urgent matters require answers and other occasions when latency is the order of the day, to sit back and listen?

JB: Yes.

AB/DK: Before signing off, we just wanted to ask you to pass comment on a few things that struck us while putting together this volume. First, we found ourselves reflecting on the identity politics of engaging with "yet another white, French, male, philosopher" to help interpret the world around us. With social theory flooded with tributes to such headline personas, why should we concern ourselves with Michel Serres?

JB: Of course, Michel Serres, like every body, is located in a social and "identity" nexus. But one of the distinctive features of his books, I think, is their ability to draw attention away from the normal, people-centric strata of analyses and toward that cosmic, elemental dimension of our lives, which is co-present with and irreducible to (albeit tangled with) the social, cultural, biographical, human-historical dimensions that get so much attention because they are easier to name and because engaging with them addresses pressing human needs, targets injustice, and (in the interest of completeness it is fair to note) produces the pleasure of expressing righteous indignation.

AB/DK: We are with you on that! For us there is an almost spontaneous simultaneity or coincidence in how Serres "pops up" in our lives. For instance, we certainly find that we can be working on topics, papers, conversations that we don't necessarily associate with Serres, only to later find that we are unknowingly working in parallel or in a knotty relationship with him. On this unwitting simultaneity, Serres talks about becoming contemporary with figures of the past and scholars from diverse backgrounds. We wonder, in your case, who became contemporary with whom? Did you become contemporary with Serres, he with you, or were you always working toward each other?

JB: I don't know. Everybody is a swarm.

AB/DK: Who else would you consider to be in your team of "contemporaries"? Who accompanies you on your travels?

JB: I feel close to Io, the character in Aeschylus's *Prometheus Bound*. She is the only human in the story, but she is also bovine (result of the unjust vengeance of Hera). Io can't stop wandering, can't rest, she's prodded always by a gadfly (her characteristic utterance is "O! O! O!"). So many among the living are my co-existents and co-thinkers; to name only some of the most recent, there is the photographer Janet Steinberg, the Baltimore eco-artist Jordan Tierney, the film and architectural theorist Guiliana Bruno. Among the dead, I think my best friends are Henry Thoreau and Gilles Deleuze, both of whom have such a good sense of humor and a refreshingly "Pagan" sensibility.

AB/DK: The connecting thread in this book is "porosity"—of disciplines, concepts, scales, and timespaces. Could you briefly comment on what porosity conjures up for you?

JB: Porous is what people are because, like any body, each of us is necessarily open to an outside, necessarily complicit with (folded into) more-than-human comings and goings, with processes of influx and efflux. "Influx and efflux" is a phrase from Walt Whitman's "Song of Myself" ("Howler and scooper of storms, capricious and dainty sea,/I am integral with you, I too am of one phase and of all phases/Partaken of influx and efflux I" [Whitman (1892) 2017, section 22]), and it was the inspiration for a book I wrote that tries to put porosity (and its allied, politically relevant capacities of sympathy, anxiety, and contagion) at the heart of a model of "self" that might better suit a world of vibrant matter, a world, that is, no longer understood as the mere context for human action but as itself a lively cauldron of co-agents including us. If *Vibrant Matter* accented the influential powers of nonhuman things and bracketed the question of subjectivity, *Influx and Efflux* turns back to the question of "I." A porous, processual body—leaky, seeping, affecting, and susceptible to influence—still retains a certain *consistency,* and there remains a felt sense of a difference between outside and inside. Serres talks about all this in *Genesis,* where he offers a brilliant account of the phases of congealment by which a body can emerge from within the more protean, less determinate field of *noise.*

Also helpful to me in exploring the implications of porosity is François Jullien's *The Great Image Has No Form, or On the Nonobject through*

Painting (2012), which contrasts the Greek/European tradition of thinking about porosity with the Chinese (primarily Taoist) tradition. Here's
Jullien:

> There are two ways in which my existence is continuously connected
> to something outside: I breathe *and* I perceive. Now, I can privilege
> the gaze and the activity of perception, the Greek choice, which
> led them to grant priority to a conception of reality as an object
> of knowledge. . . . Or I can base my conception of the world . . . on
> respiration: that is the Chinese choice. From the fact that I am alive,
> breathing in-breathing out, I deduce the principle of a regulating al-
> ternation from which the process of the world flows. . . . [T]he Chi-
> nese conceived of the primordial reality not in terms of the category
> of being and the relation of form to matter . . . , but as 'breath-energy,'
> *qi*. (134–35)

Whitman's influx and efflux also leans in that direction.

AB/DK: Finally, we want to ask, in a word or short phrase, what does it
mean to travel with Serres?

JB: In *The Birth of Physics*, Serres invokes a child's toy, a top that is kept
upright by the vortex even as it remains in motion. To travel with Serres
is to keep trying to cope well with turbulence, to become that top.

AB/DK: Thank you so much, Jane Bennett, for the conversation!

The conversations in this book are about to end or, rather, in a Serresian
gesture, perhaps they are about to begin. To begin new conversations on old
themes and continue old conversations with new partners. We appreciate the
thoughts, labor, and care that have gone into this particular conversation with
Jane Bennett as well as with all the contributors. We hope that the diverse
interactions with Serres will branch out and allow for novel engagements with
anthropological problems and beyond. One might therefore end by reflecting
on what kind of engagement Serres allows. Is he an optimist, a pessimist, a real-
ist with regard to the current state of affairs? Perhaps he is all that. Aware of the
cruelty, violence, ignorance, and quest for power found in humans. The turbulence, noise, and whirlpools of how we interact with each other and our surroundings. But Serres is also keenly invested in how newness shoots off against
decay, how novelty surpasses repetition and death. He believes in the audacity
and creative ingenuity of humans to traverse history, time, nature, structure,

identity. Perhaps the only realistic form of engagement with Serres keeps open these variations over and with bodies, senses, and thoughts. And in this it may be wise to follow Serres, when he crosses, wanders, and binds together knowledge and hope admitting, humbly and in humility, that this is just an attempt at making the world a little less stupid ([2004] 2020, 106). As such, what Serres might help us start conceiving is an anthropology that does not just operate on knowledge frequently marked by power struggles, but on comprehension. Comprehension allows a novel appreciation in and with bodies of what knowledge might imply and accordingly may allow us porously to devise better ways to inhabit our world. Serres, in *Branches* (125): "Without deadly dangers being run today . . . would we know how to transform some event into advent: would we invent a new world?"

REFERENCES

Abdussamad, H. Ahmad. 1997. "Priest Planters and Slavers of Zägé (Ethiopia), 1900–1935." *The International Journal of African Historical Studies* 29, no. 3: 543–56.

Abdussamad, H. Ahmad. 1999. "Trading in Slaves in Bela-Shangul and Gumuz, Ethiopia: Border Enclaves in History, 1897–1938." *Journal of African History* 40, no. 3: 433–46.

Abu-Lughod, Lila. 1993. *Writing Women's Worlds: Bedouin Stories.* Berkeley: University of California Press.

Ahmann, Chloe. 2018. "It's Exhausting to Create an Event Out of Nothing: Slow Violence and the Manipulation of Time." *Cultural Anthropology* 33, no. 1: 142–71.

Asad, Talal. 1973. "Introduction." In *Anthropology and the Colonial Encounter*, edited by Talal Asad, 9–20. New York: Humanity Books.

Assad, Maria L. 1991. "Michel Serres: In Search of a Tropography." In *Chaos and Order: Complex Dynamics in Literature and Science*, edited by N. K. Hayles, 278–98. Chicago: University of Chicago Press.

Assad, Maria L. 1999. *Reading with Michel Serres: An Encounter with Time.* SUNY Series, the Margins of Literature. Albany: State University of New York Press.

Assad, Maria. L. 2012. "Ulyssean Trajectories: A (New) Look at Michel Serres' Topology of Time." In *Time and History in Deleuze and Serres,* edited by Bernd Herzogenrath, 85–102. New York: Continuum.

Badone, Ellen. 1989. *The Appointed Hour: Death, Worldview, and Social Change in Brittany.* Berkeley: University of California Press.

Baldwin, James. (1955) 1963. "Notes on a Native Son." In *Notes on a Native Son,* 76–102. New York: Dial Press.

Bandak, Andreas. 2013. "Our Lady of Soufanieh: On Knowledge, Ignorance, and Indifference among Christians of Damascus." In *The Politics of Worship in the Contemporary Middle East: Sainthood in Fragile States,* 129–153. Leiden: Brill.

Bandak, Andreas. 2014a. "Making 'Sound' Analysis: From Raw Moments to Attuned Perspectives." In *Qualitative Analysis in the Making,* 176–91. New York: Routledge.

Bandak, Andreas. 2014b. "Of Rhythms and Refrains in Contemporary Damascus: Urban Space and Christian-Muslim Coexistence." In "Unity and Diversity: New Directions in the Anthropology of Christianity," edited by Joel Robbins. *Current Anthropology* 55, no. S10: S248–61.

Bandak, Andreas. 2015. "Exemplary Series and Christian Typology: Modelling on Sainthood in Damascus." *Journal of the Royal Anthropological Institute* 21, no. S1: 47–63.

Bandak, Andreas. 2022. *Exemplary Life. Modeling Sainthood in Christian Syria.* Toronto: Toronto University Press.

Bandak, Andreas, and Tom Boylston. 2014. "The 'Orthodoxy' of Orthodoxy: On Moral Imperfection, Correctness, and Deferral in Religious Worlds." *Religion and Society: Advances in Research* 5: 25–46.

Bandak, Andreas, and Simon Coleman. 2019. "Different Repetitions: Anthropological Engagements with Figures of Return, Recurrence and Redundancy." *History and Anthropology* 30, no. 2: 119–32.

Bandak, Andreas, and Manpreet Kaur Janeja, eds. 2018. *Ethnographies of Waiting: Doubt, Hope, and Uncertainty.* London: Bloomsbury.

Bandak, Andreas, and Daniella Kuzmanovic, eds. 2014. *Qualitative Analysis in the Making.* London: Routledge.

Bandak, Andreas, and Simon Stjernholm. 2022. "Genealogies, Limits, Openings: Introductory Remarks on Engaging Religion." In "Engaging Religion: Directions between Analytical Categories and Research Material in the Study of Religion," edited by Andreas Bandak and Simon Stjernholm. *Religion and Society* 13: 95–110.

Barclay, William. 1967. *The Lord's Supper.* London: SCM Press.

Barker, Timothy. 2023. Michel Serres and the Philosophy of Technology. *Theory, Culture and Society* 40, no. 6: 35–50.

Barr, Stephen. 1964. *Experiments in Topology.* New York: Dover Publications.

Barth, Fredrik. 2002. "An Anthropology of Knowledge." *Current Anthropology* 43, no. 1: 1–18.

Basso, Keith H. 1996. *Wisdom Sits in Places: Landscape and Language among the Western Apache.* Albuquerque: University of New Mexico Press.

Bateson, Gregory. (1972) 2000. *Steps to an Ecology of Mind.* New York: Ballantine.

Bateson, Gregory. (1979) 2002. *Mind and Nature: A Necessary Unity.* Cresshill, NJ: Hampton Press.

Bateson, Gregory. 1991. *A Sacred Unity: Further Steps to an Ecology of the Mind.* Edited by R. E. Donaldson. New York: HarperCollins.

Bateson, Gregory, and Mary Catherine Bateson. 1988. *Angels Fear: Towards an Epistemology of the Sacred.* London: Bantam.

Bateson, Mary Catherine. 1978. "'Daddy, Can a Scientist Be Wise?'" In *About Bateson,* edited by J. Brockman, 57–76. London: Wildwood House.

Bausinger, Tobias, Eric Bonnaire, and Johannes Preuß. 2007. "Exposure Assessment of a Burning Ground for Chemical Ammunition on the Great War Battlefields of Verdun." *Science of The Total Environment* 382, no. 2: 259–71.

Bear, Laura. 2014. "Doubt, Conflict, Mediation: The Anthropology of Modern Time." *Journal of the Royal Anthropological Institute* 20, no. S1: 3–30.

Bennett, Jane. 2010. *Vibrant Matter: A Political Ecology of Things.* Durham, NC: Duke University Press.

Bennett, Jane. 2020. *Influx and Efflux: Writing Up with Walt Whitman*. Durham, NC: Duke University Press.

Bennett, Jane, and William Connolly. 2011. "The Crumbled Handkerchief." In *Time and History in Deleuze and Serres*, edited by Bernd Herzogenrath, 153–72. London: Bloomsbury.

Ben-Yehoyada, Naor. 2012. "Dead Reckoning, or the Unintended Consequences of Clueless Navigation." *Magazin* 31, no. 16/17: Was ist ein Weg?—What is a Path? Bewegungsformen in einer globalen Welt—Forms of Movement in a Global World.

Bialecki, Jon. 2018. "Anthropology and Theology in Parallax." *Anthropology of this Century*, no. 22.

Bloch, Maurice. 1974. "Symbols, Song, Dance and Features of Articulation: Is Religion an Extreme Form of Traditional Authority?" *European Journal of Sociology* 15: 54–81.

Bloch, Maurice. 2005. "Ritual and Deference." In *Essays on Cultural Transmission*, 123–38. Oxford: Berg (LSE Monographs on Social Anthropology).

Boisvert, R. 2019. "Obituary: Michel Serres (1930–2019)." *Philosophy Now*, October/November.

Bolton, Matthew. 2010. *Foreign Aid and Landmine Clearance: Governance, Politics and Security in Afghanistan, Bosnia and Sudan*. London: I. B. Tauris.

Boylston, Tom. 2018. *The Stranger at the Feast: Prohibition and Mediation in an Ethiopian Orthodox Christian Community*. Berkeley: University of California Press.

Braverman, Richard. 1987. "Libertines and Parasites." *Restoration: Studies in English Literary Culture, 1660–1700* 11, no. 2: 73–86.

Brown, Kate. 2020. *Manual for Survival: A Chernobyl Guide to the Future*. London: Penguin Books.

Brown, Norman O. 1990. *Hermes the Thief: The Evolution of a Myth*. Great Barrington, MA: Lindisfarne Press.

Brown, Steven D. 2002. "Michel Serres: Science, Translation and the Logic of the Parasite." *Theory, Culture and Society* 19, no. 3: 1–27.

Brown, Steven D. 2013. "In Praise of the Parasite: The Dark Organizational Theory of Michel Serres." *Informática na Educação: teoria e prática* 16, no. 1: 83–100.

Brown, Steven D., L. Baczor, D. Dahill, D. King, and S. Couloigner. 2022. "The Impact of COVID-19 on Psychological Wellbeing in Occupational Contexts." Nottingham: Nottingham Trent University. https://irep.ntu.ac.uk/id/eprint/46445/.

Bryant, Rebecca. 2014. "History's Remainders: On Time and Objects after Conflict in Cyprus." *American Ethnologist* 41, no. 4: 681–97.

Bryant, Rebecca, and Daniel M. Knight. 2019. *The Anthropology of the Future*. Cambridge: Cambridge University Press.

Buckert, Walter. 1983. *Homo Necans: The Anthropology of Ancient Greek Sacrificial Ritual and Myth*. Translated by Peter Bing. Berkeley: University of California Press.

Burke, Kenneth. 1969. *A Grammar of Motives*. Berkeley: University of California Press.

Burridge, Kenelm. 1960. *Mambu: A Melanesian Millennium*. Princeton, NJ: Princeton University Press.

Caillois, Roger. 2001. *Man, Play and Games*. Urbana: University of Illinois Press.

Calasso, Roberto. 1983. *La rovina di Kasch*. Milan: Adelphi.

Calasso, Roberto. 2010. *L'Ardore*. Milan: Adelphi.

Candea, Matei. 2012. "Derrida en Corse? Hospitality as Scale-Free Abstraction." *Journal of the Royal Anthropological Association* 18, no. S1: S34–S48.

Candea, Matei. 2016. "We Have Never Been Pluralist: On Lateral and Frontal Comparisons in the Ontological Turn." In *Comparative Metaphysics: Ontology after Anthropology*, edited by Pierre Charbonnier, Gildas Salmon, and Peter Skafish, 85–106. London: Rowman and Littlefield.

Candea, Matei. 2019. *Comparison in Anthropology: The Impossible Method*. Cambridge: Cambridge University Press.

Candea, Matei. 2023. "The Anthropology of Science as an Anthropology of Ethics (and Vice Versa): Elements, Problems and Possibilities." In *Cambridge Handbook of the Anthropology of Ethics*, edited by James Laidlaw. Cambridge: Cambridge University Press.

Candea, Matei, and Giovanni da Col. 2012. "The Return to Hospitality." *Journal of the Royal Anthropological Institute* 18, no. S1: S1–S19.

Carrasco, Davíd. 1999. *City of Sacrifice: The Aztec Empire and the Role of Violence in Civilization*. Boston: Beacon Press.

Carrithers, Michael. 2012. "Seriousness, Irony and the Mission of Hyperbole." *Religion and Society: Advances in Research* 3, no. 1: 51–75.

Césaire, Aimé. 2001. *Discourse on Colonialism*. New York: Monthly Press.

Chilson, Clark. 2020. "How Wisdom Is Discovered: Discretion and Emotional Insights in Naikan Meditation in Japan." In *Capturing the Ineffable*, edited by Philip Y. Kao and Joseph S. Alter, 67–81. Toronto: University of Toronto Press.

Clark, Nigel. 2021. "Vertical Fire: For a Pyropolitics of the Subsurface." *Geoforum* 127: 364–72.

Clark, Nigel, and Bronislaw Szerszynski. 2021. *Planetary Social Thought: The Anthropocene Challenge to the Social Sciences*. Cambridge: Polity Press.

Clark, Nigel, and Kathryn Yusoff. 2017. "Geosocial Formations and the Anthropocene." *Theory, Culture and Society* 34, nos. 2–3: 3–23.

Clarke, Jennifer. 2019. "Porosity and Protection." *Cultural Anthropology* "Theorizing the Contemporary," *Fieldsights*. April 25. https://culanth.org/fieldsights/porosity-and-protection.

Clastres, Pierre. 1990. *Society against the State: Essays in Political Anthropology*. New Edition. New York: Zone Books.

Clayton, Kevin. 2012. "Time Folded and Crumpled: Time, History, Self-Organization and the Methodology of Michel Serres." In *Time and History in Deleuze and Serres*, edited by Bernd Herzogenrath, 31–50. New York: Continuum.

Colebrook, Claire. 2011. "Time and Autopoiesis: The Organism Had No Future." In *Deleuze and the Body*, edited by Laura Guillaume and Joe Hughes, 9–28. Edinburgh: Edinburgh University Press.

Coleman, Simon. 2022. *Powers of Pilgrimage. Religion in a World of Movement*. New York: New York University Press.

Colgan, Jeff D. 2018. "Climate Change and the Politics of Military Bases." *Global Environmental Politics* 18, no. 1: 33–51.

Colgan, William, Horst Machguth, Mike MacFerrin, Jeff D. Colgan, Dirk van As, and Joseph A. MacGregor. 2016. "The Abandoned Ice Sheet Base at Camp Century, Greenland, in a Warming Climate." *Geophysical Research Letters* 43, no. 15: 8091–96.

Conder, Claude. 1895. *Tentwork in Palestine*. London: A.P. Watt and Son.

Connor, Steven. 2003. "An Air That Kills: A Familiar History of Poison Gas." http://stevenconnor.com/gas.html.

Cook, Joanna. 2010. *Meditation in Modern Buddhism: Renunciation and Change in Thai Monastic Life*. Cambridge: Cambridge University Press.

Corsín Jiménez, Alberto. 2021. "Anthropological Entrapments: Ethnographic Analysis before and after Relations and Comparisons." *Social Analysis: The International Journal of Anthropology* 65, no. 3: 110–30.

Coulthard, Glen. 2014. *Red Skin, White Masks: Rejecting the Colonial Politics of Recognition*. Minneapolis: University of Minnesota Press.

Cousteau, Jacques, and Louis Malle, dir. 1956. *The Silent World*. Titanus.

Critchley, Simon. 2002. *On Humor*. London: Routledge.

Cruikshank, Julie. 2005. *Do Glaciers Listen? Local Knowledge, Colonial Encounters, and Social Imagination*. Vancouver: University of British Columbia Press.

Das, Veena. 2007. *Life and Words: Violence and the Descent into the Ordinary*. Berkeley: University of California Press.

Davies, Thom. 2019. "Slow Violence and Toxic Geographies: 'Out of Sight' to Whom?" *Environment and Planning C: Politics and Space*. April 10. https://doi.org/10.1177/2399654419841063.

Debray, Régis. 2004. *God: An Itinerary*. London: Verso.

Deleuze, Gilles. 1990. *The Logic of Sense*. New York: Columbia University Press.

Deleuze, Gilles. 2014. *Difference and Repetition: Rejecting the Colonial Politics of Recognition*. Minneapolis: University of Minnesota Press.

Deleuze, Gilles, and Félix Guattari. 1983. *Anti-Oedipus: Capitalism and Schizophrenia*. Minneapolis: University of Minnesota Press.

Deleuze, Gilles, and Félix Guattari. 1987. *A Thousand Plateaus: Capitalism and Schizophrenia*. Minneapolis: University of Minnesota Press.

Deleuze, Gilles, and Claire Parnet. 1987. *Dialogues*. New York: Columbia University Press.

Deleuze, Gilles, and Claire Parnet. 2007. *Dialogues II*. New York: Columbia University Press.

Deloria, Vine, Jr. 1972. *God Is Red: A Native View of Religion*. Ann Arbor: University of Michigan Press.

Derrida, Jacques. 1992. *Given Time: 1. Counterfeit Money*. Chicago: University of Chicago Press.

Derrida, Jacques. 2000. *Of Hospitality*. Palo Alto, CA: Stanford University Press.

Descola, Philippe. 2005. *Beyond Nature and Culture*. Chicago: University of Chicago Press.

Despret, Vinciane. 1996. *Naissance d'une Théorie Éthologique: La Danse Du Cratérope Écaillé*. Vol. Collection Les empêcheurs de penser en rond. Le Plessis-Robinson [France]: Synthélabo.

Dilley, Roy. 2007. "The Construction of Ethnographic Knowledge in a Colonial Context: The Case of Henri Gaden (1867–1939)." In *Ways of Knowing: Anthropological Approaches to Crafting Experience and Knowledge*, edited by Mark Harris, 139–57. Oxford: Berghahn.

Doane, Mary Ann. 1987. *The Desire to Desire: The Woman's Film of the 1940s*. Indianapolis: Indiana University Press.

Dolphijn, Rick, ed. 2018a. "Introduction: Michel Serres and the Times." In *Michel Serres and the Crises of the Contemporary*, 1–9. London: Bloomsbury.

Dolphijn, Rick, ed. 2018b. "The World, the Mat(t)Er of Thought." In *Michel Serres and the Crises of the Contemporary*, 127–45. London: Bloomsbury.

Douglas, Mary. 1966. *Purity and Danger: An Analysis of Concepts of Pollution*. London: Routledge.

Douglas, Mary. 1986. *Risk Acceptability According to the Social Sciences*. New York: Routledge.

Douglas, Mary. 2003. *Natural Symbols*. London: Routledge.

Douglas, Mary, and Aaron Wildavsky. 1983. *Risk and Culture: An Essay on the Selection of Technological and Environmental Dangers*. Berkeley: University of California Press.

Dresch, Paul. 1998. "Mutual Deception: Totality, Exchange, and Islam in the Middle East." In *Marcel Mauss: A Centenary Tribute*, edited by Wendy James and N. J. Allen, 111–33. New York: Berghahn Books.

Dresch, Paul. 2012. "Legalism, Anthropology, and History: A View from Part of Anthropology." In *Legalism: Anthropology and History*, edited by P. Dresch and Hannah Skoda, 1–38. Oxford: Oxford University Press.

Dulley, Iracema. 2019. *On the Emic Gesture. Difference and Ethnography in Roy Wagner*. London: Routledge.

Dunn, Elizabeth C. 2022. "Event." Anthropological Theory Commons, accessed July 27, 2023. https://www.at-commons.com/2022/06/03/event/.

Dupuy, Jean-Pierre. 2009. *On the Origins of Cognitive Science: The Mechanization of the Mind*. Cambridge, MA: MIT Press.

Ehrlich, Pippa, and James Reed, dir. 2020. *My Octopus Teacher*. Netflix Original Documentary.

Epictetus. 1989. *Discourses, Books 3–4. Fragments. The Encheiridion*. Cambridge, MA: Loeb.

Epstein, Steven. 2022. *The Quest for Sexual Health: How an Elusive Ideal Has Transformed Science, Politics, and Everyday Life*. Chicago: University of Chicago Press.

Eriksen, Thomas Hylland. 2018. "Being Irrelevant in a Relevant Way: Anthropology and Public Wisdom." *Kritisk Etnografi—Swedish Journal of Anthropology* 1, no. 1: 43–54.

Erlmann, Veit. 2004. *Hearing Cultures: Essays on Sound, Listening and Modernity*. London: Bloomsbury.

Evans-Pritchard, Edward. E. (1937) 1976. *Witchcraft, Oracles, and Magic among the Azande*. Oxford: Oxford University Press.

Fabian, Johannes. 1983. *Time and the Other: How Anthropology Makes Its Object*. New York: Columbia University Press.

Fadil, Nadia. 2009. "Managing Affects and Sensibilities: The Case of Not-Handshaking and Not-Fasting." *Social Anthropology* 17, no. 4: 439–54.

Fanon, Frantz. 1967. *Black Skin, White Masks*. New York: Grove Press.

Faubion, James D. 2020. "Of Uncertainty, Sophiology, and Governance: Zen and the Art of Scenario Planning." In *Capturing the Ineffable*, edited by Philip Y. Kao and Joseph S. Alter, 155–78. Toronto: University of Toronto Press.

FDA (US Food and Drug Administration). n.d. "Best Practices for Retail Food Stores, Restaurants, and Food Pick-Up/Delivery Services during the COVID-19 Pandemic." https://www.fda.gov/food/food-safety-during-emergencies/best-practices-retail-food -stores-restaurants-and-food-pick-updelivery-services-during-covid-19.

Ferreira da Silva, Denise. 2014. "Toward a Black Feminist Poethics: The Quest(ion) of Blackness toward the End of the World." *The Black Scholar* 44, no. 2: 81–97.

Ferreira da Silva, Denise. 2016. "On Difference without Separability. Catalogue, 32a São Paulo Art Biennial, 'Incerteza viva' (Living Uncertainty)." November 17. https://issuu .com/amilcarpacker/docs/denise_ferreira_da_silva.

Ferry, Luc. 1996. *The New Ecological Order*. Chicago: University of Chicago Press.

Figlio, Karl. 2012. "The Dread of Sameness: Social Hatred and Freud's 'Narcissism of Minor Differences.'" In *Psychoanalysis and Politics: Exclusion and the Politics of Representation. Theoretical Reflections,* edited by Lene Auested, 7–24. London: Karnac Books.

Finnegan, Ruth. 1967. *Limba Stories and Story-Telling*. Oxford: Clarendon Press.

Flyn, Cal. 2021. *Islands of Abandonment: Nature Rebounding in the Post-Human Landscape*. New York: Viking.

Fortun, Kim. 2012. "Ethnography In Late Industrialism." *Cultural Anthropology* 27, no. 3: 446–64.

Foucault, Michel. 1980. "The Confession of the Flesh." In *Power/Knowledge: Selected Interviews and Other Writings 1972-1977,* edited by C. Gordon, 194–228. New York: Pantheon Books.

Foucault, Michel. 1988. "Technologies of the Self." In *Technologies of the Self. A Seminar with Michel Foucault,* edited by L. H. Martin, 16–49. London: Tavistock.

Foucault, Michel. 2001. *Fearless Speech*. Cambridge, MA: MIT Press.

Fournier, Marcel. (1994) 2006. *Marcel Mauss: A Biography*. Princeton, NJ: Princeton University Press.

Freud, Sigmund. (1919) 1973. *The Uncanny*. Cambridge, MA: MIT Press.

Freud, Sigmund. (1930) 2010. *Civilization and Its Discontents*. Hartford, CT: Martino Fine Books.

Galvez, Paul. 2013. "Second Nature, a Conversation with Michel Serres." *Artforum,* September 2013.

Gamble, Clive. 2007. *Origins and Revolutions: Human Identity in Earliest Prehistory*. New York: Cambridge University Press.

Garbus, Liz, dir. 2021. *Becoming Cousteau*. National Geographic Documentary Films.

Geertz, Clifford. 1973. *The Interpretation of Cultures*. New York: Basic Books.

Ghosh, Amitav. 2021. *The Nutmeg's Curse: Parables for a Planet in Crisis*. Chicago: University of Chicago Press.

Gilroy, Paul. 1993. *The Black Atlantic: Modernity and Double-Consciousness*. Cambridge, MA: Harvard University Press.

Gilroy, Paul. 2014. "Lecture I: Suffering and Infrahumanity." The Tanner Lectures in Human Values. Yale University, February 21. https://tannerlectures.utah.edu/Gilroy%20 manuscript%20PDF.pdf.

Gilsenan, Michael. 1990. "Very Like a Camel: The Appearance of an Anthropologist's Middle East." In *Localising Strategies: Regional Traditions of Ethnographic Writing,* edited by Richard Fardon, 222–39. Washington, DC: Smithsonian Institution Press.

Girard, René. 1989. *The Scapegoat*. Baltimore: Johns Hopkins University Press.

Girard, René. 2010. *Battling to the End*. East Lansing: Michigan State University Press.

Glissant, Édouard. 1997. *Poetics of Relation*. Ann Arbor: University of Michigan Press.

Glowchewski, Barbara. 2021. *Réveiller les esprits de la terre*. Paris: Dehors.

Godelier, Maurice. 1999. *The Enigma of the Gift*. Chicago: University of Chicago Press.

Gomart, Emilie, and Antoine Hennion. 1999. "A Sociology of Attachment: Music Amateurs, Drug Users." *The Sociological Review* 47, no. S1: 220–47.

Gombrich, E. H. 2002. *A Preference for the Primitive: Episodes in the History of Western Taste and Art*. London: Phaidon Press, Ltd.

Gregory, Derek. 2016. "The Natures of War." *Antipode* 48, no. 1: 3–56.

Grote, Gudela. 2012. "Safety Management in Different High-Risk Domains—All the Same? *Safety Science* 50, no. 10: 1983–92.

Grove, Jairus Victor. 2019. *Savage Ecology: War and Geopolitics at the End of the World*. Durham, NC: Duke University Press.

Gusterson, Hugh. 2004. *People of the Bomb: Portraits of America's Nuclear Complex*. Minneapolis: University of Minnesota Press.

Hadot, Pierre. 1992. *La Citadelle Intérieure*. Paris: Fayard.

Hadot, Pierre. 1995. *Qu'Est-Ce Que La Philosophie Antique?* Paris: Gallimard.

Hadot, Pierre. 2003. *La Philosophie Comme Maniere de Vivre*. Paris: Livre de Poche.

Harari, Josué V., and David F. Bell. 1982. "Introduction. Journal á Plusieurs Voies." In *Hermes. Literature, Science, Philosophy*, edited by Josué V. Harari and David F. Bell, ix–xl. Baltimore: Johns Hopkins University Press.

Haraway, Donna Jeanne. 1988. "Situated Knowledges: The Science Question in Feminism and the Privilege of Partial Perspective." *Feminist Studies* 14, no. 3: 575.

Haraway, Donna Jeanne. 1989. *Primate Visions: Gender, Race and Nature in the World of Modern Science*. London: Routledge.

Haraway, Donna Jeanne. 1997. *Modest-Witness@Second-Millennium.FemaleMan-Meets-OncoMouse: Feminism and Technoscience*. New York: Routledge.

Hart, Keith. 2007. "Marcel Mauss: In Pursuit of the Whole." *Comparative Studies in Society and History* 49, no. 2: 473–85.

Hartman, Saidiya. 1997. *Scenes of Subjection: Terror, Slavery, and Self-Making in Nineteenth-Century America*. Oxford: Oxford University Press.

Hayles, N. Katherine. 1988. "Two Voices, One Channel: Equivocation in Michel Serres." *SubStance* 17, no. 3: 3–12.

Hayman, Ronald. 1987. *Sartre: A Life*. New York: Simon and Schuster.

Henig, David. 2012. "Iron in the Soil: Living with Military Waste in Bosnia-Herzegovina." *Anthropology Today* 28, no. 1: 21–23.

Henig, David. 2019. "Living on the Frontline: Indeterminacy, Value, and Military Waste in Postwar Bosnia-Herzegovina." *Anthropological Quarterly* 92, no. 1: 85–110.

Henig, David. 2020. "Emptiness and Deadly Environments. Society for Cultural Anthropology." https://culanth.org/fieldsights/emptiness-and-deadly-environments.

Hermez, Sami. 2017. *War Is Coming: Between Past and Future Violence in Lebanon*. Philadelphia: University of Pennsylvania Press.

Hertz-Ohmes, Peter. 1987. "Serres and Deleuze: Hermes and Humor." *Canadian Review of Comparative Literature* 14, no. 2: 239–50.

Herzogenrath, Bernd, ed. 2012. *Time and History in Deleuze and Serres*. London: Continuum Books.

Heywood, Paolo. 2019. "Fascism, Uncensored: Legalism and Neo-Fascist Pilgrimage in Predappio, Italy." *Terrain*, no. 72 (November). http://journals.openedition.org/terrain /18955.

High, Casey, Ann H. Kelly, and Jonathan Mair, eds. 2012. *The Anthropology of Ignorance: An Ethnographic Approach*. New York: Palgrave.

Hirsch, Eric, and Charles Stewart. 2006. "Introduction: Ethnographies of Historicity." *History and Anthropology* 16, no. 3: 261–74.

Hirschkind, Charles. 2006. *The Ethical Soundscape: Cassette Sermons and Islamic Counterpublics*. New York: Columbia University Press.

Højer, Lars, and Andreas Bandak. 2015. "Introduction: The Power of Example." *Journal of the Royal Anthropological Institute* 21, no. S1: 1–17.

Holm, Nicholas. 2017. *Humour as Politics. The Political Aesthetics of Contemporary Comedy*. London: Palgrave Macmillan.

hooks, bell. 1992. "The Oppositional Gaze: Black Female Spectators." In *Black Looks: Race and Representation*, 115–31. Boston: South End Press.

Horvath, Agnes. 2021. *Political Alchemy: Technology Unbounded*. London: Routledge.

Horvath, Agnes, and Bjørn Thomassen. 2008. "Mimetic Errors in Liminal Schismogenesis: On the Political Anthropology of the Trickster." *International Political Anthropology* 1, no. 1: 3–24.

Humphrey, Caroline. 2015. "Detachment, Difference and Separation: Lévi-Strauss at the Wedding Feast." In *Detachment: Essays on the Limits of Relational Thinking*, edited by Matei Candea, Joanna Cook, Catherine Trundle, and Thomas Yarrow, 147–67. Manchester: Manchester University Press.

Iengo, Ilenia, Panagiota Kotsila, and Ingrid L. Nelson. 2023. "Ouch! Eew! Blech! A Trialogue on Porous Technologies, Places and Embodiments." In *Contours of Feminist Political Ecology: Gender, Development and Social Change*, edited by W. Harcourt, A. Agostino, R. Elmhirst, M. Gómez, and P. Kotsila, 76–103. New York: Palgrave Macmillan.

The Independent Sage. 2020. "UK Government Messaging and Its Association with Public Understanding and Adherence to COVID-19 Mitigations: Five Principles and Recommendations for a COVID Communication Reset." https://www.independentsage.org/wp -content/uploads/2020/11/Messaging-paper-FINAL-1-1.pdf.

Ingold, Tim. 2021. *Imagining for Real: Essays on Creation, Attention and Correspondence*. London: Routledge.

Ingold, Tim. 2022. "On Not Knowing and Paying Attention: How to Walk in a Possible World." *Irish Journal of Sociology* 31, no. 1: 20–36.

Jackson, Michael. 1982. *Allegories of the Wilderness: Ethics and Ambiguity in Kuranko Narratives*. Bloomington: Indiana University Press.

Jackson, Michael. 1998. *Minima Ethnographica: Intersubjectivity and the Anthropological Project*. Chicago: University of Chicago Press.

Jackson, Michael. 2002. *The Politics of Storytelling: Violence, Transgression and Intersubjectivity*. Chicago: University of Chicago Press.

Jackson, Zakiyyah Iman. 2020. *Becoming Human: Matter and Meaning in an Antiblack World*. New York: NYU Press.

James, William. 2017. *Pragmatism*. Milpitas, CA: Jovian Press.

Johnson, Khari. 2022. "Russian Missiles and Space Debris Could Threaten Satellites." *Wired*. https://www.wired.com/story/space-debris-russia-satellites/.

Josephides, Lisette. 1991. "Metaphors, Metathemes, and the Construction of Sociality: A Critique of the New Melanesian Ethnography." *Man* 26, no. 1: 145–61.

Jovanović, Deana. 2018. "Prosperous Pollutants: Bargaining with Risks and Forging Hopes in an Industrial Town in Eastern Serbia." *Ethnos* 83, no. 3: 489–504.

Jullien, François. 2012. *The Great Image Has No Form, or On the Nonobject through Painting*. Chicago: University of Chicago Press.

Jung, Carl. G. 1972. *Four Archetypes*. London: Routledge and Kegan Paul.

Kao, Philip Y., and Joseph S. Alter, eds. 2020. *Capturing the Ineffable: An Anthropology of Wisdom*. Toronto: University of Toronto Press.

Kaufmann, Wanda Ostrowska. 1996. *The Anthropology of Wisdom Literature*. Westport, CT: Greenwood Publishing Group.

Keane, Webb. 2007. *Christian Moderns: Freedom and Fetish in the Mission Encounter*. Berkeley: University of California Press.

Khaldun, Ibn. 1967. *The Muqaddimah: An Introduction to History*. Princeton, NJ: Princeton University Press.

Khayyat, Munira. 2022. *A Landscape of War: Ecologies of Resistance and Survival in South Lebanon*. Oakland: University of California Press.

Kim, Eleana J. 2016. "Toward an Anthropology of Landmines: Rogue Infrastructure and Military Waste in the Korean DMZ." *Cultural Anthropology* 31, no. 2: 162–87.

Kirksey, Eben. 2015. *Emergent Ecologies*. Durham, NC: Duke University Press.

Kirtsoglou, Elisabeth. 2003. "The Other Then, the Other Now, the Other Within: Stereotypical Images and Narrative Captions of the Turk in Northern and Central Greece." *Journal of Mediterranean Studies* 13, no. 2: 189–213.

Kirtsoglou, Elisabeth, and Bob Simpson, eds. 2020. *The Time of Anthropology: Studies of Contemporary Chronopolitics*. London: Routledge.

Knight, Daniel M. 2012. "Cultural Proximity: Crisis, Time and Social Memory in Central Greece." *History and Anthropology* 23, no. 3: 349–74.

Knight, Daniel M. 2015. *History, Time, and Economic Crisis in Central Greece*. New York: Palgrave.

Knight, Daniel M. 2017. "Fossilized Futures: Topologies and Topographies of Crisis Experience in Central Greece." *Social Analysis* 61, no. 1: 26–40.

Knight, Daniel M. 2021. *Vertiginous Life: An Anthropology of Time and the Unforeseen*. New York: Berghahn.

Knight, Daniel M. 2022. *The Vertiginous: Temporalities and Affects of Social Vertigo: Ilinx*. Anthropological Theory Commons, June 3. https://www.at-commons.com/2022/06/03/ilinx/.

Knight, Daniel M., and Charles Stewart. 2016. "Ethnographies of Austerity: Temporality, Crisis and Affect in Southern Europe." *History and Anthropology* 27, no. 1: 1–18.

Koch, Gertrud. 1982. "Why Women Go to the Movies." *Jump Cut* 27. http://www.ejumpcut.org/archive/onlinessays/JC27folder/KochonWmSpectship.html.

Kockelman, Paul. 2010. "Enemies, Parasites, and Noise: How to Take Up Residence in a System without Becoming a Term in It." *Journal of Linguistic Anthropology* 20, no. 2: 406–21.

Krämer, Sybille. 2015. *Medium, Messenger, Transmission: An Approach to Media Philosophy*. Amsterdam: University of Amsterdam Press.

Kristeva, Julia. 1982. *The Powers of Horror: An Essay on Abjection*. New York: Columbia University Press.

Kuijt, Ian. 2009. "What Do We Really Know about Food Storage, Surplus, and Feasting in Preagricultural Communities?" *Current Anthropology* 50, no. 5: 641–44.

La Barre, Weston. 1972. *The Ghost Dance: Origins of Religion*. New York: Dell.

Lacan, Jacques. 1977. The Mirror Stage. In *Écrits: A Selection*, 1–7. New York: Norton.

La Fontaine, Jean de. 2007. *The Complete Fables of Jean de La Fontaine*. Champaign: University of Illinois Press.

Lambek, Michael. 2011. "Kinship as Gift and Theft: Acts of Succession in Mayotte and Israel." *American Ethnologist* 38, no. 1: 2–16.

Larkin, Brian. 2008. *Signal and Noise: Media, Infrastructure, and Urban Culture in Nigeria*. Durham, NC: Duke University Press.

Latour, Bruno. 1987. "The Enlightenment without the Critique: A Word on Michel Serres' Philosophy." *Royal Institute of Philosophy Lecture Series* 21: 83–97.

Latour, Bruno. 2005. *Reassembling the Social: An Introduction to Actor-Network-Theory*. Oxford: Oxford University Press.

Latour, Bruno. 2017a. "Anthropology at the Time of the Anthropocene: A Personal View of What Is to Be Studied." In *The Anthropology of Sustainability*, edited by M. Brightman and J. Lewis, 35–49. New York: Palgrave.

Latour, Bruno. 2017b. *Facing Gaia: Eight Lectures on the New Climatic Regime*. Cambridge: Polity.

Lemonnier, Pierre. 2012. "Entwined by Nature: Eels, Traps, and Ritual." In *Mundane Objects: Materiality and Non-Verbal Communication*, 45–62. Walnut Creek, CA: Left Coast Press.

Lemons, J. Derrick, ed. 2018. *Theologically Engaged Anthropology*. Oxford: Oxford University Press.

Lepselter, Susan. 2016. *The Resonance of Unseen Things: Poetics, Power, Captivity, and UFOs in the American Uncanny*. Ann Arbor: University of Michigan Press.

Lévi-Strauss, Claude. 1954. "The Mathematics of Man." *International Social Science Bulletin* 6: 581–90.

Lévi-Strauss, Claude. 1955. "The Structural Study of Myth." *Journal of American Folklore* 68, no. 270: 428–44.

Lévi-Strauss, Claude. 1963. *Structural Anthropology*. New York: Basic Books.

Lévi-Strauss, Claude. 1966. *The Savage Mind*. London: Weidenfeld and Nicolson.

Lévi-Strauss, Claude. 1969. *The Elementary Structures of Kinship*. Boston: Beacon Press.

Lévi-Strauss, Claude. 1970. *The Raw and the Cooked. Vol. 1 of Mythologiques*. New York: Harper Torchbooks.

Lévi-Strauss, Claude. 1973a. *From Honey to Ashes*. Translated by John Weightman and Doreen Weightman. London: Jonathan Cape.

Lévi-Strauss, Claude. 1973b. *From Honey to Ashes. Vol. 2, Introduction to a Science of Mythology*, translated by John Weightman and Doreen Weightman. New York: Harper and Row.

Lévi-Strauss, Claude. 1995. *The Story of Lynx*. Chicago: University of Chicago Press.

Lévi-Strauss, Claude. 2010. "The Story of Asdiwal." In *The Structural Study of Myth and Totemism*, edited by Edmund. R. Leach, 1–48. London: Routledge.

Liboiron, Max. 2021. *Pollution Is Colonialism*. Durham, NC: Duke University Press.

Loader, Ian. 2021. "Recognition and Redemption: Visions of Safety and Justice in Black Lives Matter." http://dx.doi.org/10.2139/ssrn.3840766.

Logan, Drake. 2018. "Toxic Violence: The Politics of Militarized Toxicity in Iraq and Afghanistan." *Cultural Dynamics* 30, no. 4: 253–83.

Long, Charles. 1986. *Significations: Signs, Symbols, and Images in the Interpretation of Religion*. Aurora, CO: The Davies Group.

Lowe, Celia. 2006. *Wild Profusion: Biodiversity Conservation in an Indonesian Archipelago*. Princeton, NJ: Princeton University Press.

Lowe, Celia. 2014. "Infection." *Environmental Humanities* 5, no. 1: 301–5.

Lowe, Celia, and Ursula Muenster. 2016. "The Viral Creep: Elephants and Herpes in Times of Extinction." *Environmental Humanities* 8, no. 1: 118–42.

Lucretius. 1968. *The Way Things Are: The De Rerum Natura of Titus Lucretius Carus*. Translated by Rolfe Humphries. Bloomington: Indianna University Press.

MacLeish, Kenneth, and Zoë H. Wool. 2018. "US Military Burn Pits and the Politics of Health." *Medical Anthropology Quarterly Critical Care Blog Series* (blog), 2018. https://medanthroquarterly.org/critical-care/2018/08/us-military-burn-pits-and-the-politics-of-health/.

Malara, Diego Maria. 2018. "The Alimentary Forms of Religious Life: Technologies of the Other, Lenience, and the Ethics of Ethiopian Orthodox Fasting." *Social Analysis* 62, no. 3: 21–41.

Malara, Diego Maria. 2022. "Exorcizing the Spirit of Protestantism: Ambiguity and Spirit Possession in an Ethiopian Orthodox Ritual." *Ethnos* 87, no. 4: 749–70.

Marcus Aurelius. 1889. *The Thoughts of the Emperor Marcus Aurelius Antoninus*. Boston: Little, Brown and Company.

Martin-Nielsen, Janet. 2013. "The Deepest and Most Rewarding Hole Ever Drilled: Ice Cores and the Cold War in Greenland." *Annals of Science* 70, no. 1: 47–70.

Masco, Joseph. 2015. "The Age of Fallout." *History of the Present: A Journal of Critical History* 5, no. 2: 137–68.

Masco, Joseph. 2021. *The Future of Fallout, and Other Episodes in Radioactive World-Making*. Durham, NC: Duke University Press.

Mauss, Marcel. (1925) 1967. *The Gift: Forms and Functions of Exchange in Archaic Societies*. Translated by Ian Cunnison. New York: Norton.

Mauss, Marcel. 1954. *The Gift*. Translated by Ian Cunnison. London: Cohen and West.

Mauss, Marcel, 1973. "Techniques of the Body." *Economy and Society* 2, no. 1: 70–88.

Max Planck Institute for the History of Science. 2023. *Nuclear Anthropocene*. https://www.anthropocene-curriculum.org/anthropogenic-markers/nuclear-anthropocene#read-more.

Mayblin, Maya. 2014. "People Like Us: Intimacy, Distance, and the Gender of Saints." *Current Anthropology* 55, no. S10: 271–80.

Mayblin, Maya, and Diego Maria Malara. 2018. "Introduction: Lenience in Systems of Religious Meaning and Practice." *Social Analysis* 62, no. 3: 1–20.

Mayblin, Maya, Kristin Norget, and Valentina Napolitano. 2017. "Introduction: The Anthropology of Catholicism." In *The Anthropology of Catholicism: A Reader,* edited by Kristin Norget, Valentina Napolitano, and Maya Mayblin, 1–30. Berkeley: University of California Press.

Maybury-Lewis, David. 1992. *Millennium: Tribal Wisdom and the Modern World.* New York: Viking.

Mercier, Lucie Kim-Chi. 2019. "Michel Serres' Leibnizian Structuralism." *Angelaki* 24, no. 6: 3–21.

Merleau-Ponty, Maurice. 1962. *Phenomenology of Perception.* London: Routledge and Kegan Paul.

Meyer, Birgit. 2011. "Mediation and Immediacy: Sensational Forms, Semiotic Ideologies and the Question of the Medium." *Social Anthropology* 19, no. 1: 23–39.

Millward, Jessica. 2017. "From the Ocean Floor: Death, Memory and the Atlantic Slave Trade." *Black Perspectives,* March 8. https://www.aaihs.org/from-the-ocean-floor-death -memory-and-the-atlantic-slave-trade.

Milman, Thomas. 2022. *The Insect Crisis: The Fall of the Tiny Empires That Run the World.* New York: Atlantic Books.

Milton, John. (1667) 2013. *Paradise Lost.* Cambridge: Cambridge University Press.

Mintz, Lawrence. E. 1998. "Stand-Up Comedy as Social and Cultural Mediation." In *What's So Funny? Humor in American Culture,* edited by Nancy. A. Walker, 193–203. Wilmington, DE: Scholarly Resources.

Moi, Toril. 2021. "I Came with a Sword." *London Review of Books* 43, no. 13. https://www.lrb .co.uk/the-paper/v43/n13/toril-moi/i-came-with-a-sword.

Molière. 2021. *Molière. The Complete Richard Wilbur Translations.* New York: Library of America.

Moreiras, Alberto. 2020. *Against Abstraction.* Austin: University of Texas Press.

Morelon, Claire. 2019. "Sounds of Loss: Church Bells, Place, and Time in the Habsburg Empire during the First World War." *Past and Present* 244, no. 1: 195–234.

Moreton-Robinson, Aileen. 2015. *The White Possessive: Property, Power, and Indigenous Sover-eignty.* Minneapolis: University of Minnesota Press.

Moten, Fred, and Stefano Harney. 2013. *The Undercommons: Fugitive Planning and Black Study.* New York: Minor Compositions.

Mulvey, Laura. 1973. "Visual Pleasure and Narrative Cinema." *Screen* 16, no. 3: 6–18.

Murer, Jeffrey. 2009. "Constructing the Enemy-Other: Anxiety, Trauma and Mourning in the Narratives of Political Conflict." *Psychoanalysis, Culture and Society* 14: 109–30.

Musa, Snježana, Željka Šiljković, and Dario Šakić. 2017. "Geographic Reflections of Mine Pollution in Bosnia and Herzegovina and Croatia." *Revija Za Geografijo—Journal of Geography* 12, no. 2: 53–70.

Needham, Rodney. 1971. *Rethinking Kinship and Marriage.* New York: Barnes and Noble.

Nesteroff, Kliph. 2015. *The Comedians: Drunks, Thieves, Scoundrels and the History of American Comedy.* New York: Grove Press.

Newell, Sasha. 2018. "Uncontained Accumulation: Hidden Heterotopias of Storage and Spillage." *History and Anthropology* 29, no. 1: 37–41.

Nguyen, Viet Thanh. 2016. *Nothing Ever Dies: Vietnam and the Memory of War.* Cambridge, MA: Harvard University Press.

Nielsen, Henry, Kristian Hvidtfelt Nielsen, and Heidi Flegal. 2021. *Camp Century: The Untold Story of America's Secret Arctic Military Base under the Greenland Ice.* New York: Columbia University Press.

Nielsen, Morten. 2018. "An Army of Comedy: Political Jokes and Tropic Ambiguity in the Trump Era." In *The Politics of Joking: Anthropological Engagements,* edited by Jana K. Rehak and Susanna Trnka, 152–62. London: Routledge.

Nielsen, Morten. 2019a. "Beyond the Punchline: Mythic Involution as Recursive Futural Orientation among Stand-Up Comics in New York City." *American Ethnologist* "Orientations to the Future," edited by Rebecca Bryant and Daniel M. Knight. http://americanethnologist.org/features/collections/orientations-to-the-future/beyond-the-punchline.

Nielsen, Morten. 2019b. "Comedic Lies as Transitory Truths." *Anthropology News* 60, no. 2: e94–e97.

Nietzsche, Friedrich. 1974. *The Gay Science.* New York: Random House.

Nikolovska, Astrea. Forthcoming. *Social Life of Depleted Uranium: Remembering NATO Bombing through Radioactive Glasses in Serbia.*

Nilges, Mathias. 2009. "We Need the Stars: Community, and the Absent Father in Octavia Butler's *Parable of the Sower* and *Parable of the Talents.*" *Callalo* 32, no. 4: 1332–52.

Nixon, Rob. 2013. *Slow Violence and the Environmentalism of the Poor.* Cambridge, MA: Harvard University Press.

Oring, Elliott. 2016. *Joking Asides: The Theory, Analysis, and Aesthetics of Humor.* Boulder: University Press of Colorado.

Orsi, Robert. 2017. "What Is Catholic about the Clergy Sex Abuse Crisis?" In *The Anthropology of Catholicism: A Reader,* edited by Kristin Norget, Valentina Napolitano, and Maya Mayblin, 282–92. Berkeley: University of California Press.

Parry, Jonathan. 1986. "The Gift, the Indian Gift, and 'the Indian Gift.'" *Man* 21: 453–73.

Paz, Octavio. 1985. *The Labyrinth of Solitude and Other Writings.* New York: Grove Press.

Peirce, Charles Sanders. 1994. *Collected Papers of Charles Sanders Peirce.* Cambridge, MA: Harvard University Press.

Pentzopoulos, Dimitri. 1962. *The Balkan Exchange of Minorities and Its Impact on Greece.* London: C. Hurst and Co.

Peristiany, John G. 1992. "The Sophron—a Secular Saint? Wisdom and the Wise in a Cypriot Community." In *Honor and Grace in Anthropology,* edited by John G. Peristiany and Julian Alfred Pitt-Rivers, 103–27. Cambridge: Cambridge University Press.

Peters, John Durham. 2015. *The Marvelous Clouds: Toward a Philosophy of Elemental Media.* Chicago: University of Chicago Press.

Phillips, Patti, and Catherine Phillips. 2010. "The Nature of Feminist Science Studies." *Resources for Feminist Research* 33, no. 3–4: 9. Gale Academic OneFile. https://link.gale.com/apps/doc/A257127025/AONE?u=anon~871e8207&sid=googleScholar&xid=aea2f86f.

Pigaht, Janina, and Rick Dolphin. 2018. "A New Culture to Suit the World. Interview with Michel Serres by Janina Pigaht and Rick Dolphijn." In *Michel Serres and the Crises of the Contemporary,* edited by Rick Dolphijn, 169–76. London: Bloomsbury.

Pina-Cabral, João de. 1986. *Sons of Adam, Daughters of Eve: The Peasant Worldview of the Alto Minho.* Oxford: Clarendon Press.

Piot, Charles. 2010. *Nostalgia for the Future: West Africa after the Cold War*. Chicago: University of Chicago Press.

Pipyrou, Stavroula. 2014. "Cutting *Bella Figura*: Crisis, and Secondhand Clothes in South Italy." *American Ethnologist* 41, no. 3: 532–46.

Pipyrou, Stavroula. 2016. *The Grecanici of Southern Italy: Governance, Violence, and Minority Politics*. Philadelphia: University of Pennsylvania Press.

Pipyrou, Stavroula. 2021. "On Security, Minorities, and Opportunistic Narcissism." *Journal on Ethnopolitics and Minority Issues in Europe* 20, no. 1: 24–44.

Pipyrou, Stavroula, and Antonio Sorge. 2021. "Emergent Axioms of Violence: Toward an Anthropology of Post-Liberal Modernity." *Anthropological Forum* 31, no. 3: 225–40.

Pitt-Rivers, Julian. 1977. *The Fate of Shechem or the Politics of Sex. Essays in the Anthropology of the Mediterranean*. Cambridge: Cambridge University Press.

Pitt-Rivers, Julian. 1992. "Postscript: the Place of Grace in Anthropology." In *Honor and Grace in Anthropology*, edited by Julian Pitt-Rivers and John. G. Peristiany, 215–46. Cambridge: Cambridge University Press.

Pizzorno, Alessandro. 2000. "Risposte e proposte." In *Identità, riconoscimento e scambio: Saggi in onore di Alessandro Pizzorno*, edited by Donatella Della Porta, Monica Greco, and Arpad Szakolczai, 197–245. Bari: Laterza.

Pizzorno, Alessandro. 2007. *Il velo della diversità: Studi su razionalità e riconoscimento*. Milan: Feltrinelli.

Pizzorno, Alessandro. 2008. "Rationality and Recognition." In *Approaches and Methodologies in the Social Sciences: A Pluralist Perspective*, edited by Donatella della Porta and Michael Keating, 162–74. Cambridge: Cambridge University Press.

Povinelli, Elizabeth A. 2011. *Economies of Abandonment: Social Belonging and Endurance in Late Liberalism*. Durham, NC: Duke University Press.

Povinelli, Elizabeth A. 2016. *Geontologies: A Requiem to Late Liberalism*. Durham, NC: Duke University Press.

Povinelli, Elizabeth A. 2021. *Between Gaia and Ground*. Durham, NC: Duke University Press.

Radin, Paul. 1972. *The Trickster: A Study in American Indian Mythology*. New York: Schocken.

Radin, Paul. 2017. *Primitive Man as Philosopher*. New York: New York Review Books.

Reason, James. 1997. *Managing the Risks of Organisational Accident*. Hampshire: Ashgate.

Reicher, Stephen, John Drury, Ann Pheonix, and Elizabeth Stokoe. 2021. "Government Ministers Not Wearing Masks Was Bad Enough, but Their Defence of This Position Is Even Worse." *BMJ* 375:n2682. https://doi.org/10.1136/bmj.n2682.

Reiman, Teemu, and Elina Pietikäinen. 2012. "Leading Indicators of System Safety— Monitoring and Driving the Organizational Safety Potential." *Safety Science* 50: 1993–2000.

Reno, Joshua. 2015. "Waste and Waste Management." *Annual Review of Anthropology* 44, no. 1: 557–72.

Ricoeur, Paul. 1998. *Critique and Conviction: Conversations with François Azouvi and Marc de Launay*. New York: Columbia University Press.

Riddick, Iain. 2013. *Nova: The Earth from Space*. Darrow Smithson Productions Ltd.

Robbins, Joel. 2013. "Beyond the Suffering Subject: Toward an Anthropology of the Good." *Journal of the Royal Anthropological Institute* 19, no. 3: 447–62.

Robbins, Joel. 2015. "Ritual, Value, and Example: On the Perfection of Cultural Representations: Ritual, Value, and Example." *Journal of the Royal Anthropological Institute* 21, no. S1: 18–29.

Robbins, Joel. 2020. *Theology and the Anthropology of Christian Life*. Oxford: Oxford University Press.

Rodgers, David. 2013. "The Filter Trap: Swarms, Anomalies, and the Quasi-Topology of Ikpeng Shamanism." *HAU: Journal of Ethnographic Theory* 3, no. 3: 77–105.

Rousseau, Jean-Jacques. 1962. *Discours sur l'origine de l'inégalité parmi les hommes*. Paris: Editions Garnier.

Rousseau, Katherine. 2016. "Pilgrimage, Spatial Interaction, and Memory at Three Marian Sites." PhD diss., University of Denver.

Sahlins, Marshall. 1968. "On the Sociology of Primitive Exchange." In *The Relevance of Models for Social Anthropology*, edited by Michael Banton, 139–236. London: Tavistock.

Sahlins, Marshall. 1972. *Stone Age Economics*. New Brunswick, NJ: Aldine Transaction.

Sanchez, Andrew, James G. Carrier, Christopher Gregory, James Laidlaw, Marilyn Strathern, Yunxiang Yan, and Jonathan Parry. 2017. "The Indian Gift: A Critical Debate." *History and Anthropology* 28, no. 5: 553–83.

Sartre, Jean-Paul, and Benny Levy. 2007. *Hope Now: The 1980 Interviews*. Translated by Adrian van der Hoven. Chicago: University of Chicago Press.

Scheele, Judith, and Andrew Shryock. 2019. "Introduction: On the Left Hand of Knowledge." In *The Scandal of Continuity in Middle East Anthropology: Form, Duration, Difference*, edited by Judith Scheele and Andrew Shryock, 1–26. Bloomington: Indiana University Press.

Scheper-Hughes, Nancy. 2008. "A Talent for Life: Reflections on Human Vulnerability and Resilience." *Ethnos* 73, no. 1: 25–56.

Schloss, Jeffrey. 2000. "Wisdom Traditions as Mechanisms for Organismal Integration: Evolutionary Perspectives on Homoestatic 'Laws of Life.'" In *Understanding Wisdom: Sources, Science and Society*, edited by Warren S. Brown, 154–91. Radnor, PA: Templeton Foundation Press.

Schneider, David. 1984. *A Critique of the Study of Kinship*. Ann Arbor: University of Michigan Press.

Schüll, Natasha Dow. 2012. *Addiction by Design: Machine Gambling in Las Vegas*. Princeton, NJ: Princeton University Press.

Schutz, Alfred. 1973. *Collected Papers*, Vol. 1. The Hague: Martinus Nijhoff.

Seaver, Nick. 2019. "Captivating Algorithms: Recommender Systems as Traps." *Journal of Material Culture* 24, no. 4: 421–36.

Sebald, W. G. 2003. *On the Natural History of Destruction*. New York: Modern Library.

Serres, Michel. 1968. *Hermès I: La Communication*. Paris: Éditions de Minuit.

Serres, Michel. 1980. *Hermes V: Le Passage du nord-ouest*. Paris: Editions de Minuit.

Serres, Michel. (1980) 1982. *The Parasite*. Baltimore: Johns Hopkins University Press.

Serres, Michel. (1980) 2001. *The Birth of Physics*. Manchester: Clinamen Press.

Serres, Michel. (1980) 2007. *The Parasite*. Minneapolis: University of Minnesota Press.

Serres, Michel. (1980) 2014. *Le Parasite*. Paris: Fayard.

Serres, Michel. 1982a. "The Apparition of Hermes: Dom Juan." In *Hermes. Literature, Science, Philosophy*, edited by Josué V. Harari and David F. Bell, 3–14. Baltimore: Johns Hopkins University Press.

Serres, Michel. 1982b. *Hermes. Literature, Science, Philosophy*. Baltimore: Johns Hopkins University Press.

Serres, Michel. (1982) 1995. *Genesis*. Ann Arbor: University of Michigan Press.

Serres, Michel. (1982) 1997. *Genesis*. Translated by Genevieve James and James Nielson. Revised ed. Ann Arbor: University of Michigan Press.

Serres, Michel. (1983) 1991. *Rome: The Book of Foundations*. Stanford, CA: Stanford University Press.

Serres, Michel. (1983) 2015. *Rome: The First Book of Foundations*. London: Bloomsbury.

Serres, Michel. (1985) 2008. *The Five Senses: A Philosophy of Mingled Bodies*. New York: Continuum.

Serres, Michel. (1985) 2015. *The Five Senses: A Philosophy of Mingled Bodies*. London: Bloomsbury.

Serres, Michel. 1986. *Détachement: Apologue*. Paris: Flammarion.

Serres, Michel. (1986) 1989. *Detachment*. Athens: Ohio University Press.

Serres, Michel. 1987. *L'Hermaphrodite: Sarrasine sculpteur*. Paris: Flammarion.

Serres, Michel. (1987) 2014. *Statues: The Second Book of Foundations*. London: Bloomsbury.

Serres, Michel. (1990) 1995. *The Natural Contract*. Ann Arbor: University of Michigan Press.

Serres, Michel. (1991) 1997. *The Troubadour of Knowledge*. Translated by Sheila Faria Glaser and William Paulson. Ann Arbor: University of Michigan Press.

Serres, Michel. 1992a. *Le Contrat Naturel*. Paris: Flammarion.

Serres, Michel. 1992b. "The Natural Contract." *Critical Inquiry* 19, no. 1: 1–21.

Serres, Michel. 1993. *La Légende des Anges* [Angels: A modern myth]. Paris: Flammarion.

Serres, Michel. (1993) 2017. *Geometry: The Third Book of Foundations*. London: Bloomsbury.

Serres, Michel. 1999. *Variations Sur le Corps*. Paris: Le Pommier.

Serres, Michel. (1999) 2011. *Variations on the Body*. Minneapolis: Univocal.

Serres, Michel. (1999) 2012. *Variations on the Body*. Minneapolis: University of Minnesota Press.

Serres, Michel. (2001) 2019. *Hominescence*. London: Bloomsbury.

Serres, Michel. 2003. *Jules Verne: La science et l'homme contemporain*. Paris: Le Pommier.

Serres, Michel. (2003) 2018. *The Incandescent*. London: Bloomsbury.

Serres, Michel. (2004) 2020. *Branches: A Philosophy of Time, Event and Advent*. London: Bloomsbury.

Serres, Michel. 2006. *L'Art des Ponts: Homo Pontifex*. Paris: Le Pommier.

Serres, Michel. (2008) 2011. *Malfeasance: Appropriation through Pollution?* Stanford: Stanford University Press.

Serres, Michel. (2009) 2014. *Times of Crisis: What the Financial Crisis Revealed and How to Reinvent Our Lives and Future*. London: Bloomsbury.

Serres, Michel. 2010. *Biogée* [Biogea]. Paris: Le Pommier.

Serres, Michel. 2012a. "Difference: Chaos in the History of the Science." *Environment and Planning D: Society and Space* 30: 369–80.

Serres, Michel. 2012b. *Petite Poucette*. Paris: Editions Le Pommier.

Serres, Michel. (2012) 2015. *Thumbelina: The Culture and Technology of Millennials*. London: Rowman and Littlefield.

Serres, Michel. 2015a. *Eyes*. London: Bloomsbury.

Serres, Michel. 2015b. *Le Gaucher Boiteux: Puissance de la Pensée*. Paris: Éditions le Pommier.

Serres, Michel. 2019. *Relire le Relié*. Paris: Pommier.

Serres, Michel. (2019) 2022. *Religion: Rereading What Is Bound Together*. Stanford, CA: Stanford University Press.

Serres, Michel, and Alain Badiou. 1968. *Modèle et Structure*. Produced by Dina Dreyfus. https://youtube.com/watch?v=HveDVp5XVa4&feature=sharea.

Serres, Michel, and Nayla Farouki. 1998. *Le trésor: Dictionnaire des sciences*. Paris: Flammarion.

Serres, Michel, and Peter Hallward. 2003. "The Science of Relations: An Interview." *Angelaki* 8, no. 2: 227–38.

Serres, Michel, and Bruno Latour. 1992. *Éclaircissements: Entretiens Avec Bruno Latour*. Paris: François Bourin Editeur.

Serres, Michel, and Bruno Latour. (1992) 1995. *Conversations on Science, Culture, and Time*. Ann Arbor: University of Michigan Press.

Serres, Michel, Martin Legros, and Sven Ortoli. 2016. *Pantopie Ou Le Monde de Michel Serres: De Hermès à Petite Poucette*. Paris: Éditions Le Pommier.

Sharpe, Christina. 2016. *In the Wake: On Blackness and Being*. Durham, NC: Duke University Press.

Sherzer, Joel. 2002. *Speech Play and Verbal Art*. Austin: University of Texas Press.

Shryock, Andrew. 1997. *Nationalism and the Genealogical Imagination: Oral History and Textual Authority in Tribal Jordan*. Berkeley: University of California Press.

Shryock, Andrew. 2004. "The New Jordanian Hospitality: House, Host, and Guest in the Culture of Public Display." *Comparative Studies in Society and History* 46, no. 1: 35–62.

Shryock, Andrew. 2008. "Thinking about Hospitality, with Derrida, Kant, and the Balga Bedouin." *Anthropos* 103: 405–21.

Shryock, Andrew. 2019. "Keeping to Oneself: Hospitality and the Magical Hoard in the Balga of Jordan." *History and Anthropology* 30, no. 5: 546–62.

Shryock, Andrew, and Sally Howell. 2001. "Ever a Guest in Our House: the Emir 'Abdullah, Shaykh Majid al-Adwan, and the Practice of Jordanian House Politics, as Remembered by Umm Sultan, the Widow of Majid." *International Journal of Middle East Studies* 33, no. 2: 247–69.

Shryock, Andrew, and Daniel Lord Smail. 2018a. "On Containers: A Forum. Concluding Remarks." *History and Anthropology* 29, no. 1: 49–51.

Shryock, Andrew, and Daniel Lord Smail. 2018b. "On Containers: A Forum. Introduction." *History and Anthropology* 29, no. 1: 1–6.

Silko, Leslie Marmon. 1986. "Notes on the Deer Dance." In *The Delicacy and Strength of Lace: Letters between Leslie Marmon Silko and James Wright*, edited by Anne Wright, 9–10. Saint Paul, MN: Graywolf Press.

Sloterdijk, Peter. 2009. *Terror from the Air*. Semiotext(e) Foreign Agents Series. Cambridge, MA: Semiotext(e), MIT Press.

Sosna, Daniel. 2024. "Circular Economy of Wastewater: Recirculation, Spinning, and Rolling to the Future." In *Circular Economies in an Unequal World: Waste, Renewal, and the Effects of Global Circularity*, edited by Patrick O'Hare and Dagna Rams. London: Bloomsbury.

Sosna, Daniel, David Henig, and Roman Figura. 2022. "Raven Polluters." *Anthropology News* 63, no. 3: 21–23.

Spillers, Hortense. 1987. "Mamma's Baby, Papa's Maybe: An American Grammar Book." *Diacritics* 17, no. 2: 64–81.

Steil, Carlos. 2018. "The Paths of Saint James in Brazil: Body, Spirituality and Market." In *Pilgrimage and Political Economy: Translating the Sacred*, edited by Simon Coleman and John Eade, 155–72. Oxford: Berghahn.

Stengers, Isabelle. 1993. *L'invention Des Sciences Modernes*. Paris: La Découverte.

Stengers, Isabelle. 2010. *Cosmopolitics I*. Minneapolis: University of Minnesota Press.

Stewart, Kathleen. 2011. "Atmospheric Attunements." *Environment and Planning D: Society and Space* 29, no. 3: 445–53.

Stokoe, Elizabeth, S. Simons, J. Drury, S. Michie, M. Parker, A. Phoenix, S. Reicher, B. Wardlaw, and R. West. 2022. "What Can We Learn from the Language of 'Living with COVID'?" *BMJ* 376:0575. https://doi.org/10.1136/bmj.0575.

Stoller, Paul. 2020. "Imaging Knowledge: Visual Anthropology, Storytelling and the Slow Path toward Wisdom." *Kritisk Etnografi—Swedish Journal of Anthropology* 3, no. 1: 11–20.

Strathern, Marilyn. 1980. "No Nature, No Culture: The Hagen Case." In *Nature, Culture and Gender,* edited by Carol MacCormack and Marilyn Strathern, 174–222. Cambridge: Cambridge University Press.

Strathern, Marilyn. 1996. "Cutting the Network." *Journal of the Royal Anthropological Institute* 2, no. 3: 517–35.

Subramaniam, Banu, Laura Foster, Sandra Harding, Deboleena Roy, and Kim TallBear. 2016. "Feminism, Postcolonialism, Technoscience." In *The Handbook of Science and Technology Studies*, edited by Ulrike Felt, Rayvon Fouché, Clark Miller, and Laurel Smith-Doerr, 407–33. Cambridge, MA: MIT Press.

Szakolczai, Arpad. 2022. *Post Truth Society: A Political Anthropology of Trickster Logic*. London: Routledge.

Szakolczai, Arpad, and Bjørn Thomassen. 2019. *From Anthropology to Social Theory: Rethinking the Social Sciences*. Cambridge: Cambridge University Press.

TallBear, Kim. 2019. "Caretaking Relations, Not American Dreaming." *Kalfou: A Journal of Comparative and Relational Ethnic Studies* 6, no. 1: 24–41.

Taussig, Michael. 1993. *Mimesis and Alterity: A Particular History of the Senses*. London: Routledge.

Thomas, Julia Adeney, Mark Williams, and J. A. Zalasiewicz. 2020. *The Anthropocene: A Multidisciplinary Approach*. Cambridge: Polity.

Tihut Yirgu Asfaw. 2009. *Gender, Justice and Livelihoods in the Creation and Demise of Forests in North Western Ethiopia's Zeghie Peninsula*. PhD thesis, University of British Columbia.

Todd, Zoe. 2017. "Fish, Kin and Hope: Tending to Water Violations in *amiskwaciwaskahikan* and Treaty Six Territory." *Afterall: A Journal of Art, Context, and Enquiry* 43, no, 1: 103–7.

Touhouliotis, Vasiliki. 2018. "Weak Seed and a Poisoned Land: Slow Violence and the Toxic Infrastructures of War in South Lebanon." *Environmental Humanities* 10, no. 1: 86–106.

Trumble, Ruth. 2021. "Rhythms of Crises: Slow Violence Temporalities at the Intersection of Landmines and Natural Hazards." In *A Research Agenda for Geographies of Slow Violence*, edited by Shannon O'Lear, 41–55. Northampton, MA: Edward Elgar.

Tsing, Anna L. 2000. "The Global Situation." *Cultural Anthropology* 15, no. 3: 327–60.

Tsing, Anna L. 2015. *The Mushrooms at the End of the World: On the Possibility of Life in Capitalist Ruins*. Princeton, NJ: Princeton University Press.

Tuana, Nancy. 2008. "Viscous Porosity: Witnessing Katrina." In *Material Feminisms*, edited by Stacy Alaimo and Susan Hekman, 188–213. Bloomington: Indiana University Press.

Turner, Terence. 1991. "'We Are Parrots,' 'Twins Are Birds.' Play of Tropes as Operational Structure." In *Beyond Metaphor*, edited by James. W. Fernandez, 121–58. Stanford, CA: Stanford University Press.

Turner, Terence. 2017. *The Fire of the Jaguar*. Chicago: Hau Books and University of Chicago Press.

Turner, Victor. 1985. "Experience and Performance: Towards a New Processual Anthropology." In *On the Edge of the Bush*, edited by Edith Turner, 205–66. Tucson: University of Arizona Press.

Uexküll, Jakob von. 2010. *A Foray into the Worlds of Animals and Humans: With a Theory of Meaning*. Minneapolis: University of Minnesota Press.

Van Binsbergen, Wim M. J. 2020. "Grappling with the Ineffable in Three African Situations: An Ethnographic Approach." In *Capturing the Ineffable*, edited by Philip Y. Kao and Joseph S. Alter, 179–242. Toronto: University of Toronto Press.

Van de Weyer, Robert. 2009. "The Monastic Community of Ethiopia." *Ethiopian Review*. https://www.ethiopianreview.com/index/13534.

Van Milders, Lucas. 2022. "White Hallucinations." *Postcolonial Studies* 25, no. 2: 175–91.

Vinnecombe, Patricia. 1976. *People of the Eland: Rock Paintings of the Drakensberg Bushmen as a Reflection of Their Life and Thought*. Pietermaritzburg: University of Natal Press.

Virilio, Paul. 1994. *Bunker Archeology*. New York: Princeton Architectural Press.

Viveiros de Castro, Eduardo. 1998. "Cosmological Deixis and Amerindian Perspectivism." *Journal of the Royal Anthropological Institute* 4, no. 3: 469–88.

Viveiros de Castro, Eduardo. 2012. *Cosmological Perspectivism in Amazonia and Elsewhere*. Chicago: University of Chicago Press.

Viveiros de Castro, Eduardo. 2014. *Cannibal Metaphysics*. Minneapolis: Univocal.

Vonnegut, Kurt. 2000. *Slaughterhouse-Five or The Children's Crusade: A Duty-Dance with Death*. London: Vintage.

Vorbrugg, Alexander. 2022. "Ethnographies of Slow Violence: Epistemological Alliances in Fieldwork and Narrating Ruins." *Environment and Planning C: Politics and Space*, 40, no. 2: 447–62.

Vries, Hent de. 2001. "In Media Res: Global Religion, Public Spheres, and the Task of Contemporary Comparative Religious Studies." In *Religion and Media,* edited by Hent de Vries and Samuel Weber, 3–43. Stanford, CA: Stanford University Press.

Wagner, Roy. 1979. *Lethal Speech: Daribi Myth as Symbolic Obviation*. Ithaca: Cornell University Press.

Wagner, Roy. 1981. *The Invention of Culture*. Chicago: University of Chicago Press.

Wagner, Roy. 1986. *Symbols That Stand for Themselves*. Chicago: University of Chicago Press.

Wagner, Roy. 2001. *An Anthropology of the Subject. Holographic Worldview in New Guinea and Its Meaning and Significance for the World of Anthropology*. Berkeley: University of California Press.

Wagner, Roy. 2010. *A Invenção da Cultura*. Sao Paulo: Cosac and Naify.

Watkin, Christopher. 2015. "Michel Serres' Great Story: From Biosemiotics to Econarratology." *SubStance* 44, no. 3: 171–87.

Watkin, Christopher. 2017. *French Philosophy Today: New Figures of the Human in Badiou, Meillassoux, Malabou, Serres, and Latour*. Edinburgh: University of Edinburgh Press.

Watkin, Christopher. 2020. *Michel Serres: Figures of Thought*. Edinburgh: Edinburgh University Press.

Watkin, Christopher. 2022. "Michel Serres: Philosophy and the Contemporary World." https://christopherwatkin.com/2020/10/21/interview-michel-serres-philosophy-and-the-contemporary-world/.

Watson, Janell. 2019. "Sexed or Sexist? The Androgynous Cosmocracy of Michel Serres." *SubStance* 48, no. 3: 41–44.

Webb, David. 2012. "Michel Serres: From the History of Mathematics to Critical History." In *Time and History in Deleuze and Serres,* edited by Bernd Herzogenrath, 51–68. New York: Continuum.

Webb, David. 2022. "Narrative and the Natural Contract." Presentation to Michel Serres and the Social workshop, Queens College Cambridge, June 21–22.

Weiner, Annette. 1992. *Inalienable Possessions: The Paradox of Keeping-While-Giving*. Berkeley: University of California Press.

Whitman, Walt. (1892) 2017. *Song of Myself*. Pocket Books.

Williams, Forrest. 2000. "Preface." In *An Investigation of Jean-Paul Sartre's Posthumously Published Notebooks for an Ethics,* by Gail Evelyn Linsenbard, v–viii. Lampeter: Edwin Mellen Press.

Wiseman, Boris. 2007. *Lévi-Strauss, Anthropology and Aesthetics*. Cambridge: Cambridge University Press.

Yates, Frances. 1964. *Giordano Bruno and the Hermetic Tradition*. London: Routledge.

Yates, Frances. 1975. *The Rosicrucian Enlightenment*. London: Paladine Books.

Zagzebski, Linda. 2017. *Exemplarist Moral Theory*. Oxford: Oxford University Press.

Zalasiewicz, Jan, Mark Williams, Colin N. Waters, Anthony D. Barnosky, John Palmesino, Ann-Sofi Rönnskog, and Matt Edgeworth. 2017. "Scale and Diversity of the Physical Technosphere: A Geological Perspective." *The Anthropocene Review* 4, no. 1: 9–22.

Zammito, John H. 2021. "Koselleck's Times." *History and Theory* 60, no. 2: 396–405.

Zani, Leah. 2019. *Bomb Children: Life in the Former Battlefields of Laos*. Durham, NC: Duke University Press.

Zasiadko, Yevheniia. 2022. "Polluted to Death: The Untold Environmental Consequences of the Ukraine War." May 30. https://www.ispionline.it/it/pubblicazione/polluted-death-untold-environmental-consequences-ukraine-war-35224?fbclid=IwAR0uFeIg8menDIhBewYbP-_sL-FqG9jyjOc82cox2ik385p9QjNy5GYeiqM.

Zografou, Magda, and Stavroula Pipyrou. 2011. "Dance and Difference: Toward an Individualization of the Pontian Self." *Dance Chronicle* 34, no. 3: 422–46.

Zwijnenburg, Wim. 2021. "Climate Crisis Exacerbates Military Legacy Contamination." September 21. https://www.newsecuritybeat.org/2021/09/climate-crisis-exacerbates-military-legacy-contamination/.

CONTRIBUTORS

ANDREAS BANDAK is Associate Professor in the Department for Cross-Cultural and Regional Studies at the University of Copenhagen, Denmark, specializing in temporality and religion. He has conducted research in Syria, is author of *Exemplary Life: Modelling Sainthood in Christian Syria* (2022), and has edited numerous volumes, most recently *Ethnographies of Waiting* (2018).

JANE BENNETT is Andrew W. Mellon Professor of the Humanities at Johns Hopkins University, specializing in political theory, ecological philosophy, art and politics, American political thought, political rhetoric and persuasion, and contemporary social theory. She is author of numerous books, most recently *Influx and Efflux: Writing Up with Walt Whitman* (Duke University Press, 2020) and *Vibrant Matter: A Political Ecology of Things* (Duke University Press, 2010).

TOM BOYLSTON is Senior Lecturer in Anthropology at the University of Edinburgh, Scotland, specializing in play, technology, and religion. He has conducted research in Ethiopia and the United Kingdom and is author of *The Stranger at the Feast* (2018).

STEVEN D. BROWN is Professor of Health and Organizational Psychology at Nottingham Trent University, UK. He researches service user experiences of mental health care, social remembering among vulnerable groups, and psychological well-being and safety. He is author of numerous books and volumes, most recently *Vital Memory and Affect* (with Paula Reavey, 2015) and *Psychology without Foundations* (with Paul Stenner, 2009).

MATEI CANDEA is Professor of Social Anthropology at the University of Cambridge, UK, specializing in the anthropology of interspecies relations, comparative understandings of free speech, and the history of social theory. He has conducted research in France, Britain, and Africa and is author of *Corsican Fragments* (2010) and *Comparison in Anthropology* (2018) as well as numerous edited volumes.

ALBERTO CORSÍN JIMÉNEZ is based in the Department of Anthropology at the Spanish National Research Council, specializing in science, cities, and traps. He has conducted research in Europe and Latin America and is author of *Free Culture and the City: Hackers, Commoners, and Neighbors in Madrid 1997–2017* (with Adolfo Estalella, 2023) and *An Anthropological Trompe l'Oeil for a Common World* (2013), as well as numerous edited volumes.

DAVID HENIG is Associate Professor of Cultural Anthropology at Utrecht University, Netherlands, specializing in Muslim politics, military waste and war ecologies, and conflict and coexistence. He has conducted research in West Asia and Europe and is author of *Remaking Muslim Lives* (2020) and co-editor of *Economies of Favour after Socialism* (2017).

MICHAEL JACKSON is Senior Research Fellow of World Religions at the Harvard Divinity School, specializing in existential-phenomenological thought and ethical understandings of otherness. He has conducted research in Sierra Leone and Australia and is author of numerous books, most recently *Coincidences* (2021) and *The Genealogical Imagination* (Duke University Press, 2021).

DANIEL M. KNIGHT is Reader in Social Anthropology and Director of the Centre for Cosmopolitan Studies at the University of St. Andrews, Scotland, specializing in time and temporality and philosophical anthropology. He has conducted research in Greece and is author of numerous books, most recently *Vertiginous Life: An Anthropology of Time and the Unforeseen* (2021) and *The Anthropology of the Future* (with Rebecca Bryant, 2019).

CELIA LOWE is Professor of Anthropology and International Studies at the University of Washington, specializing in postcolonial science and technology studies, human/microbial relations, political ecologies, and theories of materialism. She has conducted research in Southeast Asia and is the author of *Wild Profusion* (2006).

MORTEN NIELSEN is an anthropologist and research professor at the National Museum of Denmark, specializing in urban development, time, urban modeling, and stand-up comedy. He has conducted research in Brazil, Mozambique, Scotland, and the United States and is the co-author of *Collaborative Damage* (with Mikkel Bunkenborg and Morten Axel Pedersen, 2022) and co-editor of *The Composition of Anthropology* (with Nigel Rapport, 2018).

STAVROULA PIPYROU is Senior Lecturer in Social Anthropology and Founding Director of the Centre for Minority Research at the University of St. Andrews, Scotland. She has conducted research in Italy, Greece, and Brazil on minority politics and governance, civil society, and performance and is author of *The Grecanici of Southern Italy: Governance, Violence, and Minority Politics* (2016).

ELIZABETH A. POVINELLI is Franz Boas Professor of Anthropology at Columbia University, specializing in critical theory and settler late liberalism. She has conducted research in Australia and the United States and is author of numerous books, most recently *The Inheritance* (Duke University Press, 2021) and *Geontologies* (Duke University Press, 2016).

ANDREW SHRYOCK is Arthur F. Thurnau Professor of Anthropology at the University of Michigan, specializing in the relationship between anthropology, history, and the natural sciences. He has conducted fieldwork in Yemen, Jordan, and the United States and is author and co-author of numerous books and volumes, most recently *The Scandal of Continuity in Middle East Anthropology* (2019) and *From Hospitality to Grace* (2017).

ARPAD SZAKOLCZAI is Professor Emeritus of Sociology at University College Cork, Ireland. His recent books include *Permanent Liminality and Modernity: Analysing the Sacrificial Carnival through Novels* (2017), *From Anthropology to Social Theory: Rethinking the Social Sciences* (2019), and *Post-Truth Society: A Political Anthropology of Trickster Logic* (2022).

INDEX

comedic transubstantiation, 241–43, 245, 248–50

comedy, 233, 235; confessional, 238–40; contradiction in, 245, 247; exchange and, 248; Hermesian paradox of, 240–44; stand-up, 236; truthful, 240

Comedy Cellar, 239, 241, 245

comparative analysis, 94

comparative anthropology, 265

comparative sophiology, 255, 264–69

comparative work, 23

comprehension: angels and, 96; knowledge transformation to, 91

concepts, 5; hyphenated, 8, 222

Conder, Claude Reignier, 86

confessional comedy, 238–40

connection, 22–25, 217–21

Connolly, William, 4, 22, 277, 283

Connor, Steven, 141

consciousness, 188, 192

containment, 38–39

contemporary relevance, 281

contracts: micro, 3; social, 4, 60–61, 177–78, 183; symbiotic, 47. *See also* natural contracts

contradiction, 245

conversations, 27–28; as analysis, 164

Conversations on Science, Culture, and Time (Serres and Latour), 13–14, 27, 274–76; analogism in, 22; on nuclear weapons, 135, 140, 143; on proximity, 220; time and, 6–7; on wars, 141–42

Corcoran, John, 122

coronaviruses, 171

corpses, 159, 165–67

cosmos, 181; proximity and, 221

Cousteau, Jacques, 100, 105–10, 112–13

COVID-19 pandemic, 154, 163; safety guidance for, 155–57, 159–60

crises, 7, 13, 226; climate, 16–17, 144, 148, 283; ecological, 146, 176, 186, 283; Greek, 11–12, 26; of Neolithic, 180; temporal topologies and, 136

critical race theory, 124

Crutzen, Paul, 144

culinary triangle, 66n2

cultural contexts, wisdom and, 263

cultural exceptionalism, 229

cultural relativity, 216

curbed interactions, 184

cybernetics, 189

Cyprus, 266

da Col, Giovanni, 82

Damascus, Syria, 18

dancing, 206–7

dead bodies, 165

death, 158–63; monks and, 209

Death of Orpheus (drawing), 178

debate, 182

decolonization, 203

degrowth, 261

Deleuze, Gilles, 1, 15, 45, 103, 120, 235, 278, 286

Deloria, Vine, Jr., 132

de Martino, Ernesto, 14

democratization of knowledge, 160

denial of coevalness, 262

depleted uranium, 149

Derrida, Jacques, 1, 15, 75–76, 103, 278

Descola, Philippe, 22

Detachment (Serres), 169, 179

Dhabta, al-, 83–85

dictatorships, 12

difference, 2, 189; minority, 230; narcissism of minor, 219, 228, 230, 232

differentiation: codification and categorization and, 224; commonalities and, 228; tropic, 244–46

digital networks, 160

Diogene (ship), 106, 110

Diogenes, 169

Discours sur le Colonialism (Césaire), 124

discursive colonization, 262

dissociation, 15; analysis as, 164

distributive morality, 65

Dolphijn, Rick, 142–43

domestication of animals, 170

Don Juan (Molière), 93, 233–34, 248–49

doodling, 275

double capture, 40, 45

Douglas, Mary, 145, 182

Dreyfuss, Richard, 239, 241

Dürer, Albrecht, 178

earthly violence, 144–48

Éclaircissements (Serres), 255

eco-activism, 61

ecological crises, 146, 176, 186, 283

ecological revolution, 283

eco-movements, 283

eco-politics, 183

mimesis, 45–46

Mimesis and Alterity (Taussig), 44

mimetic culture, 44

mimetic defense, 45

Mind and Nature (Bateson, G.), 184, 187, 189–91

minority difference, 230

minority groups, 220, 225, 231

misdirection, 247–48

misery, 258

modernity: gift exchange and, 73; Greece and, 229; parasitism and, 180–81; Stoicism and, 257–58; wisdom and, 271

Molière, 93, 140, 171, 233–34, 248–50

money-market systems, 72

monitor lizards, 63

monks, 209

morality: distributive, 65; Orthodox Christianity and, 205

moral scripts, 234

mousetraps, 42

Muenster, Ursula, 111

musical instruments, 41; as traps, 42

music recommendation systems, 42

Mustakeem, Sowande, 130

mutualism, 170

My Octopus Teacher (film), 109

myth, history and, 6–7

mythic narratives, 244–45

mythistory, 7, 25

Nagasaki, 101, 140, 142–44, 149

Napoleonic Wars, 177

narcissism of minor differences, 219, 228, 230, 232

narcissistic opportunism, 230, 232

nationalism, 215, 230

Natural Contract, The (Serres), 3, 16, 27, 36, 100, 143–46, 172, 282–83; core argument of, 177–83; on law and science, 38, 47; on parasitism, 47; on reciprocity, 47

natural contracts, 34–37, 46, 60–63, 178, 281, 284; animals and, 62; recognition and, 185

natural law, 61

natural sciences, 182, 216–18

Nature, sacredness of, 191

navigation, 284

negentropy, 202

neglect, 185–86

negligence, 185

neighborhood of indiscernibility, 45

Neolithic crisis, 180

networks, 29n8

Nilges, Mathias, 126

Noah, Trevor, 248

noise, 4, 85, 274; background, 18, 40, 46; figure-ground gradation of, 40

nonlinear time, 15

"Notes on a Native Son" (Baldwin), 125

nuclear weapons, 101, 135, 142, 148–49

Nutmeg's Curse, The (Ghosh), 148

objective violence, 144, 146

obligation, 73

octopus, 109–11

Odysseus, 58

Of Hospitality (Derrida), 76

oneway relations, 77

On the Natural History of Destruction (Sebald), 135–36

opportunistic narcissism, 230, 232

oral tradition, 96

Orthodox Christianity, 203–4, 210–12; mediation and, 200; morality and, 205

orthopedic body, 128

Oswalt, Patton, 236

othering, 262

Otherness, 2, 8, 124, 126, 228

P'agumen, 200

Pali language, 267

panbiota, 169–73

pandemics, 154–55

panic, 161

panurgy, 161–62, 167

Papandreou, Andreas, 229–30

Parasite, The (Serres), 2–3, 82, 91, 93–94, 143; on collectivity, 163–64; on comedy, 233; fables and, 35; on information and energy flows, 77–78; interrupters and, 85; Molière and, 171; on parasitism and original conditions, 179; porosity and, 27, 46; symbiosis and, 28

parasites, 37

parasitic relation, 35, 79, 170

parasitism, 38–40, 46–47, 50, 58, 77–78, 227, 284; hospitality and, 82; life and, 172; modernity and, 180–81; original conditions and, 179; war and, 36

San, 61
Sartre (Hayman), 175
Sartre, Jean-Paul, 15, 26, 60
scales of analysis, 281
scaling, 21–22
scapegoat, 166
Schloss, Jeffrey, 263
scholarly exchange, 160
Schüll, Natasha, 42–43
science, 38; applied, 190; art and, 102; Christianity and, 94; cognitive, 189, 263; meanings of, 182; natural, 182, 216–18; natural contracts and, 47; nuclear weapons and, 143; philosophy of, 9; rationalism and, 176, 178; social, 2, 100, 182–83, 217–18
scientific methodology, 190
scientific optimism, 135
scientific responsibility, 143–44, 149
scuba, 105–6, 112
sea training, 127–31
Seaver, Nick, 42
Sebald, W. G., 135–36
secretly familiar, 219
self, 6, 15
self-creation, 247
senses, 225; technology influence on, 2
Serres, Michel, 1–14, 175–76, 215–16, 232, 253, 274, 285; analogy use by, 93–94; on angels, 94–95; anthropological reading of, 9; on anthropology, 93; authoritative discourse and, 254; on background noise, 40; on body, 123; climate change and, 144; on comedy, 233, 235–37, 248–49; connecting and, 22–25, 217–18; conversations and, 27–28; on cultural exceptionalism, 229; on dancing, 207; debate and, 182; on Diogenes, 169; on Don Juan, 234, 248–49; early life, 101–2, 142; on egocentric and sociocentric action, 58; epistemodicé and, 16–17; on exchange, 78–79; on gift exchange, 78; hearing and, 18; Hermes and, 49, 234; on historical circumstances and philosophical approach, 158; history as movement and, 131; on hominescence and death, 158–63; on hospitality and parasitism, 82; on hyphens, 279; intuition and, 279–80; on invariable, 275; knots and, 19; on knowledge and comprehension, 91–92; knowledge production and consumption and, 17–18; on law and society origins, 36;

on media, 94–95; metaphors and, 95; method of connections, 20; mode of inquiry, 277, 284; on natural and social contracts, 60; navigation and, 284; on negotiations with world, 172; on new human, 108; nuclear bombs and, 101, 135, 143; oneway relations idea of, 77; on parasite in milieu, 46; on parasitic relation, 35; planetary consciousness and, 112; on pollution, 145, 153–55; on pre-posing, 206; proximity variations in, 220–25; responses to, 92; on sacrifice, 65–66; on sailors as firemen, 117, 127–28; scaling and, 21–22; sea and, 100–101; ship metaphor and, 183–87; on simultaneity and connection, 220–21; structuralism on, 241; surge idea, 244; systematization and, 278; temporal topologies and, 136, 150; as theorist, 253–55; time and, 118; topology and, 217; on totalities, 100; universal humanism and, 103–4; universalism and, 119; war and violence and, 140–42; wisdom and, 255–64, 270–71; on words with pan-, 161
Sharpe, Christina, 119, 127
Shepherd and the Lion, The (La Fontaine), 34–35
ship metaphor, 183–87
Shryock, Andrew, 38–39
sibling rivalry, 52–54
Sierra Leone, 52, 54
Silent World, The (film), 106–7
Silko, Leslie Marmon, 61
simultaneity, 220–21
Sise, Bakunko, 64
situated ethics, 67n6
Slaughterhouse-Five (Vonnegut), 135–36
slave merchants, 130
Slavery at Sea (Mustakeem), 130
Sloterdijk, Peter, 141
slowness, 277
slow violence, 152n13
smallhold farming, 170
Smith, Adam, 185
social authority: of monks, 209; of Orthodox Church, 203
social contract, 4, 60–61, 177–78, 183
social distancing, 155–56, 163, 168
social existence, 60
social intelligence, 58
social media, 236
social relationality, 242, 247

social sciences, 2, 100, 182–83, 217–18

social spaces, 77

social vertigo, 26

society: archaic, 73–78, 93; origins of, 36; pre-modern, 65; self and, 6, 15

soft pollution, 153, 155

soil erosion, 147

Solitude (Serres), 141

"Song of Myself" (Whitman), 286

sophiology, 255, 264–69

sophron, 266, 268

sovereign spaces, 88

sovereignty, 77; life and, 123

space, 218

Spanish Civil War, 13–14

Speculum of the Other Woman (Irigaray), 120

spiderweb anthropologies, 42–48

stand-up comedy, 236–37, 246, 249–50; confessional, 238–40; form and content tension in, 245

Statues (Serres), 6, 65, 92, 96, 143, 164; on corpses, 159

Stengers, Isabelle, 45–46

Steps to an Ecology of the Mind (Bateson, G.), 187

Steven, Siaka, 52

Stewart, Charles, 7

stock market crashes, 11–12

Stoermer, Eugene, 144

Stoicism, 255–58, 272n3

Stone Age Economics (Sahlins), 261

stoning, 165

storage: of knowledge, 72–73; material cultures and, 72

"Story of Asdiwal, The" (Lévi-Strauss), 246

Story of Lynx, The (Lévi-Strauss), 244

streaming services, 236

Structural Anthropology (Lévi-Strauss), 244

structural consistency, 235

structuralism, 241

structure, 222

STS, feminist, 9

Stupak, Bob, 42

subcertainties, 19, 22

subjective wars, 146

surge, 244

surgeons, 130

Swiss Cheese model of safety, 154–55

symbiosis, 47, 50, 58–59, 170, 172, 284; natural contract and, 60

symbiotic contract, 47

Symbols That Stand for Themselves (Wagner), 244

Symposium (Plato), 184, 193

Syrian Christians, 18

Tartuffe (Molière), 171

Taussig, Michael, 44

Taylor, Jason deCaires, 130

"Techniques of the Body" (Mauss), 121

technology, 5–6, 162–63, 182–83; influence on senses of, 2

temporal agency, 2

temporal linearity, 7

temporal topologies, 136, 150

temporal turn, 6

terror, 135

Thoreau, Henry, 286

Thumbelina (Serres), 160, 279

Tiepolo, Giandomenico, 177

Tierney, Jordan, 286

Timaeus (Plato), 193

time, 6–7, 218; nonlinear, 15; percolating, 13; polytemporality, 13

Times of Crisis (Serres), 7, 226

Todd, Zoe, 132

Togo Hare, 51

tools, 162

topology, 135, 217, 231

totalities, 100

totemic myths, 63–64

Transjordan, 81

transubstantiation, 240, 251n2

traps, 40–41; gambling machines and, 42–44; musical instruments as, 42

travel, 24

tricksters, 62, 234, 243; animals and, 50, 56; forms of, 49–50; Mande, 50–54; as rebels, 54–58

tropic differentiation, 244–46

Troubadour of Knowledge, The (Serres), 2, 19, 24, 142, 222, 279

truth: biological, 224; minorities and, 225; risk of, 223

truthful comedy, 240

Turner, Terence, 247

Turner, Victor, 184

type-essence, 251n2

Uexküll, Jacob von, 44–45

umbilical thinking, 20

universal human, 103
universal humanism, 103–4
universalism, 280; angels and, 201; Serres and, 119
universities, 203
US Food and Drug Administration, 155–56

Van Milders, Lucas, 126
Variations on the Body (Serres), 14, 24, 26–27, 108, 123, 159, 217, 223
vaudeville, 236
vegetarianism, 208
Verne, Jules, 24
vertical binding, 173
vertiginous aesthetic, 29n2
vertiginous life, 26
Vertiginous Life (Knight), 14
vibrant matter, 224
Vibrant Matter (Serres), 286
Vicissitudes (Taylor), 130
Vinnecombe, Patrician, 61–62
violence, 2, 4, 140–42, 158; analysis and, 164; body and, 159; earthly, 144–48; objective, 144, 146; property and, 168; slow, 152n13
Virilio, Paul, 141
Viveiros de Castro, Eduardo, 39–40, 48, 61
voluntarism, 73
Vonnegut, Kurt, 135–37

Wagner, Roy, 244
war, 140–42; history and, 36; parasitism and, 36; percolation of, 148–51; pollution and, 150; socio-environmental aftermaths of, 137; subjective, 146; temporal topologies and, 136; wastes of, 148–49
Warburg, Aby, 178
War Childhood Museum, 137, 139
waste, 145; of war, 148–49
Watkin, Christopher, 21, 36, 101, 103, 111, 119, 141, 271
Webb, David, 172
Weil, Simone, 131
Where Angels Fear to Tread (Bateson, G.), 187, 189
whiteness, 35
Whitman, Walt, 286–87
Williams, Eric, 118
wisdom: comparative sophiology and, 264–69; emplacement of, 262–64; knowledge and, 267; as method, 269–72; modernity and, 271; problem with, 259–64; Serres and, 255–59
Wisdom Sits in Places (Basso), 265
Witchcraft, Oracles, and Magic among the Azande (Pritchard), 16
world, 181; negotiations with, 172; science and, 182
world-objects, 167

Yates, Francis, 182
Young, Peter, 248
Yusoff, Kathryn, 146

Zege, Ethiopia, 200, 203–4, 209–10

www.ingramcontent.com/pod-product-compliance
Lightning Source LLC
Chambersburg PA
CBHW032343280326
41935CB00008B/427